Biological Control of Plant-Parasitic Nematodes:

Progress in Biological Control

Volume 11

Published:

Volume 1
H.M.T. Hokkanen and A.E. Hajek (eds.):
Environmental Impacts of Microbial Insecticides – Need and Methods for Risk Assessment. 2004
ISBN 978-1-4020-0813-9

Volume 2
J. Eilenberg and H.M.T. Hokkanen (eds.):
An Ecological and Societal Approach to Biological Control. 2007
ISBN 978-1-4020-4320-8

Volume 3
J. Brodeur and G. Boivin (eds.):
Trophic and Guild Interactions in Biological Control. 2006
ISBN 978-1-4020-4766-4

Volume 4
J. Gould, K. Hoelmer and J. Goolsby (eds.):
Classical Biological Control of *Bemisia tabaci* in the United States. 2008
ISBN 978-1-4020-6739-6

Volume 5
J. Romeis, A.M. Shelton and G. Kennedy (eds.):
Integration of Insect-Resistant Genetically Modified Crops within IPM Programs. 2008
HB ISBN 978-1-4020-8372-3; PB ISBN 978-1-4020-8459-1

Volume 6
A.E. Hajek, T.R. Glare and M.O'Callaghan (eds.):
Use of Microbes for Control and Eradication of Invasive Arthropods. 2008
ISBN: 978-1-4020-8559-8

Volume 7
H.M.T. Hokkanen (ed.):
Relationships of Natural Enemies and Non-Prey Foods. 2008
ISBN: 978-1-4020-9234-3

Volume 8
S.S. Gnanamanickam:
Biological Control of Rice Diseases
ISBN: 978-90-481-2464-0

Volume 9
F.L. Cônsoli, J.R.P. Parra and R.A. Zucchi (eds.):
Egg Parasitoids in Agroecosystems with Emphasis on *Trichogramma*
ISBN: 978-1-4020-9109-4

Volume 10
W.J. Ravensberg:
A Roadmap to the Successful Development and Commercialization of Microbial Pest Control Products for Control of Arthropods
ISBN: 978-94-007-0436-7

For further volumes:
http://www.springer.com/series/6417

Keith Davies • Yitzhak Spiegel
Editors

Biological Control of Plant-Parasitic Nematodes:

Building Coherence between Microbial Ecology and Molecular Mechanisms

Springer

Editors
Dr. Keith G. Davies
Rothamsted Research
Department Plant Pathology &
Microbiology
AL5 2JQ Harpenden
Hertfordshire
United Kingdom
keith.davies@rothamsted.ac.uk

Prof. Yitzhak Spiegel
Agricultural Research
Organization (ARO)
The Volcani Center
Department of Nematology
PO Box 6
Bet Dagan, Israel
spiegely@volcani.agri.gov.il

ISBN 978-1-4020-9647-1 e-ISBN 978-1-4020-9648-8
DOI 10.1007/978-1-4020-9648-8
Springer Dordrecht Heidelberg London New York

Library of Congress Control Number: 2011928081

Printed on acid-free paper

Springer is part of Springer Science+Business Media (www.springer.com)

Progress in Biological Control

Series Preface

Biological control of pests, weeds, and plant and animal diseases utilising their natural antagonists is a well-established and rapidly evolving field of science. Despite its stunning successes world-wide and a steadily growing number of applications, biological control has remained grossly underexploited. Its untapped potential, however, represents the best hope to providing lasting, environmentally sound, and socially acceptable pest management. Such techniques are urgently needed for the control of an increasing number of problem pests affecting agriculture and forestry, and to suppress invasive organisms which threaten natural habitats and global biodiversity.

Based on the positive features of biological control, such as its target specificity and the lack of negative impacts on humans, it is the prime candidate in the search for reducing dependency on chemical pesticides. Replacement of chemical control by biological control – even partially as in many IPM programs – has important positive but so far neglected socio-economic, humanitarian, environmental and ethical implications. Change from chemical to biological control substantially contributes to the conservation of natural resources, and results in a considerable reduction of environmental pollution. It eliminates human exposure to toxic pesticides, improves sustainability of production systems, and enhances biodiversity. Public demand for finding solutions based on biological control is the main driving force in the increasing utilisation of natural enemies for controlling noxious organisms.

This book series is intended to accelerate these developments through exploring the progress made within the various aspects of biological control, and via documenting these advances to the benefit of fellow scientists, students, public officials, policy-makers, and the public at large. Each of the books in this series is expected to provide a comprehensive, authoritative synthesis of the topic, likely to stand the test of time.

Heikki M.T. Hokkanen, Series Editor

Preface

The need for alternative management systems for the control of plant-parasitic nematodes has increased dramatically over the last decade, mainly because of the banning of the most important nematicides. Therefore, biological control of phytonematodes has received an enhanced impetus and several attempts in the industrial/commercial sector as well as in academia, have been made to fulfill this need. The last relevant handbook on this treatise was published in 1991 and since then there has been no specific volume addressing this important topic. This book was written at a time when molecular biology as well as different 'omic' approaches, were just beginning to encroach on the subject area but were not included. Therefore, the progress that has been made in biotechnology and the new tools available for research have augmented new perspectives that help in our understanding, in areas as diverse as as aspects of mode-of-action through population dynamics to knowledge about formulation and application techniques, which have so far not been covered by any other volume.

The offered volume intends to review the biological control theme from several prospects: (1) Various ecological aspects such as: suppressive soils, organic amendments, issues related to the farming system both at present and in the future together with the role of nematodes in soil food webs, that covers application, conservation and enhancement of indigenous and introduced antagonists (Chaps. 1, 2 and 11); (2) *Caenorhabditis elegans* as a model and lessons from other natural systems (Chap. 3); (3) Exploiting advanced genomic tools to promote the understanding of biocontrol processes and thereafter helping to improve specific biological control agents (Chaps. 3, 4, 6 and 7); (4) Interaction between the plant host, nematodes' surface and microorganisms: the role of the nematode surface-coat in interactions with their host-plant and their surrounding bacteria and fungi (Chap. 5), emphasizing on the biochemical, molecular and genomic interactions of nematodes with nematode-trapping fungi (Chap. 6), and understanding the mode-of-action of various biocontrol systems such as the eggs- and cyst-parasite *Pochonia chlamydosporia* (Chap. 7) and *Trichoderma* spp. (Chap. 8). (5) Candidates for biocontrol - microorganism's applicative as well as commercial state of the art (nematode-trapping fungi, endophytes fungi, *Pochonia chlamydosporia*, *Trichoderma* sp., or *Pasteuria penetrans* (Chap. 4, Chaps. 6–10); and (6) Extrapolation of the wide knowledge existed in another systems for understanding biocontrol processes (Chap. 9).

This volume comprises a wide spectrum of topics and ideas relevant not only to biological control of plant-parasitic nematodes, but also to generic aspects of host- parasite interactions that can be used by scientists with little knowledge or experience with phytonematodes.

Hertfordshire, UK Keith G. Davies
Bet Dagan, Israel Yitzhak Spiegel

Contents

1 **Biological Control of Plant-Parasitic Nematodes: An Ecological Perspective, a Review of Progress and Opportunities for Further Research** 1
Graham R. Stirling

2 **Microbial Ecology and Nematode Control in Natural Ecosystems** 39
SofiaR. Costa, Wim H. van der Putten, and Brian R. Kerry

3 **Microbial Interactions with *Caenorhabditis elegans*: Lessons from a Model Organism** 65
Maria J. Gravato-Nobre and Jonathan Hodgkin

4 **Exploiting Genomics to Understand the Interactions Between Root-Knot Nematodes and *Pasteuria penetrans*** 91
Jenn E. Schaff, Tim H. Mauchline, Charles H. Opperman, and Keith G. Davies

5 **Plant Nematode Surfaces** 115
Rosane H.C. Curtis, John T. Jones, Keith G. Davies, Edna Sharon, and Yitzhak Spiegel

6 **Molecular Mechanisms of the Interaction Between Nematode-Trapping Fungi and Nematodes: Lessons From Genomics** 145
Anders Tunlid and Dag Ahrén

7 **Ecology of *Pochonia chlamydosporia* in the Rhizosphere at the Population, Whole Organism and Molecular Scales** 171
Brian R. Kerry and Penny R. Hirsch

8 ***Trichoderma* as a Biological Control Agent** 183
Edna Sharon, Ilan Chet, and Yitzhak Spiegel

9 New Insights on the Mode of Action of Fungal Pathogens of Invertebrates for Improving Their Biocontrol Performance 203
Jose G. Maciá-Vicente, Javier Palma-Guerrero, Sonia Gómez-Vidal, and Luis V. Lopez-Llorca

10 Endophytic Fungi 227
Johannes Hallmann and Richard A. Sikora

11 Utilization of Biological Control for Managing Plant-Parasitic Nematodes 259
Patricia Timper

12 Root Patho-Systems Nematology and Biological Control 291
Keith G. Davies and Yitzhak Spiegel

Index 305

Contributors

Dag Ahrén
Department of Microbial Ecology, Lund University,
Ecology Building, SE 224 62 Lund, Sweden
dag.ahren@biol.lu.se

Ilan Chet
Department of Microbiology and Plant Pathology, Faculty of Agricultural,
Food and Environmental Quality Sciences, Hebrew University of Jerusalem,
Rehovot 76100, Israel
chet@agri.huji.ac.il

Sofia R. Costa
Centre for Functional Ecology, Department of Life Sciences,
University of Coimbra, PO Box 3046, 3001-401 Coimbra, Portugal
sofia.costa@ci.uc.pt

Rosane H.C. Curtis
Department of Plant Pathology and Microbiology, Rothamsted Research,
Harpenden, Hertfordshire AL5 2JQ, UK
rosane.curtis@bbsrc.ac.uk

Keith G. Davies
Department of Plant Pathology and Microbiology, Rothamsted Research,
Harpenden, Hertfordshire AL5 2JQ, UK
keith.davies@rothamsted.ac.uk

Sonia Gómez-Vidal
Laboratory of Plant Pathology, Department of Marine Sciences and Applied
Biology, Multidisciplinary Institute for Environmental Studies (MIES)
Ramón Margalef, University of Alicante, Apto. 99, Alicante 03080, Spain
sonia.gomez@ua.es

Maria J. Gravato-Nobre
Department of Biochemistry, University of Oxford,
South Parks Road, Oxford OX1 3QU, UK
Maria.Gravato-Nobre@bioch.ox.ac.uk

Johannes Hallmann
Julius Kühn-Institut, Federal Research Centre for Cultivated Plants, Institute for Epidemiology and Pathogen Diagnostics, Toppheideweg 88, 48161 Münster, Germany
johannes.hallmann@jki.bund.de

Penny R. Hirsch
Nematode Interactions Unit, Plant Pathology & Microbiology Department, Rothamsted Research, Harpenden, Hertfordshire AL3 2JQ, UK
penny.hirsch@bbsrc.ac.uk

Jonathan Hodgkin
Department of Biochemistry, University of Oxford, South Parks Road, Oxford OX1 3QU, UK

John T. Jones
SCRI, Invergowrie, Dundee, DD2 5DA, Scotland, UK
John.Jones@scri.ac.uk

Brian R. Kerry
Nematode Interactions Unit, Centre for Soils and Ecosystem Function, Rothamsted Research, Harpenden, Hertfordshire AL5 2JQ, UK
and
Nematode Interactions Unit, Plant Pathology & Microbiology Department, Rothamsted Research, Harpenden, Hertfordshire AL3 2JQ, UK
brian.kerry@bbsrc.ac.uk

Luis V. Lopez-Llorca
Laboratory of Plant Pathology, Department of Marine Sciences and Applied Biology, Multidisciplinary Institute for Environmental Studies (MIES) Ramón Margalef, University of Alicante, Apto. 99, Alicante 03080, Spain
luisvilopezllorca@gmail.com

Jose G. Maciá-Vicente
Laboratory of Plant Pathology, Department of Marine Sciences and Applied Biology, Multidisciplinary Institute for Environmental Studies (MIES) Ramón Margalef, University of Alicante, Apto. 99, Alicante 03080, Spain
jgmv@ua.es

Tim H. Mauchline
Department of Plant Pathology and Microbiology, Rothamsted Research, Harpenden, Hertfordshire, AL5 2JQ, UK
tim.mauchline@rothamsted.ac.uk

Charles H. Opperman
Plant Pathology Department, North Carolina State University, Raleigh, NC 27695, USA

Javier Palma-Guerrero
Laboratory of Plant Pathology, Department of Marine Sciences and Applied Biology, Multidisciplinary Institute for Environmental Studies (MIES) Ramón Margalef, University of Alicante, Apto. 99, Alicante 03080, Spain
jpalma@ua.es

Jenn E. Schaff
Plant Pathology Department, North Carolina State University, Raleigh, NC 27695, USA
jeschaff@ncsu.edu

Edna Sharon
Division of Nematology, Institute of Plant Protection, ARO, The Volcani Center, Bet Dagan 50250, Israel
ednash2@gmail.com

Richard A. Sikora
University of Bonn, Institute for Crop Science and Ressource Conservation, Nematology in Soil Ecosystems, Nußallee 9, 53115 Bonn, Germany
rsikora@uni-bonn.de

Yitzhak Spiegel
Division of Nematology, Institute of Plant Protection, ARO, The Volcani Center, Bet Dagan 50250, Israel
spiegely@volcani.agri.gov.il

Graham R. Stirling
Biological Crop Protection Pty. Ltd., 3601 Moggill Road, Moggill, QLD 4070, Australia
graham.stirling@biolcrop.com.au

Patricia Timper
Crop Protection and Management Research Unit, USDA-ARS, P.O. Box 748, Tifton GA 31793, USA
Patricia.Timper@ars.usda.gov

Anders Tunlid
Department of Microbial Ecology, Ecology Building, SE 224 62 Lund, Sweden
anders.tunlid@mbioekol.lu.se

Wim H. van der Putten
Department of Terrestrial Ecology, Netherlands Institute of Ecology (NIOO-KNAW), P. O. Box 50, 6700 AB Wageningen, The Netherlands
and
Laboratory of Nematology, Wageningen University, NL-6709 PD Wageningen, The Netherlands
w.vanderputten@nioo.knaw.nl

Chapter 1
Biological Control of Plant-Parasitic Nematodes: An Ecological Perspective, a Review of Progress and Opportunities for Further Research

Graham R. Stirling

Abstract Plant-parasitic nematodes are important pests, causing billions of dollars damage to the world's food and fibre crops. However, from an ecological perspective, this group of nematodes is simply one component in a vast array of organisms that live in soil. All these organisms interact with nematodes and with each other, and during that process, contribute to regulatory mechanisms that maintain the stability of the soil food-web. Populations of individual species do not increase indefinitely but are subject to a constant series of checks and balances, which more or less stabilises their population densities. Thus, biological control is a normal part of a properly functioning soil ecosystem, with plant-parasitic nematodes only becoming pests when they are no longer constrained by the biological buffering mechanisms that normally keep them in check. This chapter therefore focuses on approaches that can be used to restore, maintain or enhance the natural nematode-suppressive mechanisms that should operate in all agricultural soils. The positive impact of organic matter and the negative effects of tillage, biocides, fertilisers and other management practices on suppressiveness are discussed, together with examples of suppression due to host-specific natural enemies. The problems associated with replacing soil fumigants and nematicides with biological alternatives, and the ecological issues likely to affect the efficacy of such products, are also considered.

Keywords Soil food web • Organic matter • Soil health • Organic amendments • Nematode-suppressive soil • Minimum tillage • Egg parasites • Predatory nematodes • Nematode-trapping fungi • *Pasteuria* • *Brachyphoris* • *Pochonia* • *Paecilomyces*

G.R. Stirling (✉)
Biological Crop Protection Pty. Ltd., 3601 Moggill Road, Moggill, QLD 4070 Australia
e-mail: graham.stirling@biolcrop.com.au

K. Davies and Y. Spiegel (eds.), *Biological Control of Plant-Parasitic Nematodes: Building Coherence between Microbial Ecology and Molecular Mechanisms*, Progress in Biological Control 11, DOI 10.1007/978-1-4020-9648-8_1,

1.1 Introduction

The relatively stable behaviour of animal populations in natural environments should serve as a constant reminder that in nature, all organisms are subject to a constant series of checks and balances. Populations of individual species do not increase indefinitely but are constrained by the physical environment and by the community of organisms within which they co-exist. Cyclic changes in populations will occur, but provided there is no major change in the physical or biotic environment, populations will fluctuate between certain upper and lower limits. This phenomenon, commonly referred to as 'biological balance' or the 'balance of nature', more or less stabilises animal population densities and applies to all organisms, including plant-parasitic nematodes. The action of soil organisms in maintaining nematode population densities at lower average levels than would occur in their absence is generally termed 'biological control'.

These words, which were included on the first page of my book on biological control of nematodes (Stirling 1991) define the general area of biological control, indicate that it operates wherever nematodes occur, and remind us that plant-parasitic nematodes only reach unacceptably high population densities (i.e. become pests of economic concern) when they are no longer constrained by the biological mechanisms that normally keep them in check. Phrases such as 'the balance of nature' also provide a focus for this chapter, because the aim is to discuss biological control of nematodes within an ecological framework. Thus the chapter begins with a discussion of the soil environment and the regulatory forces that operate within the soil food web and then considers how these natural regulatory mechanisms can be exploited in various farming systems to improve the level of nematode control achievable by biological means.

1.2 Fundamentals of Soil Ecology

It is only in the last few decades that ecologists have undertaken detailed studies of belowground soil processes, and this has led to a better understanding of the nature of the soil environment and the complex biological communities that live in soil. Bacteria and fungi have always been recognised as the most numerically abundant members of the soil biota, but culture-independent molecular tools are now indicating that they are far more numerous and diverse than previously thought (Coleman 2008; Buée et al. 2009a, b). Our knowledge of the feeding habits of the microfauna (e.g. protozoa), mesofauna (e.g. rotifers, nematodes, tardigrades, collembolans, mites and enchytraeids) and macrofauna (e.g. earthworms, termites and millipedes) is also improving, and this is giving us a better insight into the numerous biotic interactions that occur within the soil environment, and how these interactions influence major ecosystem processes such as organic matter turnover and nutrient cycling. These issues are only covered briefly here, but further information is available in several comprehensive textbooks in soil microbiology (e.g. Tate 2000;

Davet 2004; Sylvia et al. 2005; Paul 2007; van Elsas et al. 2007) and in recent books on soil biology and ecology (e.g. Wardle 2002; Coleman and Crossley 2003; Bardgett 2005).

1.2.1 The Soil Food Web

The reason for interest in biological control of nematodes is that some plant-feeding nematodes are important pests, causing billions of dollars damage to the world's food and fibre crops. However, from an ecological perspective, this group of nematodes is simply one component of a large community of organisms that make up what is known as the soil food web. This community is sustained by the photosynthetic activity of plants, its food supply coming from roots, root exudates and plant-derived materials that either accumulate on the soil surface or become available when roots die. The primary consumers within the food web are bacteria, fungi, plant-feeding nematodes and root-grazing insects that feed directly on living plant roots, and the bacteria and fungi that decompose detritus. However, bacteria and fungi are by far the most important component of the soil food-web: they comprise most of the living biomass in soil and are primarily responsible for breaking down and mineralising organic compounds from plant tissue.

The resources transferred from plants and detritus to primary consumers do not remain locked up for very long because these organisms soon become food and energy sources for secondary consumers. Thus bacteria are consumed by nematodes and protozoa, fungal hyphae are pierced by stylet-bearing nematodes and then plant-feeding and free-living nematodes are parasitised by fungi or eaten by predators. These secondary consumers are eventually utilised by organisms at higher levels in the soil food web, while nutrients that are defecated, excreted or contained in dead bodies are also a resource for other organisms. Thus the soil food-web contains a complex array of interacting organisms with numerous pathways that transfer energy from producers (plants) to primary and secondary consumers. Since some of the resources available to the food web are lost at each trophic interchange due to respiration, detrital food chains do not continue indefinitely. They are generally limited in length to about five members (Coleman and Crossley 2003).

1.2.2 Functions of the Soil Food Web

The two most important functions of the soil food-web are to decompose plant material that enters the soil as litter and dead roots, and to mineralise the nutrients contained within that organic matter so that they can be re-used by plants. The decomposition process is mainly the result of microbial activity, but the soil fauna plays a role by fragmenting and ingesting organic matter, thereby increasing the surface area available for microbial colonisation. As plant material is decomposed, elements are converted from organic to inorganic forms that can be taken up by plants or used by microbes. This process is of critical importance in natural ecosystems (e.g. forests and grasslands), as almost all the nutrients required to sustain primary productivity are derived from mineralisation of soil humus and indigenous biomass. The soil food web also has many other important functions, as it regulates populations of plant pests and pathogens (discussed in the following section), immobilises nutrients within microbial biomass, sequesters carbon, detoxifies pollutants and stabilises soil aggregates.

1.2.3 Biotic Interactions Within the Soil Food-Web

The soil food-web contains huge populations of innumerable species and these populations are continually interacting with each other. These interactions become more complex as the diversity within the soil food-web increases, with multiple forces exerting pressures that prevent the uncontrolled proliferation of particular populations. Interactions between populations therefore have the effect of stabilising the community that makes up the food-web.

Given the complexity of the soil food-web, it is not surprising that populations interact in many different ways. Davet (2004) gives examples of the types of interaction that can occur, and most are relevant to a discussion of biological control.

Antibiosis is the inhibition of one organism by the metabolic product of another. It usually involves interactions where the adversary is killed or inhibited but is not consumed. The metabolic products (usually soluble or volatile antibiotics) are produced in such small quantities by bacteria or fungi that it is difficult to prove conclusively that they are present in the natural environment. Nevertheless, they are known to play a role in interactions between various plant pathogens and the soil biota, with one well-studied example being inhibition of the take-all pathogen *Gaeumannomyces graminis* var. *tritici* by two antibiotics (2,4-diacetylphloroglucinol and phenazine-1-carboxylic acid) produced by fluorescent pseudomonads on wheat roots (Weller et al. 2002).

Lysis is similar to antibiosis in that its effects are manifested at a distance from the organism responsible for lytic activity, but differs in that the adversary is exploited. It occurs when an organism produces extracellular enzymes (e.g. chitinases, cellulases and glucanases) that digest the cell wall or cuticle of another organism. Sometimes the process is accompanied by the production of toxins that immobilise or kill the prey. Bacteria, and more particularly actinobacteria, are significant producers of lytic enzymes and toxins, and important agents in the lysis of fungi.

Predation is generally characterised by the consumption or assimilation of one organism (the prey) by a larger organism (the predator). It requires intimate contact between the two organisms and usually involves an active search for the prey by the predator. Protozoans, nematodes and microarthropods all have the capacity to consume other soil organisms, some feeding indiscriminately on a wide range of organisms and others having quite specific food preferences. With respect to nematodes, predators of bacteria and fungi can be differentiated from predators of organisms further along the food chain by referring to the latter as 'top predators'.

Parasitism occurs when an organism (the parasite) lives in or on another organism (the host) and obtains all or part of its nutritional resources from that host. Bacteria and viruses are known to parasitise some soil organisms (e.g. protozoans and nematodes), but fungi are probably the most important parasitic organisms in soil. Numerous fungal parasites of arthropods and nematodes are known, and mycoparasitism (parasitism of one fungus by another) is also commonly observed.

Competition between organisms occurs when the amount of an essential substrate or nutrient is insufficient to satisfy the needs of both organisms. The organism most adept at accessing the limiting element, making it inaccessible to others or eliminating those trying to obtain it, will prosper relative to its competitors. Competition is a universal phenomenon within the soil food web, but becomes particularly intense when organisms in the same ecological niche are attempting to access the same scarce resource.

The word **antagonism** is often used instead of antibiosis to describe the situation where one organism inhibits another through antibiotic production. However, the term is used in a more general sense in this chapter to cover all situations where one organism (the pest) is detrimentally affected by the actions of other organisms. Such a definition is commonly used in the literature on biological pest control, as it is useful for describing the general suppressive effects of an organism on a pest, regardless of whether the antagonist is acting through parasitism, predation, antibiosis, competition or some other process.

Although the above mechanisms depict the types of interaction that occur between organisms in the soil food web, outcomes from these interactions are not easy to predict. Environmental factors have marked effects on relationships between organisms, while the interactions between two organisms will be modified by the introduction of a third organism. Thus the structure of a microbial community is the result of environmental effects and multiple interactions that are often quite difficult to comprehend.

1.2.4 Biotic Interactions in the Root Zone

The principal means by which plant roots impact on soil food webs is through the quality and quantity of organic matter that they return to soil. These carbon inputs are derived from fine roots (which have a relatively short life span and rapid turnover times), from cells that slough off as roots move through the soil, and from root exudates. Exfoliation and exudation from roots are particularly important processes because they contribute sugars, amino acids, mucilage and other materials that are high quality nutrient sources for rhizosphere microorganisms. Thus the area in the immediate vicinity of roots is a zone of intense biological activity and complexity

(Buée et al. 2009a). Since herbivores such as arthropods, plant-parasitic nematodes and pathogenic fungi also live in this zone, their activities are most likely to be influenced by organisms that are able to establish and maintain themselves in this extremely competitive ecological niche.

The surface of the root (often referred to as the rhizoplane) is a particularly important niche for soil microorganisms. Some of these organisms thrive in regions where exudation is most intense and protective mucilage is thickest, others survive saprophytically on senescent epidermal and cortical cells, and others are endophytes, colonising root cortical tissue and living in a symbiotic association with the plant. Mycorrhizal fungi are a well-known example of the latter association, as they receive carbon substrates from the plant and provide fungal-acquired nutrients to the plant. Since ramifying mycelial filaments affect soil structure and the mycorrhizal colonisation process improves plant growth, alters root morphology, changes exudation patterns and provides some protection against root pathogens, mycorrhizae influence the biotic interactions that occur in and near roots. Other symbiotic associations also add complexity to the soil-root interface. Examples include rhizobia and other bacteria that fix nitrogen in nodules on plant roots; plant growth promoting rhizobacteria that enhance seed germination and plant growth; and endophytic fungi that deter pests from feeding on plants or improve the plant's capacity to adapt to stress conditions.

1.3 Soil Ecology and Biological Control

The preceding discussion demonstrates that plant-parasitic nematodes cannot be considered in isolation from other components of the soil biological community. Their root-feeding habit brings them into contact with a vast number of root and rhizosphere-associated microorganisms and they also interact with numerous organisms in the detritus food web (Fig. 1.1). Additionally, the activities of plant-parasitic nematodes and other soil organisms are influenced, directly and indirectly, by various soil physical and chemical properties and by environmental factors such as temperature and moisture. These ecological realities must be recognised in any discussion of biological control.

One reason for opening this chapter with a general discussion of soil biology and ecology is to make the point that biological control is a normal part of a properly functioning soil ecosystem. Numerous soil organisms interact with nematodes and with each other and in that process they contribute to the regulatory mechanisms that maintain the stability of the soil food-web. Since plant-feeding nematodes become pests when these biological buffering processes are inadequate, biological control should be thought of as maintaining, restoring or enhancing the natural suppressive mechanisms that exist in all soils. Given that it may take months or years to arrive at a new 'balance' of interactions, the difficulties involved in shifting a stabilised system to a new equilibrium should not be underestimated.

Although most nematologists have some understanding of soil ecology, many fail to view biological control from an ecological perspective. Instead, biological

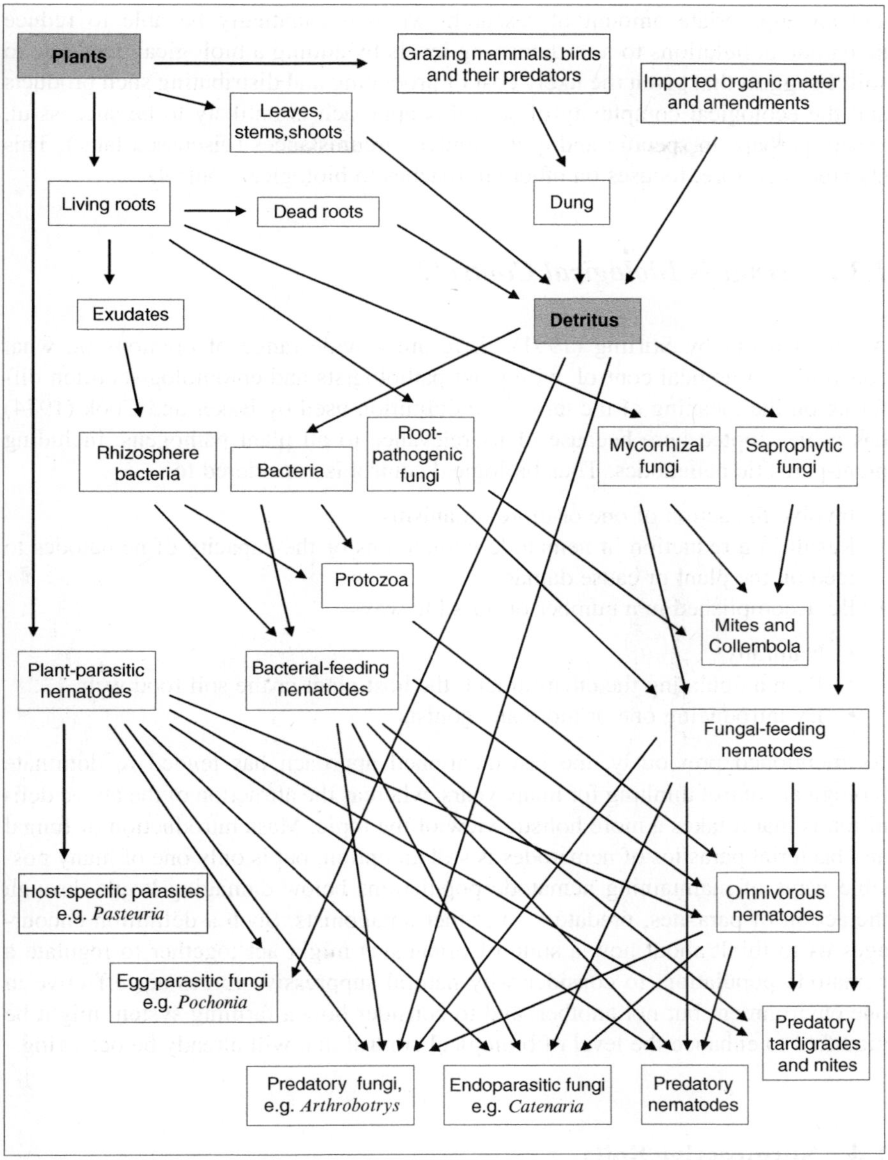

Fig. 1.1 Representation of a soil food web, showing the main interactions between plant-parasitic nematodes, some other primary consumers, and the detrital food web

control is thought of, in relatively simplistic terms, as the introduction of beneficial organisms to control a pest. Most farmers are no different. Having depended on soil fumigants and nematicides for many years, they consider that biological control is about replacing relatively toxic chemicals with safe biological products. Thus there is a common perception amongst both professionals and growers that given time and an appropriate amount of research, we will eventually be able to reduce nematode populations to non-damaging levels by adding a biological pesticide to soil. I suggest that given the likely cost of producing and distributing such products and the ecological complexity of soil, this approach is unlikely to be successful, except perhaps in specific and quite limited circumstances (discussed later). This chapter, therefore, focuses on other approaches to biological control.

1.3.1 What Is Biological Control?

As pointed out by Stirling (1991), there are a wide range of opinions on what constitutes biological control, with plant pathologists and entomologists often differing on the meaning of the term. The definition used by Baker and Cook (1974) has been adopted here because of its relevance to all plant pathogens, including plant-parasitic nematodes. Thus biological control is considered to:

- Involve the action of one or more organisms
- Result in a reduction in nematode populations or the capacity of nematodes to feed on the plant or cause damage
- Be accomplished in a number of possible ways:
  - Naturally
  - By manipulating the environment, the host plant or the soil food web
  - By introducing one or more antagonists

As mentioned previously, the last-mentioned approach has tended to dominate biological control thinking for many years, whereas the attraction of the above definition is that it takes a more holistic view of the topic. Mass introduction of fungal and bacterial parasites of nematodes is still an option, but is only one of many possible ways of maintaining nematode populations below damaging levels through the action of parasites, predators and other antagonists. Such a definition encourages us to think about how a suite of organisms might act together to regulate a nematode population, to consider why natural suppressive forces are effective in one environment but not another, and to consider how a farming system might be modified to enhance the level of biological control that will already be occurring.

1.4 Suppressive Soils

Soilborne pathogens debilitate roots or cause wilt, root-rot and damping-off diseases in most of the world's crops. Although these pathogens are widely distributed, there are situations where disease severity is lower than expected, given the prevailing environment and the level of disease in surrounding areas. In some of these cases, the indigenous microflora is the reason plants are effectively protected from the pathogen, a phenomenon that is known as disease-suppression. Books by Baker and

Cook (1974), Cook and Baker (1983), Hornby (1990) and Stirling (1991) summarise much of the early work in this area and discuss many examples of suppressiveness to nematodes and other soilborne pathogens.

Two types of disease suppressiveness can occur in agricultural soils. The most common (often referred to as 'general' or 'non-specific' suppressiveness) is found in all soils and provides varying degrees of biological buffering against most soil-borne pests and pathogens. Since the level of suppressive activity is broadly related to total soil microbial biomass and is therefore enhanced by practices that conserve or enhance soil organic matter, the term 'organic matter-mediated general suppression' is also commonly used (Hoitink and Boehm 1999; Stone et al. 2004). This type of suppression can be removed by sterilising the soil and is due to the combined effects of numerous soil organisms.

A second form of suppression (usually known as 'specific' suppressiveness) is also eliminated by sterilisation and other biocidal treatments but differs from general suppressiveness in that it results from the action of a limited number of antagonists. This type of suppression relies on the activity of relatively host-specific pathogens and can be transferred by adding small amounts of the suppressive soil to a conducive soil (Westphal 2005). Since specific suppression operates against a background of general suppressiveness (Cook and Baker 1983), the actual level of suppressiveness in a soil will depend on the combined effects of both forms of suppression.

1.4.1 Broad-Spectrum, Organic Matter-Mediated Suppression

The role of organic matter in enhancing suppression of soilborne diseases caused by fungi, Oomycetes, bacteria and nematodes has been known for many years and there are now well-documented examples in many quite different agricultural systems. These include suppression of *Pythium* in Mexican fields following the application of large quantities of organic matter over many years (Lumsden et al. 1987); broad-spectrum control of *Pythium*, *Phytophthora* and *Rhizoctonia* in peat and compost-based soilless container media (Hoitink and Boehm 1999); the use of cover crops, organic amendments and mulches to suppress *Phytophthora* root rot of avocado in Australia (Broadbent and Baker 1974; Malajczuk 1983; You and Sivasithamparan 1994, 1995); suppression of the same disease with eucalyptus mulch in California, USA (Downer et al. 2001); the management of a fungal, bacterial and nematode-induced root disease complex of potato in Canada with chicken, swine and cattle manures (Conn and Lazarovits 1999; Lazarovits et al. 1999, 2001), and the use of crop residues, animal manures and organic waste materials to reduce damage caused by plant-parasitic nematodes (reviewed by Muller and Gooch 1982; Stirling 1991; Akhtar and Malik 2000; Oka 2010).

It is obvious from the above examples that a wide range of types and sources of organic matter can be used to enhance suppressiveness and that they are effective in many different situations. However, studies (summarised by Hoitink and Boehm

1999 and Stone et al. 2004) in relatively simple nursery potting media have given us a much better understanding of the mechanisms involved. Suppression is generated soon after an amendment is added to soil and is associated with the activity of indigenous microorganisms that colonise organic material during the decomposition process. Development of suppression is associated with high levels of microbial activity, with many studies showing that the rate of hydrolysis of fluorescein diacetate (FDA) is a relatively good indicator of suppressiveness. Since microbial activity must remain high to maintain suppressiveness, the quantity and quality of the organic inputs have a major impact on the duration of suppressiveness. The labile constituents of organic matter (e.g. sugars, proteins and hemicelluloses) are degraded relatively quickly and suppression is then sustained by the subsequent decomposition of more recalcitrant materials in the coarse and mid-sized particulate fraction (Stone et al. 2001).

Perhaps the most important feature of organic-matter mediated general suppression is its capacity to act against most, if not all, major soilborne pathogens of food and fibre crops. Since root disease problems in the field rarely involve a single pathogen, enhancing the suppressive potential of a soil with organic matter is one of the only non-chemical techniques available to control a suite of pathogens. This does not mean that manipulating organic matter to manage several pathogens is a simple matter. When pathogens which are good primary saprophytes but poor competitors are involved (e.g. *Pythium* and *Fusarium*), the fact that they may multiply on fresh organic matter before being suppressed must be taken into account when designing application strategies. In the case of *Rhizoctonia*, which has a high competitive saprophytic ability due to its capacity to degrade cellulose as well as simple sugars, organic-matter mediated general suppression is often insufficient to achieve control and specific antagonists may also be required (Stone et al. 2004).

1.4.2 Suppressing Nematodes with Organic Amendments

It has been known for many years that animal manures, oil-cakes, residues from leguminous crops and other materials with a low C/N ratio can be added to soil to control plant-parasitic nematodes (see reviews by Muller and Gooch 1982; Rodriguez-Kabana 1986; Stirling 1991). Although there is some evidence that such amendments increase populations of microorganisms antagonistic to nematodes, the main mechanism is thought to be the release of nematicidal compounds such as ammonia during the decomposition process. Since relatively high concentrations of ammonia are needed to achieve control, there is a direct relationship between the amount of N in an amendment and its effectiveness (Rodriguez-Kabana 1986). Thus amendments with N contents greater than 2% are usually used and application rates are typically greater than 10 t/ha.

Although the nematicidal effects of ammonia are well established (Eno et al. 1955; Rodriguez-Kabana et al. 1982; Oka and Pivonia 2002; Tenuta and Ferris 2004) and lethal concentrations are achievable with nitrogenous amendments, the commercial use of such amendments is limited by cost and by concerns about the environmental impact of large quantities of nitrogen. Most recent studies have therefore sought to achieve efficacy at lower application rates. One successful approach involved adding a nitrification inhibitor (nitrpyrin) with the amendment to slow the oxidation of ammonia to nitrite and nitrate, therefore allowing ammonia concentrations to build up for an extended period. When the inhibitor was applied with chitin or cottonseed amendments, ammonia levels were higher for longer periods than in amended soils without the inhibitor, and this was associated with reduced egg production and galling from *Meloidogyne javanica* (Oka and Pivonia 2002). Alkaline additives have also improved the effectiveness of nitrogenous amendments by increasing soil pH and therefore shifting the equilibrium between the NH_4^+ and NH_3 to the latter form, which is nematicidal (Oka et al. 2006a).

Other work in the USA and Israel has shown that specially formulated organic amendments can cause nematode mortality through mechanisms other than ammonia production. De-watered municipal biosolids applied at 1.1% w/w did not affect *Heterodera glycines*, but the nematode was killed when the biosolids were stabilised with alkaline materials such as cement kiln dust, fly ash or quicklime (Zasada 2005). Nematode mortality was associated with a rapid increase in the pH of the soil solution (to a $pH > 10$), and this occurred when CaO in the amendment reacted with water to form $Ca(OH)_2$ (Zasada and Tenuta 2004; Zasada 2005). The contribution of ammonia production to the nematicidal effect was unclear in the American studies, but work with similar products in Israel suggested that it was important there (Oka et al. 2006b). However, the mechanism is clearly chemical rather than biological, as experiments with autoclaved materials indicated that microbes associated with the amendment were not involved (Zasada 2005).

Whether it will eventually be possible to use nitrogenous amendments in nematode management programs remains a moot point. Enormous quantities of organic and inorganic wastes and industrial by-products are available in most countries and there is a need to find uses for them as commercial fertilisers and soil conditioners. Alkaline-stabilised organic amendments are effective against plant-parasitic nematodes, but fine tuning will be needed before they can be used routinely in nematode management. Thus there is a need to determine the application rates required to achieve consistent nematode control; develop methodologies to prevent overproduction of ammonia and ensure that pH does not increase excessively; understand the long-term effects of these amendments on soil physical properties, soil chemistry and soil microbial ecology; and find ways of integrating the practice into the soil and crop management programs used for specific nematode-susceptible crops.

Although most recent research on organic amendments for nematode control has focused on nitrogenous materials, the possibility of using materials with a much higher C/N ratio has also received attention. McSorley and Gallaher (1995) used a composted mixture of sticks, leaves, branches, grass clippings and wood chips from the urban environment (C/N ratio = 36) as an amendment or mulch and found that it had little effect on plant-parasitic nematodes in vegetable crops planted immediately after the amendment was applied. However, in another study that continued for 3 years, population densities of plant-parasitic nematodes on maize were reduced in the third season, once the woody compost material had broken down and levels of soil organic matter had increased in amended plots (McSorley and Gallaher 1996).

Three studies in Australia have also shown that amendments with minimal amounts of N have suppressive effects on nematodes. In the first of these studies, apple trees mulched with sawdust for 5 years had much lower populations of *Pratylenchus jordanensis* in years 2–5 than non-mulched trees or trees growing in fumigated or nematicide-treated plots. In years 4 and 5, yields from mulched trees were as good as those obtained with methyl bromide fumigation (Stirling et al. 1995). A second study in which tomato was planted into field plots that had been amended over the previous 2 years with sawdust and urea showed that the amended soil was highly suppressive to *M. javanica* and that the level of nematode control was significantly better than that obtained with the nematicide fenamiphos. Plants in amended plots were almost free of galls, whereas the untreated controls were heavily galled (Vawdrey and Stirling 1997). The third study involved an amendment of sugarcane residue (the tops and leaves remaining in the field after sugarcane is mechanically harvested). Sugarcane was planted 23 weeks after the residue was incorporated into soil, and 24 weeks after planting there were 95% fewer lesion nematodes (*Pratylenchus zeae*) in roots growing in amended soil than in roots from the non-amended control (Stirling et al. 2005).

Results from these studies indicate that amendments with a high C/N ratio act much more slowly than nitrogenous amendments. When soil is amended with materials such as yard waste, sawdust or sugarcane residue, suppressiveness may take months or years to develop whereas it develops almost immediately when the amendment has a relatively high N content. Interestingly however, suppressiveness is soon lost with nitrogenous amendments. Thus when soil was assessed 4 and 7 months after it was amended with N-rich materials (e.g. lucerne hay, feedlot manure, poultry manure, chitin and a waste product from sugar mills known as mill mud), it was not suppressive to *M. javanica* or *P. zeae* (Stirling et al. 2003). In contrast, materials with a much higher C/N ratio (e.g. sawdust, sugarcane residue and grass hay) were suppressive to both nematodes.

Although the suppressiveness generated by high C/N amendments has not been studied in detail, the evidence currently available suggests that physical or biological rather than chemical mechanisms are responsible. Relatively large predators (e.g. nematodes and arthropods) may be able to operate more effectively when soil structure is improved with organic matter, while in the Australian studies discussed previously, one common observation was that fungi appeared to be associated in some way with suppressiveness. For example, a suppressive, sawdust-amended soil had high numbers of fungal-feeding nematodes (Vawdrey and Stirling 1997), while low concentrations of nitrate nitrogen, a fungal-dominant soil biology and high numbers of omnivorous nematodes were associated with suppression in one of the other experiments (Stirling et al. 2003). In an experiment where *P. zeae* was suppressed after soil was amended with sugarcane residue, an unidentified predatory fungus was found in the amended but not the non-amended soil (Stirling et al. 2005). It is therefore possible that fungal predation on nematodes was responsible for these suppressive effects. The predatory hyphomycetes and several genera of wood-decaying basidiomycetes are commonly found in habitats that are rich in cellulose and lignin and are thought to have evolved the capacity to scavenge for additional N in low N environments by preying on nematodes (Barron 1992; Tzean and Liou 1993). Thus when high C/N amendments are added to soil, these fungi may utilise free-living nematodes as a food source and coincidently capture plant-parasitic species.

1.4.3 Farming Systems to Enhance General Suppressiveness

Although amending soil with high rates of organic matter can generate suppressiveness to nematodes and other soilborne pathogens and maintain it for some time after the amendment is applied, it is important to recognise that this approach to disease control is likely to be most useful in high value horticultural production systems. Nurseries where plants are grown in containers, glasshouses producing vegetable or ornamental crops and intensive in-field production of crops with a high monetary value are perhaps the only situations where it is realistic to use amendments in this way to manage nematodes. In all other agricultural systems, applying organic matter at rates of 10–100 t/ha/annum is never likely to be economically feasible. Importation of organic matter will generally be expensive relative to the income derived from most crops, largely because transportation costs are high and non-agricultural markets compete for the resource. Since high application rates are required to achieve the desired effects, there is also the potential for environmental problems from the nitrogen, heavy metals and other potential pollutants that may be present in the amendment. Thus for all the world's staple food and fibre crops, organic-matter mediated general suppression will mainly be achieved by developing farming systems that increase C inputs and conserve soil organic matter. Almost all soil and crop management practices affect the levels of soil organic matter, but perhaps the most important are crop rotation, cover cropping, crop residue management, organic amendments and tillage (Magdoff and Weil 2004). They are therefore the main tools that can be used to improve a soil's physical, chemical and biological status and therefore influence its capacity to suppress soilborne pests and pathogens.

Since accumulation of organic matter is directly related to C inputs (Paustian et al. 1997), reducing the frequency and duration of bare fallow periods and including perennial forages, high residue crops and cover crops within the farming system are the most practical ways of minimising the decline in soil organic C that occurs in all cropping systems. Careful management of above and below-ground plant residues also has a place, particularly in cropping systems where most of the above-ground material is harvested. Organic amendments may also be useful, but successive inputs at low application rates are likely to be more economically, agronomically and environmentally desirable than occasional inputs at high application rates. When used collectively, these practices are the first step towards increasing levels of soil organic matter and enhancing the suppressiveness of field soils to nematodes and soilborne diseases.

The second step involves reducing tillage, as conventional tillage arguably causes greater losses of soil organic matter than any other farm management practice (Magdoff and Weil 2004). In comparison to cultivated soils, non-tilled soils are less susceptible to erosion losses caused by water or wind, and decomposition also proceeds more slowly because crop residues remain on the soil surface rather than being mixed with the soil. Non-tilled soils are also cooler and subject to less pronounced wetting and drying cycles, both of which reduce rates of microbial respiration and organic matter decomposition. A compilation of studies from the literature (Franzluebbers 2004) indicates that soil under no tillage accumulates organic C to a greater extent than under inversion tillage, and that this effect is seen for both particulate organic matter and the more labile C fractions on which heterotrophic soil organisms depend. It is therefore not surprising that reducing tillage produces profound changes in the detritus food web, the most obvious impact being favourable effects on larger organisms such as predatory and omnivorous nematodes, mites, enchytraeids, earthworms, beetles and spiders (Wardle 1995).

When appropriate crop rotations, reduced tillage, residue retention, more frequent cover cropping and regular inputs of animal manures and organic wastes are integrated into a farming system, they are a powerful combination of practices that will result in improved soil and ecosystem health. Their widespread adoption in many industries in recent years is testimony to the benefits obtained. Although enhanced suppression of plant-parasitic nematodes will never be the primary reason for such improvements to a farming system, recent work on sugarcane in Australia suggests that it is one of the benefits that will accrue. Damage caused by *M. javanica* and *P. zeae*, the most important nematode pests of sugarcane, has been reduced by introducing a rotation crop and implementing residue retention and minimum tillage to enhance natural biological control mechanisms that suppress these pests (Stirling 2008). Although such suppressiveness is likely to take years to reach its full potential, particularly in farming systems where biomass production is limited by low rainfall, it is nevertheless worth pursuing because it comes with numerous other soil health benefits (e.g. improved nutrient cycling, better soil structure, increased water and nutrient holding capacity and broad-spectrum disease suppression) that are crucial for the long-term sustainability of a cropping system (Weil and Magdoff 2004). From the perspective of nematodes, future studies within improved farming systems should concentrate on establishing the levels of soil organic matter required to achieve suppression, understanding the regulatory mechanisms involved, and determining how the quality, quantity and timing of organic inputs influences the development of suppressiveness.

1.4.4 Specific Suppression of Soilborne Pathogens

There are many situations where soilborne diseases caused by fungi, bacteria or nematodes are suppressed by pathogen-specific agents. Historically, the best-documented examples for nematodes are the suppression of *Heterodera avenae* in a cereal monoculture by two fungi, *Nematophthora gynophila* and *Pochonia chlamydosporia*, and the multiplication of *Pasteuria penetrans* in some cropping systems to levels that suppress root-knot nematodes. Both examples were discussed in detail by Stirling (1991).

In the last two decades, other examples of natural suppression due to *P. penetrans* have been reported (Weibelzahl-Fulton et al. 1996) and suppressiveness has been transferred from one field to another (Kariuki and Dickson 2007). The role of other *Pasteuria* species as suppressive agents has also been recognized, with Noel et al. (2010) demonstrating that when *P. nishizawae* is introduced into a non-suppressive field soil, it induces suppressiveness to soybean cyst nematode (*H. glycines*).

Another important contribution to our understanding of nematode-suppressive soils has been a decade-long investigation (reviewed by Borneman et al. 2004 and Borneman and Becker 2007) on the development of suppressiveness to *H. schachtii* in a field that had been cropped intensively with hosts of the nematode. After a period when populations of *H. schachtii* were high and disease incidence was severe, nematode populations declined to such an extent that studies commenced on the causes of the phenomenon. Work with various biocides (Westphal and Becker 1999) and experiments in which suppression was transferred to a conducive soil using either soil or cysts (Westphal and Becker 2000, 2001) showed that the suppressiveness was biological in nature and prompted studies of the microflora associated with nematode cysts and eggs. This work showed that eggs from the field were frequently parasitised by fungi and that *Brachyphoris* (syn. *Dactylella*) *oviparasitica*, *Fusarium oxysporum*, other *Fusarium* spp., *Paecilomyces lilacinus* and various unidentified fungi could be isolated on agar media (Westphal and Becker 2001).

The above investigations showed that fungi were associated with suppressiveness and subsequent studies demonstrated that modern technologies were useful for identifying the key suppressive organisms. Soils with different levels of suppressiveness were created with biocides or by combining different amounts of suppressive and conducive soil and oligonucleotide fingerprinting of rRNA genes (OFRG) was used to identify the main fungal phylotypes associated with different levels of suppression (Yin et al. 2003). The main phylotype in the most suppressive treatments had high sequence identity to rRNA genes from various nematode destroying fungi. Subsequent analyses indicated that the fungus represented by this phylotype was most closely related to *Brachyphoris oviparasitica*, a parasite of *Meloidogyne* eggs that had previously been found to suppress this nematode in California peach orchards (Stirling and Mankau 1978; Stirling et al. 1979). A second phase of the study validated this result, with sequence-selective quantitative PCR assays showing that the largest amounts of *B. oviparasitica* PCR product came from soils possessing the highest levels of suppressiveness to *H. schachtii* (Yin et al. 2003). In phase three of the study, *B. oviparasitica* was added to fumigated soil and produced the same high level and long-term suppressiveness that was observed in the naturally suppressive soil (Olatinwo et al. 2006a, b, c).

Other recent studies indicate that when field soils are surveyed systematically for suppression using appropriate techniques, examples of specific suppressiveness to plant-parasitic nematodes are often found. Thus suppressiveness to *Rotylenchulus reniformis* was detected in cotton fields in Texas and Louisiana USA (Robinson et al. 2008), while another study showed that one of six California soils was suppressive to *M. incognita* (Bent et al. 2008). In the latter work, a negative correlation between *P. chlamydosporia* rRNA genes and nematode population densities suggested that this fungus may have been one of the major factors responsible for suppressiveness.

There are many examples of specific suppressiveness to fungal and bacterial pathogens, and work in this area has been discussed by numerous authors, including Baker and Cook 1974; Cook and Baker 1983; Hornby 1990; Schippers 1992; Whipps 1997; Alabouvette 1999; Weller et al. 2002 and Mazzola 2004, 2007. As in the examples cited for nematodes, suppression develops in situations where a pathogen increases to high population densities, causes severe disease and then declines spontaneously to levels that do not cause damage. Take-all decline of wheat is perhaps the best-known example and it is encouraging to note that after years of research on the microbial antagonists involved, there are now situations where growers can be confident that suppressiveness will be maintained and disease losses will be negligible (Weller et al. 2002).

1.5 Mass Release of Biological Control Agents

The possibility of introducing mass-produced antagonists into soil or establishing them on seeds or roots has been a major component of research on biological control of soilborne pathogens for several decades. However, any objective review of that research would have to conclude that there have been relatively few practical outcomes. By 2005, only nine bacteria and five fungi were registered with the United States Environment Protection Agency for control of soilborne diseases (Fravel 2005). Of these organisms, strains of *Agrobacterium* are known to be effective against crown gall, but it is not known whether the others are efficacious in the hands of the consumer. Worldwide, the number of biological products is greater, but as in the United States, most are formulations of the fungi *Gliocladium* and *Trichoderma* or the bacteria *Pseudomonas* and *Bacillus*, and many are marketed as plant growth promoters, plant strengtheners or soil conditioners rather than as biocontrol agents (Paulitz and Belanger 2001). The only organism listed by Fravel (2005) as registered in the USA for nematode control was a non-biological product consisting of killed mycelium and fermentation materials from *Myrothecium verrucaria*. However, since that time, a strain of *Paecilomyces lilacinus* (Melancon®, Bioact®) has been commercialised in the USA, some parts of Europe and several other countries for use against cyst and root-knot nematodes.

In considering the types of organism most likely to have potential for development as biocontrol agents, Deacon (1991) pointed out that host specificity and the capacity to operate in the same ecological niche as the target pathogen were attributes that were required to achieve success. With regard to antagonists of nematodes, endospore-forming bacteria in the genus *Pasteuria* fit these criteria, as they are specific parasites of most economically important plant-parasitic nematodes (Sturhan 1988; Sayre and Starr 1988; Ciancio et al. 1994; Chen and Dickson 1998). Recent advances in the *in vitro* culture of some members of the genus (Hewlett et al. 2004; Gerber et al. 2006) indicate that mass production by liquid fermentation is possible, thereby opening opportunities for commercial exploitation of the parasite. Initial work is being done with Candidatus *Pasteuria usgae* (Giblin-Davis et al. 2003), a parasite of sting nematode (*Belonolaimus longicaudatus*) and is focused on control of the nematode on golf courses and athletic fields in south-eastern USA (Hewlett et al. 2008).

Although host specificity is clearly advantageous to a biological control agent and also limits impacts on non-target organisms, it remains to be seen whether the extreme specificity of *Pasteuria* will limit its commercial usefulness. Thus in *P. penetrans*, for example, spores do not attach to all populations of the *Meloidogyne* species from which they are obtained, indicating that host preference is determined at a population rather than species level (Stirling 1985). Later studies have shown that *P. penetrans* produces heterogeneous sub-populations of endospores that show preferences for particular nematode populations (Davies et al. 1994; Davies and Redden 1997). Thus variability in *P. penetrans* may be a host-adaptive process that allows endospores to attach to and infect the nematodes present in a given environment. Any biological control program involving the mass culture of *P. penetrans* will therefore have to consider host specificity issues when deciding which bacterial strains are to be produced for a particular market.

Pasteuria clearly has potential as a mass-produced biological control agent but another host-related issue that requires consideration is whether it will be equally effective against all its known hosts. Examples of long-term natural suppression due *Pasteuria* have largely been confined to root-knot and cyst nematodes, presumably because millions of spores are produced in saccate females and inputs from these infected nematodes are sufficient to maintain relatively high spore concentrations in an environment where losses are always occurring due to predation and percolation. Thus, when sedentary endoparasitic nematodes are being targeted, spore populations should increase naturally, and this will limit the number mass-produced spores needed, or the number of applications required, to achieve satisfactory control. However, the same level of natural increase may not occur with vermiform nematodes, as fewer spores are produced and this limits the spore population densities achievable in soil. Low spore production in infected nematodes may have been one of the reasons that *P. usgae* did not always suppress populations of sting nematode (*Belonolaimus longicaudatus*) to acceptable levels, despite relatively high levels of parasitism (Giblin-Davis et al. 1990). Thus, when *Pasteuria* is used against ectoparasitic and migratory endoparasitic nematodes, it may be necessary to regularly supplement natural populations of the parasite with spores produced *in vitro*.

One way of ensuring that a biocontrol agent is capable of operating in the same ecological niche as the target nematode is to concentrate on organisms that naturally inhabit the rhizosphere. Initial studies with rhizosphere-inhabiting bacteria showed that some isolates were antagonistic to plant-parasitic nematodes (Becker et al. 1988; Kloepper et al. 1991, 1992; Kluepfel et al. 1993; Oka et al. 1993), while more recent work has focused on fluorescent pseudomonads and strains of *Bacillus* that have the capacity to enhance plant growth and induce disease resistance (Weller et al. 2002; Haas and Keel 2003; Kloepper et al. 2004). Some of these bacteria have given broad-spectrum protection against soilborne pathogens (Jetiyanon et al. 2003) and in tests on vegetable crops, they consistently increased plant growth and sometimes reduced galling caused by root-knot nematode (Kokalis-Burelle et al. 2002a, b). Since bacteria are relatively easy to apply to transplants, they may eventually find a place in the vegetable industry within integrated management programs for nematodes and other soilborne pathogens.

Given that fungi capable of parasitising females and eggs of endoparasitic nematodes must come into intimate contact with the target nematode to utilise them as a food source, it is hardly surprising that they are commonly found in the rhizosphere (Stirling 1979; Bourne et al. 1996). However, there have been relatively few behavioural studies of this group of fungi in this intensely competitive environment. In the case of *P. chlamydosporia*, root colonising ability is known to be important in bringing the fungus in contact with nematode eggs (DeLeij and Kerry 1991), but a capacity to colonise sites where nematodes are present may be an even more important attribute. *P. chlamydosporia* is abundant on roots infected by root-knot nematodes (De Leij et al. 1992; Bourne et al. 1996; Atkins et al. 2009) and populations increase markedly when egg masses are extruded on the galled root surface (Bourne et al. 1996), suggesting that specificity towards sedentary endoparasitic nematodes is associated in some way with a capacity to recognise the quantitative and qualitative changes in root exudation patterns that occur following nematode infection (Wang and Bergeson 1974).

Although Gaspard and Mankau (1986) were able to isolate several species of nematode-trapping fungi from the root surface, little is known about the capacity of these fungi to form traps and prey on nematodes in the rhizosphere. Persson and Jansson (1999) found that differences in the root colonising ability of nematode-trapping fungi did not explain differences in their capacity to reduce damage caused by root-knot nematode. However, it may be premature to conclude that these fungi do not prey on nematodes in the rhizosphere, as some species probably perform better in this environment than others. Also, we know little about where traps are produced in relation to the root surface, and we lack the tools required to monitor the intensity and timing of trap production. If we are to ever understand the predatory behaviour of this group of fungi at the soil/root interface, these issues must be addressed.

Given the difficulties involved in establishing an introduced organism in the extremely competitive rhizosphere environment, one approach that has received increasing attention in recent years is the possibility of using endophytic organisms for biocontrol purposes. The advantage of endophytes is that they occur in the same ecological niche as endoparasitic nematodes but are not subject to competition from microorganisms in the soil and rhizosphere. With regard to endophytic organisms for nematode control, most recent work with fungi has focused on strains of *Fusarium oxysporum* that reduce infection and reproduction of *Radopholus similis* (Athman et al. 2007) and *M. incognita* (Hallman and Sikora 1994; Dababat and Sikora 2007). Endophytic bacteria have received less attention but are of interest because they act in much the same way as the plant growth-promoting rhizobacteria mentioned previously (Compant et al. 2005). Their suppressive mechanisms have not been fully elucidated, but those commonly proposed include competition with the pathogen for an ecological niche or substrate, production of inhibitory allelochemicals and induction of systemic resistance (Hallman and Sikora 1996; Compant et al. 2005; Vu et al. 2006; Franco et al. 2007).

Since endophytic microorganisms enable plants to adapt to stress conditions and are a potential source of metabolites for the pharmaceutical industry (Maheshwari 2006), they will be the subject of increasing attention in coming years. From the perspective of biological control of nematodes, endophytes should be relatively easy to apply as inoculants to seed or seedlings and can therefore be established in the root system before nematodes are attracted to roots and begin to feed. The future challenge is to find strains that are active against nematodes, show that these organisms can be established in appropriate niches within roots, and demonstrate that they are efficacious enough to warrant inclusion in integrated management programs for nematodes.

Mononchids and stylet-bearing dorylaimids are often observed in the soil environment, but their usefulness as mass-produced biological control agents is limited by their long life cycles and low fecundity, and an inability to culture them in large quantities. Predatory nematodes in the Diplogastrida are not seen as frequently but may be a better alternative. Diplogastrid predators are much easier to mass produce, they show some specificity towards their prey and can survive periods of low prey density by feeding on bacteria (Bilgrami et al. 2005); all useful attributes for a biological control agent. Recent work with two diplogastrids in the genus *Mononchoides* has shown that *M. fortidens* reduces damage caused by root-knot nematode in pots while *M. gaugleri* decreases total populations of plant-parasitic nematodes in turf grass (Khan and Kim 2005; Bilgrami et al. 2008).

1.6 Directions for Future Research

It should be apparent from the preceding discussion that our understanding of biological control systems as they apply to nematodes has improved markedly in the last 30 years. However, biological control has still not taken the step from 'potentially useful management option' to 'reliable and effective control measure'. The following is a personal opinion of what needs to be done to ensure that in future, biological control contributes in a much greater way to integrated management systems for nematodes.

1.6.1 Developing More Sustainable Farming Systems

One of the biggest changes to world agriculture in the last 30 years has been the development of no-till farming. Various forms of conservation tillage are now applied to many millions of hectares of cropped land, and when combined with practices such as crop rotation and cover cropping, it has resulted in farming systems that are much more profitable and sustainable than they were in the past. One of the benefits from this change will be an increase in the suppressiveness of soils to soilborne disease (Stone et al. 2004).

Given the economics of broad-scale agriculture and the cropping area involved, enhancing general suppressiveness through the farming system is probably the only realistic way of improving the level of biological control in most of the world's agricultural land. The role of farming systems in enhancing suppressiveness should therefore be a major focus of future research. We need to know how the main soil management practices available to farmers (e.g. tillage, fallowing, rotation crops, cover cropping and organic inputs from crop residues and amendments) influence the physical, chemical and biological properties of soil and in turn affect the development of suppressiveness to various pathogens, including nematodes.

A comprehensive review by Wardle (1995) demonstrates that tillage practices have a major impact on the detritus food web and could therefore be expected to affect the processes that regulate populations of plant-parasitic nematodes. The quantity of C and N (the resource base for the detritus food web) is usually lower under conventional tillage than no-tillage, microbial biomass and the ratio of microbial biomass to organic C tends to decline when soil is tilled and bacteria tend to be favoured over fungi. The larger soil organisms (predatory and omnivorous nematodes, springtails and mites) are particularly vulnerable to tillage and all tend to respond positively when tillage is reduced. Given that fungi, predatory nematodes and microarthropods are the main predators of nematodes and tillage is detrimental to all of them, a move from conventional to minimum tillage could be expected to enhance the general suppressiveness of soil to plant-parasitic nematodes. Observations on cereals and sugarcane in Australia (Rovira 1990; Stirling 2008) and results from long-term tillage experiments with soybean in the USA (Westphal et al. 2008; Seyb et al. 2008) indicate that populations of several plant-parasitic nematodes are lower in soils under minimum tillage than in cultivated soils. Although this effect is not necessarily due entirely to enhanced suppressiveness, detailed ecological studies of these and other no-till systems are clearly warranted.

In addition to reducing the frequency and intensity of tillage, practices such as crop rotation, cover cropping, more careful residue management and greater organic inputs from amendments can also be used by farmers to improve levels of soil organic matter and thereby influence the biological status of soil and its general suppressiveness to nematodes. The role of organic matter in enhancing suppressiveness is discussed in the following section, but from a farmer's perspective, the challenge is to integrate these practices into a farming system that is profitable and sustainable. The way this is done will depend on factors such as climate, soil type and the principal crops involved, but results of a recent research program in Australia provide an example of what is achievable.

In the early 1990s, the Australian sugar industry was facing an uncertain future because productivity was declining due to a problem known as yield decline. At that time, sugarcane was grown on beds 1.5 m apart, machinery wheel spacings did not match crop row spacings and the crop residues remaining after harvest were often burnt rather than retained. After a plant and 2–4 ratoon crops, an expensive program of ripping and cultivation was required to remove the old crop, alleviate compaction caused by farm machinery and then replant the field to sugarcane. A multidisciplinary research team was established to develop solutions to the problem and its initial studies showed that soils under long-term sugarcane monoculture were physically and chemically degraded, while large yield responses to soil fumigation and nematicides indicated that biological constraints were also limiting productivity. A 12-year research program (summarised by Garside et al. 2005; Stirling 2008) resulted in the development of a new farming system based on residue retention, minimum tillage, a leguminous rotation crop and controlled traffic using global positioning system guidance. This system is now being adopted by growers because it increases sugar yields, reduces costs, improves soil health and provides additional income from rotation crops such as soybean and peanut. From a nematological perspective, losses from *P. zeae* and *M. javanica* have been reduced because (1) the introduction of a rotation crop has reduced nematode population densities at planting, (2) damage thresholds have increased as soil health has improved and (3) suppressive mechanisms of biological control are now operating more effectively.

Economic pressures and the entrenched attitude of some growers will always make it difficult to make major changes to a farming system. However, the fact that the Australian sugar industry was able to make such a change and in the process overcome obstacles that were initially perceived as insurmountable, indicates that the task is achievable. Reducing losses from nematodes and other soil-borne pathogens may not be the primary reason for embarking on such a process, but is likely to be one of the outcomes.

Globally, the farming system that is perhaps in most need of urgent attention from a farming systems perspective is the plasticulture system commonly used for vegetable production. In many countries, vegetable crops are grown intensively on beds mulched with plastic film; water, nutrients and pesticides are delivered to soil via trickle irrigation tubing; double or multiple cropping is common; soil is bare-fallowed between crops; there is limited crop rotation; organic inputs from cover crops and amendments are rare; and soil is routinely fumigated. This farming system treats the soil as an inert medium to support the plant, and in the absence of any biological buffering, it is not surprising that root-knot nematode and other soil-borne pathogens re-establish following fumigation and quickly build up to high population densities (Desaeger and Csinos 2006). It is therefore disappointing that over the last decade or so, much of the money allocated to finding alternatives to methyl bromide was spent on testing alternative fumigants rather than on developing more sustainable vegetable farming systems. There are production systems that warrant further testing (e.g. Stirling 2008; Stirling and Eden 2008; Bhan et al. 2010), but until the vegetable industry is prepared to take a long-term view, invest in research on alternative farming systems and then persist with those alternatives for 5–10 years, the status quo will remain.

1.6.2 Understanding the Impact of Soil Organic Matter on Suppressiveness

Organic matter has profound effects on many important soil physical and chemical properties (e.g. soil aggregation, soil water availability and nutrient cycling); it promotes biological activity and diversity through affects on the detritus food web; and it plays a key role in developing healthy soils and enhancing their suppressiveness to plant pathogens and pathogenic nematodes (Weil and Magdoff 2004; Magdoff and Weil 2004). Since levels of soil organic matter gradually decline when plant biomass is continually removed as harvested product rather than being returned to the soil, measures which provide additional C inputs and minimise C losses due to microbial respiration and erosion must always be a component of management programs for cropped soils. Retention of crop residues that would otherwise be burnt or taken off-farm, crop rotation, cover cropping, organic amendments and minimum tillage are the main options available, and where practicable, they should be used together to increase the amount of soil organic matter, improve soil health and reduce the impact of soilborne diseases (Stone et al. 2004).

Since Linford's initial work in the 1930s, there have been numerous studies on the role of organic inputs in enhancing suppressiveness to plant-parasitic nematodes (see reviews by Muller and Gooch 1982; Stirling 1991; Akhtar and Malik 2000; Widmer et al. 2002). However, the results of many of these studies cannot be readily extrapolated to the field because they focused on the relatively short-term effects of amendments when applied at rates that are unrealistically high for broadscale agriculture. There is therefore an urgent need to study the medium and long-term biological changes that take place when soil organic matter is conserved and enhanced in ways that are feasible to introduce into a farming system, and understand how they affect the development of suppressiveness. We need to measure parameters such as total and labile C, microbial activity and biological diversity and relate them to suppressiveness; identity the key groups of organisms involved in suppression; understand how they are affected by the quantity and quality of C inputs; and then use the information to find better ways of manipulating organic matter within a farming system to enhance suppressiveness.

Since plant-parasitic nematodes are particularly damaging when populations are high during crop establishment, understanding the temporal effects of crop and soil management practices on the development of suppression is important, as it may then be possible to maximise suppressiveness during the period when crops are being planted. This could perhaps be achieved by altering tillage practices or by selecting rotation crops on the basis of the C/N ratio of their residues or the relative proportion of labile to more recalcitrant compounds in the plant material. In situations where it is possible to include practices known to be beneficial to the soil biology within a cropping system (e.g. an undisturbed pasture ley), the management practices used during the transition back to cropping are likely to have a major impact on whether suppressiveness is maintained or lost.

The main energy channels within the detritus food web are either bacterial or fungal, and soil ecologists suggest that the dominant channel in natural systems is largely determined by litter quality and the environment (Bardgett 2005; Wardle 2005). However, in agroecosystems, soil management practices also determine whether energy flow occurs rapidly through bacterial channels or more slowly through fungal channels. Tillage and nitrogen fertilisation practices have particularly profound effects on the soil food web, stimulating bacterial activity to such an extent that bacterial rather than fungal decomposers predominate in many farming systems. Changes in the proportions of bacteria to fungi and their flow-on effects to other components of the soil food web, together with the detrimental effects of tillage (Wardle 1995) and nitrogen (Tenuta and Ferris 2004) on some predators may explain why soils tend to lose their natural suppressiveness once they are cropped. These issues need to be further explored, but they also raise questions about how agricultural soils should be managed to maintain suppressiveness. Do biological mechanisms of suppression operate in highly-disturbed, nitrogen-enriched and bacterially-dominant soils, and if so, how can their activity be enhanced? What is the impact of N fertilisation practices on various parasites and predators of nematodes, and is it possible to provide a crop with adequate N without detrimental effects on the organisms responsible for suppression? Will minimum tillage and surface mulching increase the activity of fungi and other organisms that parasitise or prey on nematodes? Will the effects of such practices be apparent only in surface layers or will they also occur at depth?

Because of a paucity of research on the biological processes that operate within the root zone of agricultural crops, the list of unanswered questions about interactions between soil and crop management practices, organic matter status and parasitism and predation on plant-parasitic nematodes is almost endless. Unfortunately, the ecological literature provides few answers. Nematode ecologists and soil biologists often refer to the 'top down' or predatory processes that regulate nematode populations but usually use the term in a general sense and rarely attempt to identify the organisms responsible. When predation is specifically mentioned, mononchid and Dorylaimid nematodes are often considered to be the main predators of nematodes, and other natural enemies (e.g. nematophagous fungi and arthropods) are usually ignored. Another problem is that predator-prey relationships in soil are poorly understood, particularly in situations where predators have a range of food options available to them (Small 1987). Thus we do not know, for example, whether mononchid predators can be sustained in their natural habitat by ingesting bacteria, protozoans and other soil organisms, or whether they live mainly on certain groups of nematodes. Studies in simple microcosms provide useful information on feeding habits (e.g. Bilgrami and Gaugler 2005; Bilgrami et al. 2005) but we also need to know what predators eat when given a choice in their natural environment. Some dorylaimids are known to consume nematode eggs in agar culture, but are eggs an important food source in the natural environment? Entomologists are using molecular techniques to identify organisms in the gut contents of predators (Symondson 2002; King et al. 2008) and similar approaches could perhaps be used to elucidate the food preferences of nematophagous organisms.

Fungi are perhaps the most important parasites and predators of nematodes, but the impact of organic matter on their predatory activity is poorly understood. Jaffee's recent work with nematode-trapping fungi is therefore an important contribution because it sought to clarify whether these fungi are associated with suppression in organically-amended soils. Briefly, these studies showed that suppression of root-knot nematode was positively correlated to microbial biomass but was not related to management system (organic v. conventional) or to fungal population density (Jaffee et al. 1998). Also, the two fungi tested (*Arthrobotrys oligospora* and *Dactylellina haptotyla*) did not necessarily respond in the same the same way to organic amendments (Jaffee 2004). One of these species (*A. oligospora*) responded to the addition of substrates with relatively low C:N ratios and high N contents but failed to trap nematodes, (Jaffee 2003, 2004; Jaffee et al. 2007; Nguyen Vi et al. 2007), raising questions as to why it invests resources in producing specialised hyphae capable of capturing nematodes.

The role of organic matter in influencing the trapping behaviour of nematode-trapping fungi has perplexed nematologists for many years, and advances in this area would provide vital clues to how organic matter can be better managed to enhance biological control. Currently, there are two models to explain how organic matter stimulates predatory activity (Jaffee et al. 1998). The numerical response model presupposes that nematode-trapping fungi are obligate parasites that are dependent on nematodes for carbon, nitrogen and energy. They therefore respond to the addition of organic matter by consuming the free-living nematodes which multiply on the microorganisms involved in the decomposition process. The supplemental nitrogen model presupposes that the fungi are facultative parasites and obtain nitrogen from nematodes to enable them to compete for energy in carbon-rich/ nitrogen-poor plant litter. Although both nutritional models probably occur within the nematode-trapping fungi, it would be useful to know which model predominates in particular soil types, cropping systems or environments, as this would provide clues to how organic inputs could be managed to maximise trapping activity.

Clearly, there is much more to be learnt about the ecology of the nematode-trapping fungi. However, ecological studies are difficult to undertake because procedures for quantifying these fungi are tedious, their detection efficiency is largely unknown, trap production cannot be quantified and there is not always a consistent relationship between fungal population density and trapping activity (Jaffee 2003). Techniques that could be used to monitor traps would therefore be particularly useful in ecological studies, and could possibly be developed by targeting genes or gene products involved in trap production (Ahrén et al. 2005). When such technologies are combined with the sequencing and genomic techniques being used to study fungal plant pathogens (Xu et al. 2006) and methods likely to become available in the field of transcriptomics, it may eventually be possible to understand the factors which cause nematode-trapping fungi to switch from a saprophytic to parasitic mode of nutrition.

1.6.3 *Identifying Nematode-Suppressive Soils*

Most agricultural soils are highly disturbed and their organic matter status has declined following many years of cultivation. Since organic matter is the resource that sustains the soil food web, the biodiversity of most soils has been depleted, often to such an extent that the mechanisms regulating populations of plant-parasitic nematodes are not effective enough to prevent them from becoming pests. This lack of biological complexity is manifested in the fact that the nematode community in many agricultural soils is dominated by nematodes with short generation times and relatively high reproductive rates (i.e. plant-parasitic and microbivorous species). Such nematode assemblages are indicative of relatively simple, non-structured food webs, whereas suppressive mechanisms are most likely to operate in soils that have complex food webs with long food chains and many trophic links (Jaffee et al. 1998; Berkelmans et al. 2003; Sánchez-Moreno and Ferris 2007). Suppressiveness is associated with the prevalence of omnivorous and predatory nematodes, but the ratio of predators to prey is also important (Sánchez-Moreno and Ferris 2007). Thus the best way of finding soils likely to be generally suppressive to plant-parasitic nematodes is to identify situations where there is a structured nematode community containing a range of omnivorous and predatory species. Nematode communities of this nature are most likely to be found in farming systems where crops (particularly perennials) are grown continually; there is no disturbance due to tillage; broad-spectrum biocides are not used; and inputs of synthetic fertilisers are not excessive.

Although the presence of omnivorous and predatory nematodes can be used as an indicator of general suppressiveness, this does not necessarily mean that they are the primary suppressive agents. Their presence simply indicates that a relatively complex soil food web is present and that it is likely to contain a range of nematophagous fungi, arthropods and other organisms that will also be contributing to regulatory processes. A challenge of the future is find better ways of quantifying these organisms and monitoring their predatory activities in both suppressive and conducive soils.

Plant nematologists usually focus on areas where nematodes cause problems, but locating suppressive soils requires a different mindset. Field observations must be made with the intention of finding situations where the environment is suitable for a particular nematode but population densities remain low in the presence of a susceptible host. Such situations may be quite localised and difficult to find, but could possibly be identified more easily using techniques in precision agriculture to generate data on biomass or yield variability within fields (Melakeberhan 2002; Srinivasan 2006). Such data could then be linked to high throughput, DNA-based systems for quantifying nematode populations (Ophel-Keller et al. 2008).

A recent study (Robinson et al. 2008) provides and good example of how previously unrecognised suppressiveness can be detected. Comprehensive surveys of cotton fields in the USA had previously demonstrated that reniform nematode (*Rotylenchulus reniformis*) occurred at relatively high population densities in most fields. However, some fields had inexplicably low nematode populations while others had much lower population densities in surface soils that expected. Results of assays in pots then showed that there was a biological reason for these differences in nematode distribution, raising questions about why suppressiveness built up in some soil types or environments, or whether it was enhanced by particular farming practices.

Once candidate soils are identified, a range of techniques can be used to verify suppressiveness and confirm its biological nature (Westphal 2005). One of the most common is to treat the soil with a biocide (often a fumigant or heat), re-inoculate with the nematode and check for differences in nematode multiplication rates in biocide-treated and untreated soil. Another frequently-used method, which is most useful when the suppressive agent(s) have relatively short life cycles and therefore multiply readily, is to transfer small quantities of the test soil to heat-treated or fumigated soil and demonstrate that the transfer reduces nematode multiplication or results in high levels of parasitism or predation on nematodes. However, a weakness of such methods is that plants often do not grow as well in field soil as in partially or fully-sterilized soils and a reduction in the number of feeding sites may confound detection of suppressiveness. An alternative approach is to eliminate plants from the test system. Heated and untreated soil is inoculated with a nematode that is not present naturally in the test soil and nematode mortality is measured following incubation in the laboratory (Jaffee et al. 1998; Pyrowolakis et al. 2002; Sánchez-Moreno and Ferris 2007). Although this assay avoids problems caused by the use of plants, it focuses exclusively on suppressive forces that affect the migratory stages of a nematode's life cycle. Thus the best way of confirming suppressiveness is to demonstrate its occurrence using more than one method.

The organisms associated with suppression have traditionally been determined using cultural, biochemical and microscopic methods, but molecular techniques and other technologies will become increasingly important in future. When used to characterise soil microbial communities, these tools provide new insights into the identity, diversity and functional capacities of microorganisms involved in suppressing soil-borne pathogens (Weller et al. 2002; Mazzola 2004; Garbeva et al. 2004; van Elsas et al. 2008). With regard to nematodes, molecular technologies have contributed to our understanding of specific suppression and will eventually be used to quantify and track both the nematode and its suppressive agents (Borneman and Becker 2007). Ultimately, this will allow us to understand how agronomic practices influence the development of suppressiveness.

1.6.4 Maintenance of Suppressiveness

Once soils suppressive to a particular nematode pest have been identified and characterised, the next challenge is to understand how they are best managed to maintain suppressiveness. Specific suppression is dependent on the presence of the host nematode and usually manifests itself in situations where nematode populations have remained at high levels for many years. Thus the use of nematode-resistant varieties, fallowing and other practices that reduce nematode populations to very low levels may be an impediment to the development and maintenance of specific suppressiveness. Cropping sequences that use tolerant or partially resistant cultivars may be a better option, as they would minimise crop damage while maintaining a food source for the target nematode and its suppressive agents. A comparable strategy for maintaining specific suppressiveness in perennial cropping systems would be to grow nematode-susceptible cover crops in situations where nematode-resistant or tolerant rootstocks are available.

Given the economic importance of the genera *Meloidogyne* and *Heterodera* and the fact that parasitism is relatively easy to study because infective stages of these nematodes are sedentary and their eggs are aggregated, it is not surprising that most of the research on specific suppressiveness has focused on this group of nematodes. Obligate parasites of females (e.g. *Pasteuria penetrans* and *Nematophthora gynophila*) and saprophytic fungi with a relatively specialised capacity to utilise nematode eggs as a food source (e.g. *Pochonia chlamydosporia* and *Brachyphoris oviparasitica*) sometimes parasitise a large proportion of the females or eggs, but future research needs to focus on why these suppressive forces are active in some situations and not others. The factors most likely to be involved include the continuity of supply of host nematodes; particular soil physical, chemical or environmental conditions; soil organic matter status and the genetic makeup of the suppressive agent.

Since root-knot and cyst nematodes have relatively short generation times, high reproductive capacities and relatively low damage thresholds, one shortcoming of host-specific parasites is that populations of the target pest are only reduced when levels of parasitism are high, largely because the nematodes killed by the parasite are often in excess of those required to maintain high population densities. Another potential deficiency is that some host-specific parasites (e.g. *Pasteuria*) do not always prevent the nematode from feeding, while others only act after feeding has occurred (e.g. egg-parasitic fungi). These parasites may therefore have little or no impact on crop damage. In such situations, the challenge is to find ways of maintaining high levels of specific suppressiveness while integrating other management tactics into the farming system.

The key to maintaining general suppressiveness is to sustain a soil food web with enough activity and complexity to prevent plant-parasitic nematodes from becoming predominant. Techniques for analysing nematode assemblages are now readily available (Neher and Darby 2009; Ferris and Bongers 2009) and can be used to indicate whether a soil food web is complex enough to provide the desired suppressive services. Sanchez-Moreno and Ferris (2007) provided an example of how this might be done when they showed that suppressiveness was related to the prevalence of omnivores and predators. Given that the composition of the soil food web is dependent on the quality and quality of C inputs, another way of addressing this issue might be to improve our understanding of the relationship between soil C and suppressiveness. By measuring one or more of the many forms of C in soil, it may be possible to define, in a particular soil type and environment, the soil C status required to achieve adequate suppressiveness.

With the move towards minimum till farming systems in the last 20–30 years, one area that requires more research is the role of organic matter that is retained on the soil surface rather than incorporated, in enhancing suppressiveness. Mulches and surface residues from previous crops not only improve the environment for roots and soil organisms by minimising moisture and temperature fluctuations, but also provide the C inputs required to maintain a suppressive soil food web. Soil mulched with residues from a sugarcane crop was more suppressive to plant-parasitic nematodes than non-mulched soil (Stirling 2008), while C inputs from decomposing residues appeared to be the main reason that sugarcane roots immediately below the trash blanket were healthier and had fewer plant-parasitic nematodes than roots further down the profile (Stirling et al. 2011). The next step is to determine whether the level of suppressiveness is related to the quantity or quality of the organic matter left behind after a crop is harvested.

1.6.5 Monitoring Biocontrol Agents in Soil

Many different groups of organisms are known to parasitise or prey on nematodes, but one of the main problems in working with biological control systems is the difficulty of detecting and quantifying some of these groups in soil. This applies particularly to the predatory and parasitic fungi. Nematode-trapping fungi can be quantified using sprinkle plates and soil dilution plates, but these time-consuming methods tend to detect species that grow well in culture and their efficacy is affected by the bait nematode used, soil type, moisture content and laboratory conditions. Also, estimates of fungal population density do not always correlate with trapping activity (Jaffee 2003; Smith and Jaffee 2009). The situation is even worse with parasitic fungi. Thus with *Hirsutella rhossiliensis*, for example, nematodes must be extracted from soil and examined on agar plates for signs of parasitism (Jaffee et al. 1991) or a suitable assay nematode must be found and checked for adhering conidia (McInnis and Jaffee 1989). This means that nematologists generally report only those predators that can be readily recovered from soil. Although suppressiveness to plant-parasitic nematode is positively related to the prevalence of omnivore and predator species, the lack of a strong relationship between these groups of nematodes suggests that other components of the soil food web are contributing to the regulatory process (Sánchez-Moreno and Ferris 2007).

In recent years the phylogeny and systematics of the Orbiliales (the group of ascomycetes containing most of the nematode-trapping fungi) has been revised using molecular techniques (Ahrén et al. 1998; Hagedorn and Scholler 1999; Scholler et al. 1999; Li et al. 2005; Chen et al. 2007a, b, c). A recent paper by Smith and Jaffee (2009) demonstrates that such techniques are also useful for ecological studies in soil and other substrates. Orbiliales-specific PCR primers for the ITS and 28 rDNA detected many uncultured Orbiliales that were closely related to nematode-trapping fungi and fungal parasites of nematode eggs, suggesting that molecular methods will provide a fuller picture of the nematophagous fungal community than culture-based methods alone. However, there were discrepancies between the results of molecular and culture-based studies that need to be followed up, and this process is likely to improve our understanding of the ecological role of this diverse group of parasitic and predaceous fungi.

In the case of bacterial parasites in the genus *Pasteuria*, the number of spores attached to the host nematode has been used as an indicator of spore concentration in soil (Stirling et al. 1990). However, such bioassays are not entirely satisfactory, as the relationship between spore concentration and the number of attached spores is affected by factors that affect nematode motility. Molecular assays to detect and quantify endospores in soil (Atibalentja et al. 2008) offer the opportunity to directly monitor the parasite in soil and could also be used to improve our understanding of its population dynamics.

One area where it is particularly important to monitor populations of fungal and bacterial parasites and predators is when they are mass-produced and used as biological control agents. The capacity of the introduced organism to come into contact with its target nematode and also survive in the extremely competitive soil environment is vital information from an ecological perspective. In future, data of this nature will largely be obtained using molecular technologies. One recent example is the use of species-specific primers to detect *Paecilomyces lilacinus* in soil and estimate the proportion of eggs infected by the fungus (Atkins et al. 2005). A real-time PCR primer and probe set also provided a method of detecting populations of *P. lilacinus* as low as 10 spores/g soil. These and other similar methodologies provide a new set of tools to assess the impact of various environmental and crop management factors on the spatial and temporal population dynamics of particular biological control agents and it is important that they are now employed in ecological studies.

1.6.6 Developing Biocontrol Products for Targeted Markets

As pointed out by many who have worked on biological control of soilborne pathogens, two major factors limit the potential of inoculants as a control strategy: (1) in most cropping systems, it is uneconomic to mass produce an organism and add it to soil in amounts sufficient to control a pathogen and (2) the buffering effect of the microbial community (which is responsible for the general suppressiveness of soils to pathogens) operates against a biological control agent once it is introduced into soil. These economic and ecological realities must therefore be recognised when deciding whether mass production and release of a biological control agent is a realistic nematode management strategy.

From an economic perspective, it is unreasonable to expect a biological pesticide to be cheaper than a chemical product. Biological control agents cannot be mass produced without a fermentation facility and an appropriate substrate; formulation costs are likely to be relatively high and the specificity of most biocontrol agents limits economies of scale within the production, distribution and marketing process. Furthermore, some organisms require controlled conditions during transport and storage, and this imposes additional costs. Thus applying a biological control agent to soil for nematode control is only likely to be feasible in situations where nematicides are currently the main control tactic. It will never be a realistic option for broad-acre crops (e.g. cereals, grains, oilseeds, cotton and sugarcane), and for most tree and vine crops. Future research should therefore focus on situations where monetary losses from nematodes are high enough to justify the use of a mass-produced biological product.

Root-knot nematode is a major pest of crops grown in glasshouses and other protective structures, and is an obvious target of such research for a number of reasons. First, the nematode causes problems on a global scale; the crops grown in glasshouses are relatively high in value, and the cost of nematode control with fumigants and nematicides is already an accepted component of production costs. Second, the soil environment (particularly moisture and temperature) can be reasonably well controlled; while the highly modified state of glasshouse soils (due to practices such as fumigation and intensive tillage) may mean that they are amenable to maintaining an introduced organism in the root zone throughout the life of the crop. Third, biological products can be applied within protective structures in a number of relatively simple ways (e.g. as a seed inoculants, seedling dips, soil drenches or additives to transplant mixes). I therefore suggest that this cropping system should be used as a test case by nematologists to see whether inundative biological control can be developed to the point where it is a realistic alternative to chemical control. What is needed is a coordinated effort to put our current knowledge of biological control into practice. About 30 years ago, the International Meloidogyne Program made a major contribution to worldwide knowledge of root-knot nematodes (Sasser and Carter 1985; Barker et al. 1985), and a similar program on biological control of these nematodes in glasshouse crops would provide an opportunity to move biological control from the laboratory to the market place. Numerous potentially useful biocontrol agents are available (*Paecilomyces lilacinus*, *Pochonia chlamydosporia*, *Pasteuria penetrans*, various nematode-trapping fungi, a number of readily-cultured diplogastrid predators, a range of plant growth-promoting rhizobacteria and several endophytes), and the aim would be to apply them in an integrated manner to achieve reliable and effective nematode control.

Whether the above research program is initiated or not, one disadvantage of mass production and release as a biological control strategy is that once an organism is applied to roots or soil, it is subject to the rigours of the environment. Efficacy of biological products will therefore be much more subject to environmental influences than the chemical nematicides that they are intended to replace. Thus research teams working with biological control agents should not consider that their job is done when a commercial partner is found and a formulated product is placed on the market. Many years of follow-up research is likely to be required to define the situations where a product will give reliable and effective control. Lists of registered products tend to imply that progress is being made, but the ultimate criterion for success is consistent results and widespread acceptance in the target market.

1.7 Concluding Remarks

Although it is easy to be disheartened by the lack of practical outcomes from biological control research in the last 30 years, it would be wrong to conclude that biological control cannot be developed to the point where it makes a significant contribution to integrated management programs for nematodes. Robust and durable systems of natural suppression are almost certainly operating in some fields, but they need to be sought out and the contributing factors identified, so that farming systems can then be modified to enhance suppressive mechanisms. Nematologists are contributing significantly to our understanding of the soil ecosystem, and since biological control is little more than applied soil biology and microbial ecology, these strengths must now be utilised to better understand the forces that regulate nematode populations and how they can be better deployed against plant-parasitic nematodes.

Research managers and individual scientists also need to recognise that not all impediments to progress are technical. The fragmentation of science into disciplines means that is often difficult for nematologists, plant pathologists, soil ecologists, molecular biologists and agronomists to work together, even though inputs from specialists in all these areas are required to better understand biotic interactions in the root zone, and to apply our collective knowledge to enhancing the suppressive potential of agricultural soils. The recent shift in resources from traditional areas of science into biotechnology continues a long-term trend towards specialisation that must be handled carefully. Intractable problems in biological control need to be tackled with new technologies, but without ecological and agronomic input, the desired outcomes are not likely to be achieved.

Acknowledgements I thank Gregor Yeates and Keith Davies for their comments on the manuscript, and my wife Marcelle for her constant support over many years.

References

Ahrén D, Ursing BM, Tunlid A (1998) Phylogeny of nematode-trapping fungi based on 18 S rDNA sequences. FEMS Microbiol Lett 158:179–184

Ahrén D, Tholander M, Fekete C et al (2005) Comparison of gene expression in trap cells and vegetative hyphae of the nematophagous fungus *Monacrosporium haptotylum*. Microbiology 151:789–803

Akhtar M, Malik A (2000) Role of organic soil amendments and soil organisms in the biological control of plant-parasitic nematodes. Bioresour Technol 74:35–47

Alabouvette C (1999) Fusarium wilt suppressive soils: an example of disease-suppressive soils. Australas Plant Pathol 28:57–64

Athman SY, Dubois T, Coyne D et al (2007) Effect of endophytic *Fusarium oxysporum* on root penetration and reproduction of *Radopholus similis* in tissue culture-derived banana (*Musa* sp.) plants. Nematology 9:599–607

Atibalentja N, Babadoost M, Noel GR (2008) A real-time PCR assay for the detection of *Pasteuria nishizawae* in soil. Phytopathology 98(Supplement):S15

Atkins SD, Clark IM, Pande S et al (2005) The use of real-time PCR and species-specific primers for the identification and monitoring of *Paecilomyces lilacinus*. FEMS Microbiol Ecol 51:257–264

Atkins SD, Peteira B, Clark IM et al (2009) Use of real-time quantitiative PCR to investigate root and gall colonisation by co-inoculated isolates of the nematophagous fungus *Pochonia chlamydosporia*. Ann Appl Biol 155:143–152

Baker KF, Cook RJ (1974) Biological control of plant pathogens. WH Freeman & Co., San Francisco

Bardgett R (2005) The biology of soil: a community and ecosystem approach. Oxford University Press, Oxford

Barker KR, Carter CC, Sasser JN (1985) An advanced treatise on *Meloidogyne*: volume 2, Methodology. North Carolina State University Graphics, Raleigh

Barron GL (1992) Lignolytic and cellulolytic fungi as predators and parasites. In: Carroll GC, Wicklow DT (eds) The fungal community, its organization and role in the ecosystem. Marcel-Decker, New York

Becker JO, Zavaleta-Mejia E, Colbert SF et al (1988) Effects of rhizobacteria on root-knot nematodes and gall formation. Phytopathology 78:1466–1469

Bent E, Loffredo A, McKenry MV et al (2008) Detection and investigation of soil biological activity against *Meloidogyne incognita*. J Nematol 40:109–118

Berkelmans R, Ferris H, Tenuta M et al (2003) Effects of long-term crop management on nematode trophic levels other than plant feeders disappear after 1 year of disruptive soil management. Appl Soil Ecol 23:223–235

Bhan M, McSorley R, Chase CA (2010) Effect of cropping system complexity on plant-parasitic nematodes associated with organically grown vegetables in Florida. Nematropica 40:53–70

Bilgrami AL, Gaugler R (2005) Feeding behaviour of the predatory nematodes *Laimydorus baldus* and *Discolaimus major* (Nematoda: Dorylaimida). Nematology 7:11–20

Bilgrami AL, Gaugler R, Brey C (2005) Prey preference and feeding behaviour of the diplogastrid predator *Mononchoides gaugleri* (Nematoda: Diplogastrida). Nematology 7:333–342

Bilgrami AL, Brey C, Gaugler R (2008) First field release of a predatory nematode, *Mononchoides gaugleri* (Nematoda: Diplogastrida), to control plant-parasitic nematodes. Nematology 10:143–146

Borneman J, Becker JO (2007) Identifying microorganisms involved in specific pathogen suppression in soil. Annu Rev Phytopathol 45:153–172

Borneman J, Olatinwo R, Yin B et al (2004) An experimental approach for identifying microorganisms involved in specified functions: utilisation for understanding a nematode suppressive soil. Australas Plant Pathol 33:151–155

Bourne JM, Kerry BR, De Leij FAAM (1996) The importance of the host plant on the interaction between root-knot nematodes (*Meloidogyne* spp.) and the nematophagous fungus *Verticillium chlamydosporium* Goddard. Biocontrol Sci Technol 6:539–548

Broadbent P, Baker KF (1974) Behaviour of *Phytophthora cinnamomi* in soils suppressive and conducive to root rot. Aust J Agric Res 25:121–137

Buée M, de Boer W, Martin F et al (2009a) The rhizosphere zoo: an overview of plant-associated communities of microorganisms, including phages, bacteria, archea, and fungi, and some of their structuring factors. Plant Soil 321:189–212

Buée M, Reich M, Murat C et al (2009b) 454 pyrosequencing analyses of forest soils reveal an unexpectedly high fungal diversity. New Phytol 184:449–456

Chen ZX, Dickson DW (1998) Review of *Pasteuria penetrans*: biology, ecology and biological control potential. J Nematol 30:313–340

Chen J, Xu L-L, Liu B et al (2007a) Taxonomy of *Dactylella* complex and Vermispora I. Generic concepts based on morphology and ITS sequences data. Fungal Divers 26:73–83

Chen J, Xu L-L, Liu B et al (2007b) Taxonomy of *Dactylella* complex and Vermispora II. The genus Dactylella. Fungal Divers 26:85–126

Chen J, Xu L-L, Liu B et al (2007c) Taxonomy of *Dactylella* complex and Vermispora III. A new Brachyphoris and revision of Vermispora. Fungal Divers 26:127–142

Ciancio A, Bonsignore R, Vovlas N et al (1994) Host records and spore morphometrics of *Pasteuria penetrans* group parasites of nematodes. J Invertebr Pathol 63:260–267

Coleman DC (2008) From peds to paradoxes: linkages between soil biota and their influences on soil ecological processes. Soil Biol Biochem 40:271–289

Coleman DC, Crossley DA (2003) Fundamentals of soil ecology. Academic, Burlington

Compant S, Duffy B, Nowak J et al (2005) Use of plant growth-promoting bacteria for biocontrol of plant diseases: principles, mechanisms of action, and future prospects. Appl Environ Microbiol 71:4951–4959

Conn KL, Lazarovits G (1999) Impact of animal manures on *Verticillium* wilt, potato scab, and soil microbial populations. Can J Plant Pathol 21:81–92

Cook RJ, Baker KF (1983) The nature and practice of biological control of plant pathogens. American Phytopathological Society, St. Paul

Dababat AEA, Sikora RA (2007) Induced resistance by the mutualistic endophyte, *Fusarium oxysporum* strain 162, toward *Meloidogyne incognita* on tomato. Biocontrol Sci Technol 17:969–975

Davet P (2004) Microbial ecology of the soil and plant growth. Science Publishers Inc., Enfield

Davies KG, Redden M (1997) Diversity and partial characterization of putative virulence determinants of *Pasteuria penetrans*, the hyperparasitic bacterium of root-knot nematodes (*Meloidogyne* spp.). J Appl Microbiol 83:227–235

Davies KG, Redden M, Pearson TK (1994) Endospore heterogeneity in *Pasteuria penetrans* related to adhesion to plant-parasitic nematodes. Lett Appl Microbiol 19:370–383

De Leij FAAM, Kerry BR, Dennehy JA (1992) The effect of fungal application rate and nematode density on the effectiveness of *Verticillium chlamydosporium* as a biological control agent for *Meloidogyne incognita*. Nematologica 38:112–122

Deacon JW (1991) Significance of ecology in the development of biocontrol agents against soil-borne diseases. Biocontrol Sci Technol 1:5–20

DeLeij FAAM, Kerry BR (1991) The nematophagous fungus *Verticillium chlamydosporium* as a potential biological control agent for *Meloidogyne arenaria*. Rev Nematol 14:157–164

Desaeger JA, Csinos AS (2006) Root-knot nematode management in double-cropped plasticulture vegetables. J Nematol 38:59–67

Downer AJ, Menge JA, Pond E (2001) Association of cellulytic enzyme activities in *Eucalyptus mulches* with biological control of *Phytophthora cinnamomi*. Phytopathology 91:847–855

Eno CF, Blue WG, Good JM Jr (1955) The effect of anhydrous ammonia on nematodes, fungi, bacteria, and nitrification in some Florida soils. Proc Soil Sci Soc Am 19:55–58

Ferris H, Bongers T (2009) Indices developed specifically for analysis of nematode assemblages. In: Wilson MJ, Kakouli-Duarte T (eds) Nematodes as environmental indicators. CAB International, Wallingford, pp 124–145

Franco C, Michelsen P, Perry N et al (2007) Actinobacterial endophytes for improved crop performance. Australas Plant Pathol 36:524–531

Franzluebbers AJ (2004) Tillage and management effects on soil organic matter. In: Magdoff F, Weil RR (eds) Soil organic matter in sustainable agriculture. CRC Press, Boca Raton, pp 227–268

Fravel DR (2005) Commercialization and implementation of biocontrol. Annu Rev Phytopathol 43:337–359

Garbeva P, van Veen JA, van Elsas JD (2004) Microbial diversity in soil: selection of microbial populations by plant and soil type and implications for disease suppressiveness. Annu Rev Phytopathol 42:243–270

Garside AL, Bell MJ, Robotham BG et al (2005) Managing yield decline in sugarcane cropping systems. Int Sugar J 107:16–26

Gaspard JT, Mankau R (1986) Nematophagous fungi associated with *Tylenchulus semipenetrans* and the citrus rhizosphere. Nematologica 32:359–363

Gerber JF, Hewlett TE, Smith KS et al (2006) Materials and methods for in vitro production of bacteria.US Patent 7,067,299 B2

Giblin-Davis RM, McDaniel LL, Bilz FG (1990) Isolates of the *Pasteuria penetrans* group from phytoparasitic nematodes in Bermudagrass turf. J Nematol 22(supplement):750–762

Giblin-Davis RM, Williams DS, Bekal S et al (2003) '*Candidatus* Pasteuria usgae' sp. nov., an obligate endoparasite of the phytoparasitic nematode *Belonolaimus longicaudatus*. Int J Syst Evol Microbiol 53:197–200

Haas D, Keel C (2003) Regulation of antibiotic production in root-colonising *Pseudomonas* spp. and relevance for biological control of plant disease. Annu Rev Phytopathol 41:117–153

Hagedorn G, Scholler M (1999) A reevaluation of predatory orbiliaceous fungi. 1. Phylogenetic analysis using rDNA sequence data. Sydowia 51:27–48

Hallman J, Sikora RA (1994) Influence of *Fusarium oxysporum*, a mutualistic fungal endophyte, on *Meloidogyne incognita* infection of tomato. J Plant Dis Prot 101:475–481

Hallman J, Sikora RA (1996) Toxicity of fungal endophyte secondary metabolites to plant-parasitic nematodes and soil-borne plant pathogenic fungi. Eur J Plant Pathol 102:155–162

Hewlett TE, Gerber JF, Smith KS (2004) In vitro culture of *Pasteuria penetrans*. Nematol Monogr Perspect 2:175–185

Hewlett TE, Waters JP, Luc JE et al (2008) Field studies using in vitro produced *Pasteuria* endospores to control sting nematodes on turf. In: Abstracts, Fifth International Congress of Nematology, vol 180, Brisbane, 2008

Hoitink HAJ, Boehm MJ (1999) Biocontrol within the context of soil microbial communities: a substrate-dependent phenomenon. Annu Rev Phytopathol 37:427–446

Hornby D (1990) Biological control of soil-borne plant pathogens. CAB International, Wallingford

Jaffee BA (2003) Correlations between most probable number and activity of nematode-trapping fungi. Phytopathology 93:1599–1605

Jaffee BA (2004) Do organic amendments enhance the nematode-trapping fungi *Dactylellina haptotyla* and *Arthrobotrys oligospora*? J Nematol 36:267–275

Jaffee BA, Muldoon AE, Anderson CE et al (1991) Detection of the nematophagous fungus *Hirsutella rhossiliensis* in California sugar beet fields. Biol Control 1:63–67

Jaffee BA, Barstow JL, Strong DR (2007) Suppression of nematodes in a coastal grassland soil. Biol Fertil Soils 44: 19–26

Jaffee BA, Ferris H, Scow KM (1998) Nematode-trapping fungi in organic and conventional cropping systems. Phytopathology 88:344–350

Jetiyanon K, Fowler WD, Kloepper JW (2003) Broad-spectrum protection against several pathogens by PGPR mixtures under field conditions in Thailand. Plant Dis 87:1390–1394

Kariuki GM, Dickson DW (2007) Transfer and development of *Pasteuria penetrans*. J Nematol 39:55–61

Khan Z, Kim YH (2005) The predatory nematode, *Mononchoides fortidens* (Nematoda: Diplogastrida), suppresses the root-knot nematode, *Meloidogyne arenaria*, in potted field soil. Biol Control 35:78–82

King RA, Read DS, Traugott M, et al. (2008) Molecular analysis of predation: a review of best practice for DNA-based approaches. Mol Ecol 17:947–963

Kloepper JW, Rodríguez-Kábana R, McInroy JA et al (1991) Analysis of populations and physiological characterization of microorganisms in rhizospheres of plants with antagonistic properties to phytopathogenic nematodes. Plant Soil 136:95–102

Kloepper JW, Rodríguez-Kábana R, McInroy JA et al (1992) Rhizosphere bacteria antagonistic to soybean cyst (*Heterodera glycines*) and root knot (*Meloidogyne incognita*) nematodes: identification by fatty acid analysis and frequency of biological control activity. Plant Soil 139:75–84

Kloepper JW, Ryu C-M, Zhang S (2004) Induced systemic resistance and promotion of plant growth by *Bacillus* spp. Phytopathology 94:1259–1266

Kluepfel DA, McInnis TM, Zehr EI (1993) Involvement of root-colonizing bacteria in peach soils suppressive of the nematode *Criconemella xenoplax*. Phytopathology 83:1240–1245

Kokalis-Burelle N, Martinez-Ochoa N, Rodriguez-Kabana R et al (2002a) Development of multicomponent transplant mixes for suppression of *Meloidogyne incognita* on tomato (*Lycopersicon esculentum*). J Nematol 34:362–369

Kokalis-Burelle N, Vavrina CS, Rosskopf EN et al (2002b) Field evaluation of plant growth-promoting rhizobacteria amended transplant mixes and soil solarization for tomato and pepper production in Florida. Plant Soil 238:257–266

Lazarovits G, Conn KL, Potter J (1999) Reduction of potato scab, verticillium wilt, and nematodes by soymeal and meat and bone meal in two Ontario potato fields. Can J Plant Pathol 21:345–353

Lazarovits G, Tenuta M, Conn KL (2001) Organic amendments as a disease control strategy for soilborne diseases of high-value agricultural crops. Australas Plant Pathol 30:111–117

Li Y, Hyde KD, Jeewon R et al (2005) Phylogenetics and evolution of nematode-trapping fungi (Orbiliales) estimated from nuclear and protein encoding genes. Mycologia 97:1034–1046

Lumsden RD, Garcia ER, Lewis JA et al (1987) Suppression of damping-off caused by *Pythium* spp. in soil from the indigenous Chinampa agricultural system. Soil Biol Biochem 19:501–508

Magdoff F, Weil RR (2004) Soil organic matter management strategies. In: Magdoff F, Weil RR (eds) Soil organic matter in sustainable agriculture. CRC Press, Boca Raton, pp 45–65

Maheshwari R (2006) What is an endophytic fungus? Curr Sci 90:1309

Malajczuk N (1983) Microbial antagonism of Phytophthora. In: Erwin DC, Bartnicki-Garcia S, Tsao PH (eds) Phytophthora: its biology, taxonomy, ecology and pathology. American Phytopathological Society, St. Paul

Mazzola M (2004) Assessment and management of soil microbial community structure for disease suppression. Annu Rev Phytopathol 42:35–59

Mazzola M (2007) Manipulation of rhizosphere bacterial communities to induce suppressive soils. J Nematol 39:213–220

McInnis TM, Jaffee BA (1989) An assay for *Hirsutella rhossiliensis* spores and the importance of phialides for nematode inoculation. J Nematol 21:229–234

McSorley R, Gallaher RN (1995) Effect of yard waste compost on plant-parasitic nematode densities in vegetable crops. J Nematol 27:545–549

McSorley R, Gallaher RN (1996) Effect of yard waste compost on nematode densities and maize yield. J Nematol 28:655–660

Melakeberhan H (2002) Embracing the emerging precision agriculture technologies for site-specific management of yield-limiting factors. J Nematol 34:185–188

Muller R, Gooch PS (1982) Organic amendments in nematode control. An examination of the literature. Nematropica 12:319–326

Neher DA, Darby BJ (2009) General community indices that can be used for analysis of nematode assemblages. In: Wilson MJ, Kakouli-Duarte T (eds) Nematodes as environmental indicators. CAB International, Wallingford, pp 107–123

Nguyen Vi L, Bastow JL, Jaffee BA et al (2007) Response of nematode-trapping fungi to organic substrates in a coastal grassland soil. Mycol Res 111:856–862

Noel GR, Atibalentja N, Bauer SJ (2010) Suppression of *Heterodera glycines* in a soybean field artificially infested with *Pasteuria nishizawae*. Nematropica 40:41–52

Oka Y (2010) Mechanism of nematode suppression by organic soil amendments – a review. Appl Soil Ecol 44:101–115

Oka Y, Pivonia S (2002) Effect of a nitrification inhibitor on nematicidal activity of organic and inorganic ammonia-releasing compounds against the root-knot nematode *Meloidogyne javanica*. Nematology 5:505–513

Oka Y, Chet I, Spiegel Y (1993) Control of root-knot nematode, *Meloidogyne javanica* by *Bacillus cereus*. Biocontrol Sci Technol 3:115–126

Oka Y, Tkachi N, Shuker S et al (2006a) Laboratory studies on the enhancement of nematicidal activity of ammonia-releasing fertilisers by alkaline amendments. Nematology 8:335–346

Oka Y, Tkachi N, Shuker S et al (2006b) Field studies on the enhancement of nematicidal activity of ammonia-releasing fertilisers by alkaline amendments. Nematology 8:881–893

Olatinwo R, Borneman J, Becker JO (2006a) Suppression of *Heterodera schachtii* populations by *Dactylella oviparasitica* in four soils. J Nematol 38:345–348

Olatinwo R, Borneman J, Becker JO (2006b) Induction of beet-cyst nematode suppressiveness by the fungi *Dactylella oviparasitica* and *Fusarium oxysporum* in field microplots. Phytopathology 96:855–859

Olatinwo R, Yin B, Becker JO et al (2006c) Suppression of the plant-parasitic nematode *Heterodera schachtii* by the fungus *Dactylella oviparasitica*. Phytopathology 96:111–114

Ophel-Keller K, McKay A, Hartley D et al (2008) Development of a routine DNA-based testing service for soilborne diseases in Australia. Australas Plant Pathol 37:243–253

Paul EA (ed) (2007) Soil microbiology, ecology and biochemistry. Academic, Burlington

Paulitz TC, Belanger RR (2001) Biological control in greenhouse systems. Annu Rev Phytopathol 39:103–133

Paustian K, Collins HP, Paul EA (1997) Management control on soil carbon. In: Paul EA et al (eds) Soil organic matter in temperate agroecosystems. CRC Press, Boca Raton

Persson C, Jansson H-B (1999) Rhizosphere colonization and control of *Meloidogyne* spp. by nematode-trapping fungi. J Nematol 31:164–171

Pyrowolakis A, Westphal A, Sikora RA (2002) Identification of root-knot nematode suppressive soils. Appl Soil Ecol 19:51–56

Robinson AF, Westphal A, Overstreet C et al (2008) Detection of suppressiveness against *Rotylenchulus reniformis* in soil from cotton (*Gossypium hirsutum*) fields in Texas and Louisiana. J Nematol 40:35–38

Rodriguez-Kabana R (1986) Organic and inorganic nitrogen amendments to soil as nematode suppressants. J Nematol 18:129–135

Rodriguez-Kabana R, Shelby RA, King PS et al (1982) Combinations of anhydrous ammonia and 1, 3-dichloropropenes for control of root-knot nematodes in soybean. Nematropica 12:61–69

Rovira A (1990) Ecology, epidemiology and control of take-all, Rhizoctonia bare patch and cereal cyst nematode. Australas Plant Pathol 19:101–111

Sánchez-Moreno S, Ferris H (2007) Suppressive service of the soil food web: effects of environmental management. Agric Ecosyst Environ 119:75–87

Sasser JN, Carter CC (1985) An advanced treatise on *Meloidogyne*, vol 1, Biology and control. North Carolina State University Graphics, Raleigh

Sayre RM, Starr MP (1988) Bacterial diseases and antagonisms in nematodes. In: Poinar GO Jr, Jansson H-B (eds) Diseases of nematodes. CRC Press, Boca Raton

Schippers B (1992) Prospects for management of natural suppressiveness to control soilborne pathogens. In: Tjamos EC, Papavizas GC, Cook RJ (eds) Biological control of plant diseases. Plenum, New York

Scholler M, Hagedorn G, Rubner A (1999) A reevaluation of predatory orbiliaceous fungi. II. A new generic concept. Sydowia 51:89–113

Seyb A, Xing LJ, Vyn TJ et al (2008) Effect of tillage on population levels of *Heterodera glycines* in a crop sequence of corn and a nematode-susceptible or -resistant cultivar of soybean (Abstr.). Phytopathology 98:S204

Small RW (1987) A review of the prey of predatory soil nematodes. Pedobiologia 30:179–206

Smith ME, Jaffee BA (2009) PCR primers with enhanced specificity for nematode-trapping fungi (Orbiliales). Microb Ecol 58:117–128

Srinivasan A (ed) (2006) Handbook of precision agriculture principles and applications. Haworth, New York

Stirling GR (1979) Techniques for detecting *Dactylella oviparasitica* and evaluating its significance in field soils. J Nematol 11:99–100

Stirling GR (1985) Host specificity of *Pasteuria penetrans* within the genus *Meloidogyne*. Nematologica 31:203–209

Stirling GR (1991) Biological control of plant-parasitic nematodes: progress, problems and prospects. CAB International, Wallingford

Stirling GR (2008) The impact of farming systems on soil biology and soilborne diseases: examples from the Australian sugar and vegetable industries – the case for better integration of sugarcane and vegetable production and implications for future research. Australas Plant Pathol 37:1–18

Stirling GR, Eden LM (2008) The impact of organic amendments, mulching and tillage on plant nutrition, Pythium root rot, root-knot nematode and other pests and diseases of capsicum in a subtropical environment, and implications for the development of more sustainable vegetable farming systems. Australas Plant Pathol 37:123–131

Stirling GR, Mankau R (1978) Parasitism of *Meloidogyne* eggs by a new fungal parasite. J Nematol 10:236–240

Stirling GR, McKenry MV, Mankau R (1979) Biological control of root-knot nematodes (*Meloidogyne* spp.) on peach. Phytopathology 69:806–809

Stirling GR, Sharma RD, Perry J (1990) Attachment of *Pasteuria penetrans* spores to *Meloidogyne javanica* and its effects on infectivity of the nematode. Nematologica 36:246–252

Stirling GR, Dullahide SR, Nikulin A (1995) Management of lesion nematode (*Pratylenchus jordanensis*) on replanted apple trees. Aust J Exp Agric 35:247–258

Stirling GR, Wilson EJ, Stirling AM et al (2003) Organic amendments enhance biological suppression of plant-parasitic nematodes in sugarcane soils. In: Proceedings of the Australian Society of Sugar Cane Technologists 25: (CD ROM)

Stirling GR, Wilson EJ, Stirling AM et al (2005) Amendments of sugarcane trash induce suppressiveness to plant-parasitic nematodes in sugarcane soil. Australas Plant Pathol 34: 203–211

Stirling GR, Halpin NV, Bell MJ (2011) A surface mulch of crop residues enhances suppressiveness to plant-parasitic nematodes in sugarcane soils. Nematropica 41 (in press)

Stone AG, Traina SJ, Hoitink HAJ (2001) Particulate organic matter composition and Pythium damping-off of cucumber. Soil Sci Soc Am J 65:761–770

Stone AG, Scheuerell SJ, Darby HM (2004) Suppression of soilborne diseases in field agricultural systems: organic matter management, cover cropping, and other cultural practices. In: Magdoff F, Weil RR (eds) Soil organic matter in sustainable agriculture. CRC Press, Boca Raton, pp 131–177

Sturhan D (1988) New host and geographical records of nematode-parasitic bacteria of the *Pasteuria penetrans* group. Nematologica 34:350–356

Sylvia DM, Fuhrmann JJ, Hartel PG et al (2005) Principles and applications of soil microbiology. Pearson Prentice Hall, Upper Saddle River

Symondson WOC (2002) Molecular identification of prey in predator diets. Mol Ecol 11:627–641

Tate RL (2000) Soil microbiology, 2nd edn. Wiley, New York

Tenuta M, Ferris H (2004) Sensitivity of nematode life-history groups to ions and osmotic tensions of nitrogenous solutions. J Nematol 36:85–94

Tzean SS, Liou JY (1993) Nematophagous resupinate basidiomycetous fungi. Phytopathology 83:1015–1020

Van Elsas JD, Jansson JK, Trevors JT (eds) (2007) Modern Soil Microbiology, 2nd edn. CRC Press, Boca Raton

Van Elsas JD, Speksnijder AJ, van Overbeek LS (2008) A procedure for the metagenomics exploration of disease-suppressive soils. J Microbiol Meth 75:515–522

Vawdrey LL, Stirling GR (1997) Control of root-knot nematode (*Meloidogyne javanica*) on tomato with molasses and other organic amendments. Australas Plant Pathol 26:179–187

Vu T, Hauschild R, Sikora RA (2006) *Fusarium oxysporum* endophytes induced systemic resistance against *Radopholus similis* on banana. Nematology 8:847–852

Wang ELH, Bergeson GB (1974) Biochemical changes in root exudates and xylem sap of tomato plants infected with *Meloidogyne incognita*. J Nematol 6:194–202

Wardle DA (1995) Impacts of disturbance on detritus food webs in agro-ecosystems of contrasting tillage and weed management practices. Adv Ecol Res 26:105–183

Wardle DA (2002) Communities and ecosystems: linking the aboveground and belowground components. Monographs in population biology 34. Princeton University Press, Princeton

Wardle DA (2005) How plant communities influence decomposer communities. In: Bardgett RD, Usher MB, Hopkins DW (eds) Biological diversity and function in soils. Cambridge University Press, Cambridge

Weibelzahl-Fulton E, Dickson DW, Whitty EB (1996) Suppression of *Meloidogyne incognita* and *M. javanica* by *Pasteuria penetrans* in field soil. J Nematol 28:43–49

Weil RR, Magdoff F (2004) Significance of soil organic matter to soil quality and health. In: Magdoff F, Weil RR (eds) Soil organic matter in sustainable agriculture. CRC Press, Boca Raton, pp 1–43

Weller DM, Raaijmakers JM, McSpadden Gardner BB et al (2002) Microbial populations responsible for specific soil suppressiveness to plant pathogens. Annu Rev Phytopathol 40:309–348

Westphal A (2005) Detection and description of soils with specific nematode suppressiveness. J Nematol 37:121–130

Westphal A, Becker JO (1999) Biological suppression and natural population decline of *Heterodera schachtii* in a California field. Phytopathology 89:434–440

Westphal A, Becker JO (2000) Transfer of biological soil suppressiveness against *Heterodera schachtii*. Phytopathology 90:401–406

Westphal A, Becker JO (2001) Components of soil suppressiveness against *Heterodera schachtii*. Soil Biol Biochem 33:9–16

Westphal A, Mehl H, Seyb A et al (2008) Consequences of tillage intensity on population densities of *Heterodera glycines* and severity of sudden death syndrome in corn-soybean sequence (Abstr.). Phytopathology 98:S169

Whipps JM (1997) Developments in the biological control of soil-borne plant pathogens. Adv Bot Res 26:1–134

Widmer TL, Mitkowski NA, Abawi GS (2002) Soil organic matter management of plant-parasitic nematodes. J Nematol 34:289–295

Xu J-R, Peng Y-L, Dickman MB et al (2006) The dawn of fungal pathogen genomics. Annu Rev Phytopathol 44:337–366

Yin B, Valinsky L, Gao X et al (2003) Identification of fungal rDNA associated with soil suppressiveness against *Heterodera schachtii* using oligonucleotide fingerprinting of ribosomal RNA genes. Phytopathology 93:1006–1013

You MP, Sivasithamparan K (1994) Hydrolysis of fluorescein diacetate in an avocado plantation mulch suppressive to *Phytophthora cinnamomi* and its relationship with certain biotic and abiotic factors. Soil Biol Biochem 26:1355–1361

You MP, Sivasithamparan K (1995) Changes in microbial populations of an avocado plantation mulch suppressive to *Phytophthora cinnamomi*. Appl Soil Ecol 2:33–43

Zasada IA (2005) Factors affecting the suppression of *Heterodera glycines* by N-Viro soil. J Nematol 37:220–225

Zasada IA, Tenuta M (2004) Chemical-mediated toxicity of N-Viro soil to *Heterodera glycines* and *Meloidogyne incognita*. J Nematol 36:297–302

Chapter 2
Microbial Ecology and Nematode Control in Natural Ecosystems

Sofia R. Costa, Wim H. van der Putten, and Brian R. Kerry

Abstract Plant-parasitic nematodes have traditionally been studied in agricultural systems, where they can be pests of importance on a wide range of crops. Nevertheless, nematode ecology in natural ecosystems is receiving increasing interest because of the role of nematodes in soil food webs, nutrient cycling, influences on vegetation composition, and because of their indicator value. In natural ecosystems, plant-parasitic nematode populations can be controlled by bottom-up, horizontal and top-down mechanisms, with more than one mechanism acting upon a given population. Moreover, in natural ecosystems soil nematodes inhabit probably more heterogeneous environment than in agricultural soils. New breakthroughs are to be expected when new molecular-based methods can be used for nematode research in natural ecosystems. Thus far, nematode ecology has strongly relied on coupling conventional abundance and diversity measurements with conceptual population ecology. Biochemical and molecular methods are changing our understanding of naturally co-evolved multitrophic plant-nematode-antagonist interactions in nature, the inter-connections within the soil food web and the extent to which nematodes are involved in many, disparate, soil processes. We foresee that finer

S.R. Costa (✉)
Centre for Functional Ecology, Department of Life Sciences,
University of Coimbra, PO Box 3046, 3001-401 Coimbra, Portugal
e-mail: sofia.costa@ci.uc.pt

W.H. van der Putten
Department of Terrestrial Ecology, Netherlands Institute of Ecology (NIOO-KNAW),
P. O. Box 50, 6700 AB Wageningen, The Netherlands
and
Laboratory of Nematology, Wageningen University, NL-6709 PD Wageningen,
The Netherlands
e-mail: w.vanderputten@nioo.knaw.nl

B.R. Kerry
Nematode Interactions Unit Centre for Soils and Ecosystem Function,
Rothamsted Research Harpenden, Hertfordshire AL5 2JQ, UK
e-mail: brian.kerry@bbsrc.ac.uk

K. Davies and Y. Spiegel (eds.), *Biological Control of Plant-Parasitic Nematodes: Building Coherence between Microbial Ecology and Molecular Mechanisms*,
Progress in Biological Control 11, DOI 10.1007/978-1-4020-9648-8_2,

nematode interactions that lead to their management and control can only be fully understood through the joint effort of different research disciplines that investigate such interactions from the molecular to the ecosystem level.

2.1 Introduction

Plant-parasitic nematodes have traditionally been studied in agricultural systems, where they can be pests of importance on a wide range of crops. Much research has focused on cultural, biological and chemical methods of regulating their populations. As the most abundant and diverse metazoans, nematodes are becoming of increasing interest to ecologists, particularly after research in soil ecology has become more prominent. However, it was only in the early 2000s that agricultural scientists joined efforts with ecologists in order to understand how nematodes are controlled in nature (van der Putten et al. 2006).

Plant-parasitic nematodes in agro-ecosystems have a direct economic impact in reducing crop yield or its marketability and therefore attract the attention of both fundamental and applied scientists. Yet cropping systems cover only about 10.9% of land area worldwide (FAOSTAT 2009). The remainder of plant-parasitic nematodes live, feed, reproduce and die in other types of ecosystems. We consider that systems of low human intervention or disturbance may hold vital clues on how plant-parasitic nematode populations affect and are affected by their natural, co-evolved, plant hosts and other soil biota. This new perspective on the interactions between plant-parasitic nematodes and their biotic and abiotic environment is yielding new and exciting information that may ultimately be translated back to how natural control mechanisms may have been lost or altered by plant breeding and agricultural practices (van der Putten et al. 2006).

Plants, nematodes, soil bacteria and fungi all communicate below ground and are interconnected by trophic interactions, resulting in both direct and indirect effects. In this chapter we describe the physical settings, biological and functional components of these interactions and how they are believed to provide nematode control in natural ecosystems. We further discuss how the advances in molecular tools are helping to gain insight into particular mechanisms of nematode control.

2.2 The Living Soil

Soil-dwelling nematodes are distributed in the microhabitats formed by water films on soil particles, in a complex three dimensional matrix composed of a gaseous, a liquid, and a solid phase, and interact closely with each other and also with a vast array of other organisms. In this section, we review existing knowledge on the soil environment as a driver of nematode diversity and distribution and attempt to illustrate its complexity, which partially results from and certainly prompts a unique set of potential interactions with other organisms.

2.2.1 A Patchy Environment

Soil is a harsh environment, a fragmented habitat with heterogeneous physical and chemical properties, which can be inter-related, and that vary at the regional level, but also at local levels, and even at the microscopical scale within soil cores (Ferris et al. 1990; Ettema et al. 2000; Hodge 2006). Soil characteristics are not only spatially, but also temporally variable (Ettema et al. 2000). Soil moisture and temperature are considered important factors in nematode population dynamics and these conditions vary greatly over time (Bell and Watson 2001). Some nematodes can survive long dry periods in a dormant, anhydrobiotic state, but rapidly become active when soil moisture levels increase (Freckman et al. 1975; Freckman and Mankau 1986; Liang and Steinberger 2001). Soil organisms colonise the soil environment when both water and organic matter are present, and the higher their availability, the more microhabitats can be formed (Pen-Mouratov and Steinberger 2005). Because they move in water films, nematode population size and dynamics are also influenced not only by water availability, but also by the soil hydraulic properties that vary both spatially and temporally (Avendano et al. 2004). Nematode movement is inhibited above a moisture tension of 4.45 pF and they collapse below 4.2 pF, a tension which would also cause permanent wilting of plants. But even at this point, the relative humidity in soil pores rarely drops below 98% (Jones and Jones 1964).

Nematodes, like other soil-dwelling organisms are sensitive to chemical soil properties such as pH, water content, ion content, oxygen levels and nutrient concentrations and their population dynamics are also related to physical properties such as soil texture and structure (Goralczyk 1998; McSorley and Frederick 2004). Soil texture is simply a measure of particle size, and perhaps of more importance to nematodes is soil structure: the spatial distribution of such particles, the formation and size of pores, their arrangement and continuity (Avendano et al. 2004). The diameter of soil pores can alone determine the size of the organisms that live and move within them and how plant roots are arranged, by their size exclusion limit (Watt et al. 2006). Physical and chemical corridors are also thought to form in soil, and can theoretically help communication and contact of organisms that are otherwise isolated from each others (Rantalainen et al. 2004, 2006, 2008).

Roots explore the heterogeneous soil environment in order to acquire water and nutrients. Upon finding nutrient-rich patches in soil, they exploit them through architectural changes, morphological and physiological plasticity (Hodge 2006; Watt et al. 2006), whilst avoiding plant intraspecific competition by a biochemical mechanism of self-recognition (Gruntman and Novoplansky 2004). Plant roots are the main driver of the rhizosphere dynamics, but are also affected by the soil organisms in a multitude of often complex feedback mechanisms (Wardle et al. 2004). As root apexes grow through soil, they encounter and interact with other soil organisms, which may have a mutualistic, neutral or a pathogenic role (Watt et al. 2006). Both symbionts and pathogens affect plant performance and development, and the root can develop new structures in response to these organisms (galls, nodules, mycorrhizae), the results of a biochemical interaction that has been tuned through co-evolution and horizontal gene transfer for thousands of years (Abad et al. 2008; Bauer and Mathesius 2004; Mathesius 2003; Scholl et al. 2003; Opperman et al. 2008).

Plant primary production is the sole food source for plant-parasitic nematodes and it forms the basic input into soil food webs (De Ruiter et al. 1995). Plant-parasitic nematode populations are very responsive to changes in vegetation (Korthals et al. 2001). As different plant-parasitic nematodes may have different levels of specificity to their hosts, plant identity, rather than plant diversity, is a main driver of plant-parasitic nematode diversity and abundance (De Deyn et al. 2004; Viketoft et al. 2005; Wardle et al. 2003; Yeates 1987). To understand how the ecosystem functions it is important not just to quantify different groups of nematodes, but also to know where they are situated relative to each other (Ettema and Yeates 2003). Nematodes are patchily distributed in soil and their diversity can be high at the scale of soil-cores; both its drivers and its function are still not clearly understood (Ettema et al. 1998).

For ecological purposes a key question arises: what do these nematodes do? Soil nematodes are usually classified into functional groups that reflect their feeding habit, such as bacterial-feeders, fungal-feeders, omnivores, plant-parasitic and predators (Bongers and Bongers 1998; Yeates et al. 1993). Nematodes can also be classified through their coloniser-persister strategy that permits the calculation of the maturity index, a measure of ecosystem disturbance. The coloniser-persister assessment aims at determining the extent to which a nematode species is adapted for rapid multiplication and short life-cycles to exploit rapidly changing optimal conditions (coloniser or *r*-strategist), or to tolerate and survive variable, sometimes harsh conditions (persister or *k*-strategist) (Bongers 1990).

Any classification system based on a specific trait is arguably an artificial one, since different results can be produced depending on the trait of choice. Also traits such as the coloniser-persister role are often not static and immutable, but plastic, or adaptable in response to environmental conditions. The nematode *Caenorhabditis elegans*, for example, is known to be able to switch from an *r*-strategy to a *k*-strategy when exploiting different food resource availability (Lee 2002). Incidentally, trait plasticity together with niche partitioning, small scale disturbance and parasite burden/predation, is thought to promote species coexistence, by reducing population sizes and inter-specific competition (Ettema 1998). In other words, the biological aspects of the environment, and the way organisms interact in soil, could largely determine their diversity and abundance.

2.2.2 The Soil Ecosystem

Soil biomass in the below-ground subsystem can be structured through food chains that originate either from the primary production of plant roots (grazing food chains), or from labile or recalcitrant litter and debris (the decomposer food chains). We review ecology theory on how such structures are inevitably interlinked, which theoretically may lead to nematode control by natural mechanisms; we also give practical examples of the functional involvement of nematodes.

2.2.2.1 Food Chains and Energy Channels

According to the Green World Hypothesis (GWH), plants are abundant because herbivores are top-down controlled by their predators and parasites, whereas plants themselves are bottom-up controlled by resource availability. In regulating the herbivore populations, predator and parasite populations are also resource-limited (Hairston et al. 1960). Should the GWH apply to plant-parasitic nematodes in three-level food chains, a given plant parasite would not only control its host plant (by reducing its primary production), but also be controlled by a natural enemy, e.g. a fungal or bacterial parasite. An interesting analogy of the GWH is the Brown Ground Hypothesis (BGH), in which essentially the same regulatory processes are applied to decomposition in ecosystems, or 'why there is so much carbon in soil' (Allison 2006). Organic matter accumulates in soil because there is a large amount of input of dead material (notably of plant origin) and the microbial organisms involved in its decomposition are top-down controlled.

A sometimes large proportion of roots in the rhizosphere can be inactive. In grassland ecosystems, for example, a layer of dead roots frequently accumulates in the most superficial layers of soil (Watt et al. 2006). Yet organic matter in soil in the form of dead organisms, leaf litter and root deposition is not a blind alley for energy and biomass in the soil ecosystem. Although soil is rich in carbon compounds, nitrogen is generally a limiting factor (Ingham et al. 1985). Therefore, dead material is a major input of organic matter into the system, a food source for decomposers that eventually mineralize these nutrients and make them again available to the plants. Exudates and leachates from living plant roots, for example, also support communities of decomposer microorganisms, which are thought to be selected by their interaction with the plants: they specialise in the decomposition of the plant exudates and leachates and promptly make nutrients available back to the plant (Grayston et al. 1998).

By bringing together concepts taken from the GWH and the BGH, two parallel energy chains can be identified in soil food webs, both culminating at the top-level consumers: one starting from biomass originating from plant primary production, and one starting from dead organic matter (Fig. 2.1). Predators and parasites in soil acquire energy from both the grazing and the decomposer chains and therefore predation/parasitism on the primary consumers can be driven by the decomposer chain (Moore et al. 2004). Energy channels are defined as a group of species consuming biomass that originates from the same primary energy source (Moore and Hunt 1988). In soil food-webs, the dead biomass based energy chain is developed along two lines, one starting from bacteria and one from fungi. Therefore, there are three parallel energy channels in the soil food-web: the root channel, a primary production channel based on the plant and following on to its herbivores and their predators and parasites; the bacterial channel, a decomposition channel based on

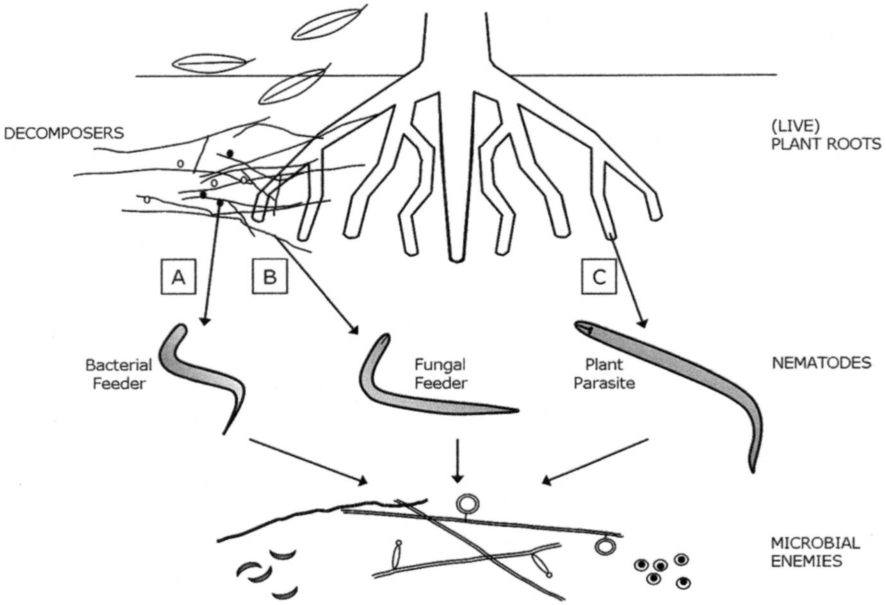

Fig. 2.1 The involvement of plants, bacterial-feeding, fungal-feeding, plant-parasitic nematodes and their microbial enemies in the three parallel energy channels in the soil food-web: A – the bacterial energy channel, originating on labile organic matter, B – the fungal energy channel, based on the decomposition of recalcitrant organic matter and C – the grazing channel, using plant primary production as an energy source. All three channels are joined together at the top consumer level (Moore and Hunt 1988), the microbial enemies of nematodes and therefore a generalist microbial enemy could obtain energy from the three channels

high quality, N-rich debris; and the fungal channel, based on low-quality, C-rich compounds (Fig. 2.1).

In analyses of soil decomposer food-webs in grassland ecosystems, bacterial-feeding nematodes have only been associated with the bacterial-based energy channel and fungal-feeding nematodes with the fungal-based energy channel. Plant-parasitic nematodes have exclusively been allocated to the root channel, as they depend solely on primary production as their food source. However, all three channels are not separate: predatory nematodes were associated not only with the bacterial-based energy channel (89.9%), but also with the fungal-based energy channel (10.6%), and (weakly) with the root channel (0.4%). The association of predatory nematodes with the primary production channel seems to imply that there may be a weak trophic link to plant-parasitic nematodes (De Ruiter et al. 1995).

Although conceptually useful, in nature, organisms are not simply organised in food chains, but rather in food-webs, with complex and indirect interactions between them, regardless of their trophic level. The sum of indirect effects that result from food-webs can easily overshadow the biomass/energy transfer of the three-level food chain, which could be demoted from the designation 'trophic cascade' to the more modest 'trophic trickle' (Strong et al. 1999). Indeed, three-level food chains are thought to be unstable and tend to chaotic dynamics (Hastings and Powell 1991), which certainly is not expected to express the natural functioning of the soil ecosystem. To understand the ecology of soil, we need to consider how food-webs, and not simply food chains, work.

2.2.2.2 Food-Web Effects and Interactions

The Santa Rosalia theory proposes that ecological interactions such as competition have a major role in the maintenance of biodiversity (Hutchinson 1959). This theory aims to explain 'why there are so many species of organisms' and was put forward after the observation of several species of plankton inhabiting a small pond by the Santa Rosalia caves. The number of species in a community and also their functional differences increase food-web complexity, which seems to promote coexistence. Long food chain loops with weak links that form in complex multitrophic interaction webs may also be responsible for the stability of ecosystems (Neutel et al. 2002). If there is a high degree of functional differences between species, then inter-specific facilitation as opposed to competition can occur; this mechanism is thought to be involved in driving decomposition processes in soil (Heemsbergen et al. 2004) and has also been shown to support the coexistence of different plant species in coastal dune systems (Stubbs and Wilson 2004).

Plant identity, as substantiated before in this chapter, is thought to be a driver of soil food-webs, leading to changes in the soil community both among and within trophic groups (Wardle et al. 2003). In a biodiversity field experiment, the soil food-webs of plant individuals were most similar within the same plant community. Individual plant soil food-webs varied between plant communities and between plant species; this variation could be detected even between plant individuals (Bezemer et al. 2010). Soil communities are also known to feed back to their host plants (Bever 1994). Soil biota, therefore, also determine the abundance of plant species, as the most abundant species have strong positive feedbacks with their own soil and rare species have a negative feedback effect (Klironomos 2002). Soil community feedbacks can also maintain the coexistence of competitor plants, where otherwise one would exclude the other (Bever 2003).

Negative feedbacks caused by soil-borne disease complexes composed of fungal pathogens and plant-parasitic nematodes have been correlated to both degeneration and successional replacement of marram grass *Ammophila arenaria* in coastal sand dunes (Van der Putten and Peters 1997; Van der Putten et al. 1993). However, if plants are released, even if only partially or temporally from their own natural enemies, they will have an increased competitive advantage and may outcompete other plants if they remain constrained by their natural enemy community (DeWalt et al. 2004). Also above-ground studies suggest that plant coexistence can be maintained by such indirect effects when parasites disproportionately repress the population density of the dominant host plant species (Yorozuya 2006).

Indirect effects encompass a wide range of interactions and can be defined as occurring when the impact of one species on another requires the presence of a third (Strauss 1991; Wooton 1994). Tritrophic interactions in which plants can communicate with the enemies of their enemies, giving indirect control, have been the object of much study, and are a good example of indirect effects (Price et al. 1980). The four most studied types of indirect effects are apparent competition (the sizes of two different populations being mediated by a shared predator), indirect facilitation (a population benefiting from the predation of another), exploitative competition (two different populations being limited by the same resource) and the above-mentioned trophic cascades (van Veen et al. 2006; White et al. 2006). Indirect effects comprise not only density-mediated effects, but also trait-mediated ones, including life-history traits and plastic or evolutionary adaptations of populations (Luttbeg et al. 2003).

2.3 Nematode Control Mechanisms in Natural Ecosystems

Plant-parasitic nematode populations can be controlled by a range of mechanisms that are active during interactions both within and among different trophic levels; these include bottom-up, horizontal and top-down interactions. Such interactions can be mediated by organisms that do not affect the nematode populations themselves, but cause indirect effects through food-web links. In Fig. 2.2, we summarize the mechanisms that are thought to contribute to plant-parasitic nematode control in the rhizosphere of *Ammophila arenaria* (marram grass) in coastal sand dunes and represent knowledge gained from the EU-EcoTrain Project (2002–2006). It is important to note that a particular nematode population can be controlled through more than one mechanism.

2.3.1 Bottom-Up Control

Bottom-up control occurs when a nematode population size is kept below a certain level by resource limitation, i.e., food availability. What would initially appear to be a simple concept potentially involves several mechanisms. Plant-parasitic nematodes are obligate parasites which have co-evolved with their host plants; during this process, both nematodes and plants have also interacted and co-evolved with a range of other rhizosphere organisms. The outcomes of this coevolving network are still not clearly understood, and some potential processes are described below. Although the existence of partial nematode resistance in some crop varieties is common knowledge, crop plants have not been naturally coevolved with their nematode parasites and other soil biota. Therefore natural systems represent a key opportunity to investigate such ecological interactions.

In a large population study of *Heterodera arenaria* parasitizing marram grass in sand dunes, the nematodes were found to be early colonisers of newly-developing roots.

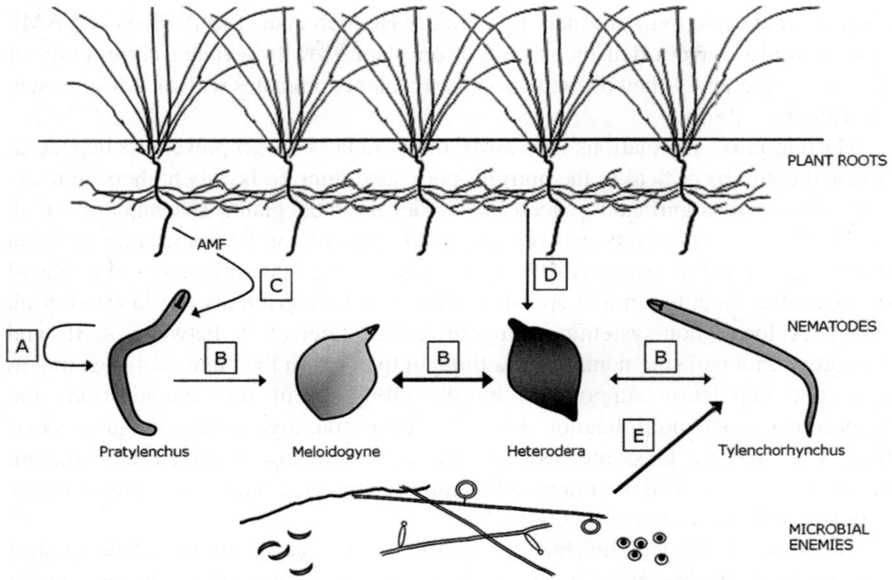

Fig. 2.2 Mechanisms of plant-parasitic nematode control in the rhizosphere of marram grass (*Ammophila arenaria*) in European coastal sand dunes, a natural ecosystem: A – Horizontal control through intraspecific competition, B – Horizontal control through inter-specific competition, C – Bottom-up control by an indirect effect, via Arbuscular Mycorrhizal Fungi (AMF) associations with the plant root, D – Bottom-up control through resource limitation, E – Top-down control

In the recently developed root layers, *Heterodera arenaria* populations increased to a level where they became resource-limited, (Fig. 2.2) whilst in deeper (older) root zones, when the nematode populations are established, they were affected by other parameters, such as resource quality (Van der Stoel et al. 2006). These sedentary parasites were considered mostly harmless in the coastal sand dunes under study, but sedentary endoparasites together with migratory endoparasites are the main nematode groups involved in disease complexes. They develop synergistic or additive effects on disease incidence and severity by association with plant-pathogenic bacteria or fungi (Hillocks 2001). A disease complex of such plant-parasitic nematodes (*H. arenaria, Meloidogyne maritima and Pratylenchus* spp.) and fungal plant-pathogens has been suggested to be involved in the decline of *Ammophila arenaria* (marram-grass) in coastal sand dunes (Van der Putten et al. 1993). These natural systems provide a unique opportunity for studies on the ecology and natural control of these nematodes.

Plant mutualists, such as mycorrhizal fungi and rhizobia are widespread and are thought to maintain the structure and diversity of natural communities. Many studies suggest the importance of mutualisms in improving plant nutrition and health, but there is little evidence for community-level impacts of mutualists (Christian 2001). The presence of arbuscular mycorrhizal fungi (AMF) can increase plant diversity and ecosystem productivity (van der Heijden et al. 1998). However, AMF fungi can also have a detrimental effect on plant growth: a richer community of these fungi increases plant diversity because no plant dominates with all AMF present (Klironomos 2003).

Marram grass associations with AMF might delay or even prevent its degeneration and could be critical in the nutrient-poor sand dune soils, where their numbers were shown to significantly decrease in degenerated plants (Kowalchuk et al. 2002). However, this study was observational. The role of the association between marram grass and its native AMF populations has been investigated in more detail in sequential inoculation and split-root glasshouse experiments (de la Pena et al. 2006). A local, non-systemic, competition-like interaction between AMF and migratory endoparasitic nematodes is thought to occur in the plant roots, leading to nematode population suppression by the inhibition of root colonisation, and reduced nematode multiplication (Fig. 2.2). Arbuscular mycorrhizal fungal associations with marram grass are also thought to be critical for plant establishment, because they can lead to improved plant growth, especially in younger plants (Rodriguez-Echeverria et al. 2004).

The role of the legume-rhizobia symbiotic interaction in nematode control appears to have idiosyncratic effects, being highly dependent on the interacting species identity. Some studies suggest that plant-parasitic nematodes may reduce nodule formation (Duponnois et al. 2000; Villenave and Cadet 1998). On the other hand, some rhizobia strains have been shown to elicit induced resistance in the plant against plant-parasitic nematodes (Mitra et al. 2004; Reitz et al. 2000). Plant-parasitic nematodes and rhizobia interact in the rhizosphere, and there is evidence of horizontal gene transfer between them (Scholl et al. 2003), but the outcomes of their interactions for plants are still not clear. Recent studies using the model legume *Medicago truncatula* have shown that rhizobial nodulation suppresses root galling by the endoparasitic nematode *Meloidogyne javanica*, which in turn increases nodulation (Costa et al. 2008).

Colonisation of land by vascular plants dates back an estimated 400 million years (Signor 1994). Throughout this time, plants have interacted with their herbivores, parasites and pathogens, and this has led to a coevolution process that is responsible for the development of plant chemical defence (Ehrlich and Raven 1964). Plants may not be vulnerable to herbivore attack, as is suggested by the GWH, but constantly release primary production compounds (CO_2, sugars) and also secondary metabolites through root exudations and leaf volatiles, which are indicative of their physiological state. These can act as cues for their herbivores, which can be attracted or repelled, and also for natural enemies of these herbivores (Price et al. 1980; Rasmann et al. 2005).

Some plant species may produce secondary metabolites with nematotoxic effects (Gommers 1981), but such effects have, to our knowledge, not been assessed in natural systems. *Tagetes* plants have been studied extensively for their effects on nematode suppression and various nematicidal polythienyl compounds were isolated from them (Uhlenbroek and Bijloo 1958). Endoroot bacterial isolates of *Tagetes erecta* and of *T. patula* have a role in this effect, which could be transferred to potato

Solanum tuberosum plants, resulting in a decrease in nematode populations without affecting the potato yield (Sturz and Kimpinski 2004). Rhizosphere bacteria also have shown activity against fungal pathogens, with effects being influenced by soil type, root morphology, root exudation and plant identity (Berg et al. 2006; Lee et al. 2005).

Plant-parasitic nematode management strategies in agricultural systems should be developed taking into account and exploiting the role of the plants as an interacting organism in the food-web.

2.3.2 Horizontal Control

The logistic model of population growth (Lewis and Taylor 1967) can be used for nematode populations such as *Pratylenchus* and *Tylenchorhynchus* that reproduce continuously and have overlapping generations in the rhizosphere (Van Den Berg and Rossing 2005). This model assumes that the carrying capacity (or maximum density in a host) of each population reflects the food source limitation as the populations grow and intraspecificic competition takes place between the nematodes (McSorley and Duncan 2004). When inter-specific competition interactions are considered, the (partial) niche overlap between the two competing populations, a proportion (depending on niche overlap) of each population can be seen as equivalent to the other, and therefore contributes to their density when carrying capacity is being considered (Lewis and Taylor 1967). Therefore, not only the competing populations of nematodes themselves, but also the host plant is a main player in horizontal control.

To evaluate the possible role of horizontal control of nematodes that are involved in the decline of marram grass in coastal sand dunes, mesocosm experiments were performed using combinations of sand burial and inoculation with *Meloidogyne maritima*, *Heterodera arenaria*, and *Pratylenchus penetrans*, alone or in combinations (Brinkman et al. 2005c). Plant biomass was only found to be reduced by one of the nematode species, *M. maritima*, and additive effects between the three plant-parasitic nematodes could not be found. Indeed, this experiment revealed that the addition of the three species of nematodes led to a decrease in the negative effect of *M. maritima* on plant biomass. *Heterodera arenaria* and *P. penetrans* were thought to interfere with the *M. maritima* life-cycle by shifting its reproductive stage to later in the season, when it takes place in sub-optimal conditions. We anticipate that the application of a specific biological control agent in agricultural systems to reduce a given nematode population could benefit its competitors, with a corresponding increase in their population size. However, the extent to which a nematode population would need to be reduced in order to produce a population outbreak of its competitors is unclear and this threshold may not be reached through biological control.

Pot studies on competition effects (horizontal control) between the three species of endoparasitic nematodes, and also with the ectoparasitic *Tylenchorhynchus ventralis*, indicate that *P. penetrans* is limited by intraspecific, but not by inter-specific competition. Moreover, *P. penetrans* is a stronger competitor than *H. arenaria* and *M. maritima* (Fig. 2.2). The sedentary endoparasites were equally strong competitors and were only weakly affected by the *T. ventralis* population (Brinkman et al. 2005a, b). Importantly such competitive interactions are mediated by the host plant, whose tolerance and attractiveness to the nematode populations, as altered by the interactions with those populations, is influenced by the carrying capacities for the nematodes (Brinkman et al. 2008; de la Pena et al. 2008).

2.3.3 Top-Down Control

Research on top-down control of plant-parasitic nematodes by soil micro-organisms has traditionally been done on agricultural ecosystems, to tentatively develop applications of biological control agents (Whipps and Davies 2000). However, the diversity and distribution of nematode microbial enemies in natural ecosystems are still mostly unknown. Nematode antagonists would need to occupy the same soil pores as nematodes, and survive the variable chemical and physical characteristics described in Sect. 2.2.1. These prerequisites limit the groups of organisms that can predate and parasitise soil-dwelling nematodes (De Ruiter et al. 1995). We restrict our review to the microbial enemies of nematodes (fungi and bacteria), as they putatively have a larger effect on nematode populations than predatory nematodes, protozoans and soil microarthropods (Piskiewicz et al. 2008; Rodriguez-Kabana 1991).

The Red Queen Hypothesis (RQH) was originally formulated as a species extinction law (Van Valen 1973), and has since been developed and expanded to include a range of ecological aspects of host-parasite interactions. The hypothesis is based on the Red Queen character of the Lewis Carroll book ‘Through the Looking Glass’, saying to Alice ‘here, you see, it takes all the running you can do to keep in the same place’. The RQH attempted to reconcile the biotic (and genetic) aspects of interactions between organisms with the environmental parameters that result in natural selection and evolution (Van Valen 1975). In order to avoid (local) extinction, the organisms at loss must evolve rapidly to improve their fitness, and this process is occurring continuously (Van Valen 1973, 1976). As the host is also the physical environment of the parasite, at least for part of its life-cycle, the RQH effect would be more pronounced if the organisms in question were a host improving in fitness, and a parasite therefore reducing its fitness.

The coevolution of parasites and their hosts is driven by and leads to a dynamic balance in which the populations interact to regulate their biological and ecological parameters, resulting in an inter-regulation of host and parasite population size (Anderson and May 1981). Host-parasite coevolution is mainly driven by virulence, a product of the host-parasite interaction. Hosts should evolve to decrease virulence, whereas parasites should evolve to maintain virulence at an optimal level, which would allow infection and multiplication of the parasite without detrimental effects on the host (the cost of parasitism) (Ebert and Hamilton 1996).

It has been shown mathematically that, through coevolution, parasites that are able to parasitise different hosts can evolve divergently, generating subpopulations or races with different host preferences. Such a heterogeneous population would be favoured to a homogeneous generalist population, in that each of the subpopulations can co-evolve faster with its host, than a generalist population can co-evolve with different and variable hosts (Kawecki 1998). Some practical examples seem to support this theory: although trapping fungi would appear to be generalists, hyphal development varies within different nematode hosts and trapping fungi have different specificities, with some nematodes remaining unaffected (Barron 1977); they also have different specificities towards surface mutants of *Caenorhabditis elegans* (de Gives et al. 1999). Molecular methods have demonstrated the existence of different host preferences in biotypes of the nematophagous fungus *Pochonia chlamydosporia* (Mauchline et al. 2004). Finally, immunodetection methods have shown that populations of *Pasteuria* spp. attacking different hosts have different endospore surface immunological properties. In a coastal sand dune, endospores from a (possibly multi-species) population of these putatively highly specific parasites were found attached to *Pratylenchus* spp., *Tylenchorhynchus* spp. and omnivorous Dorylaimid nematodes (Costa et al. 2006). The presence of *Pasteuria* spp. (sub) populations attacking phylogenetically and functionally different nematode hosts at a given site resulted in an apparent generalist role for these bacteria.

The microbial enemy community diversity and population dynamics can be influenced not only by nematode identity, but also indirectly by the host plant of the nematode, in a complex trophic interaction involving the three trophic levels (Kerry and Hominick 2002). Studies on *Pochonia chlamydosporia* suggest that this facultative nematode parasite can colonise the rhizosphere more or less extensively depending on the (agricultural) plant species. This fungus generally provides more efficient control of root-knot nematodes (*Meloidogyne* spp.) when plant susceptibility to the nematodes is moderate, because more nematode egg masses are exposed and vulnerable to fungal colonisation on the root surface. Also, the nutrition of the fungus may affect its transition from a saprotroph to a parasite (see Chap. 7).

Natural enemies of nematodes could link energy channels; plant ectoparasitic nematodes (root channel) feeding on marram grass seem to be top-down controlled by putatively generalist fungal parasites that increase their population size by feeding on bacterial-feeding and omnivorous nematodes (bacterial channel) (Piskiewicz et al. 2007) (see Fig. 2.2). In fact, by selective addition experiments, most of the plant-parasitic nematode species commonly found in the rhizosphere of marram grass could be controlled, to variable extents, by microorganisms present in soil filtrates (Piskiewicz et al. 2008). Given a choice in a Y-tube type experiment, the ectoparasites *Tylenchorhynchus ventralis* migrate towards roots that are free of such microorganisms, putatively detecting their presence and hence avoiding rhizosphere areas that have their natural enemies (Piskiewicz et al. 2009a). The microorganisms are thought to actively parasitise the nematodes in a local interaction, and not just suppress them systemically (Piskiewicz et al. 2009b).

However, nematode microbial enemies abundance and virulence are not only restricted by biological, coevolutionary factors. *Catenaria anguillulae*, the most common endoparasitic fungus attacking nematodes, is considered a generalist feeder, and its ubiquity in soil is thought to be related to not being constrained by host availability (Costa 2006). The fungus spends most of its life-cycle protected inside the host but is highly dependent on soil water content for infection. Its infective propagules are zoospores that need to move through water films following nematode exudations to find their hosts (Barron 1977; Deacon and Saxena 1997). Therefore, the soil physical, chemical and temporal heterogeneity are major factors determining the abundance of this natural enemy of nematodes. Conversely, the infective endospores of the bacterial parasites *Pasteuria* spp. are a resistance structure, which allows them to survive harsh environmental conditions. The resistance to soil abiotic conditions for infection may be a key factor for the development of large population densities of the bacteria. Interestingly, in natural coastal sand dunes, population densities of *Pasteuria* spp. have been found comparable to those that would promote biological control in agricultural systems (Costa et al. 2006).

Although nematode-trapping fungi can colonise the rhizosphere feeding on organic matter, their sensitivity to environmental changes makes them poor competitors in soil (Barron 2003; Siddiqui and Mahmood 1996). Their parasitic phase feeding on nematodes is thought to be the norm in soil conditions, yet trap formation is controlled by numerous factors (Jaffee et al. 1992) with gene expression patterns that differ from those in the saprophytic phase (Ahren and Tunlid 2003). Although the abundance of facultative parasites is only partially dependent on their nematode host density, their parasitic phase can be very influenced by it (Jaffee and Strong 2005), with such fungi being frequently found associated to large numbers of their hosts (Farrell et al. 2006).

2.3.4 Indirect Effects

In grassland ecosystems, different plants and functional groups of nematodes can affect each others' population levels and nutrient mineralization through food-web links between the root, bacterial, and fungal energy channels, as has been revealed by Phospholipid Fatty Acid (PLFA) profiling (see Sect. 2.4). Low levels of parasitism by the specific *Heterodera trifolii* on the legume *Trifolium repens* increases root leakage, releasing N and C that lead to an increase in the soil microbial biomass, involved in the mineralization of such compounds (Bardgett et al. 1999b). The interactions between the host plants and nematodes can also lead to alterations in root exudation, morphology and architecture (Haase et al. 2007). The increased microbial activity leads to an increased bacterial feeding nematode activity, and both promote net mineralization and nutrient cycling. The nutrients are then made available not only to the attacked plants but also to the neighbouring ryegrass *Lolium perenne*. Both plant-parasitic and bacterial-feeding nematode populations were shown to affect the rate and direction of nutrient fluxes in this ecosystem, which ultimately affects plant competition and thereby alters plant community structure (Bardgett et al. 1999a, b).

Different bacterial-feeding nematode species have different feeding preferences. Therefore, the species composition of bacterial-feeding nematode populations can have a significant impact in structuring the bacterial decomposer community, through preferential feeding on different bacterial populations (De Mesel et al. 2004). The transfer of nitrogen from the parasitized plants to their neighbours was found to be dependent on the density of root infestation (Dromph et al. 2006). Under high grazing pressure, nematodes with high feeding specificity altered the diversity of bacteria growing on detritus (De Mesel et al. 2004). Such effects in turn can indirectly lead to changes on nutrient mineralization rates, and consequently on plant nutrient uptake (Laakso et al. 2000). In grassland ecosystems, these bacterial channel interactions with bacterial-feeding nematodes were found to be highly species-specific; and interestingly, these indirect effects can involve higher trophic levels, as the bacterial-feeding nematode populations were also strongly regulated by top-down control (Bardgett et al. 1999a). Changes in the quantity and the quality of plant-root leachates may not only be caused by nematode feeding on roots but also by above-ground herbivory, that indirectly affect decomposition and soil processes (Bardgett and Wardle 2003).

Natural enemy recruitment, or indirect defence, was described in detail for the interaction between maize plants, their lepidopteran above-ground herbivores, and their parasitoid wasps. Upon seedling attack by lepidopteran larvae, maize plants emit a mixture of volatile compounds that are highly attractive to a range of parasitic wasps, natural enemies of the lepidopterans. This is achieved by the herbivory-induced and transcript-regulated gene expression of an enzyme, terpene synthase TPS10, that forms (*E*)-β-farnesene, (*E*)-α-bergamotene, and other herbivory-induced sesquiterpene hydrocarbons (Schnee et al. 2006).

As with insects, it is likely that nematodes respond to a range of volatile and non-volatile signals at a range of different scales (Jones and Jones 1964). Because nematodes move through water films in soil pores that are also filled with air, both volatile and water-soluble compounds could be involved in attracting nematodes to roots, but volatile compounds can potentially travel faster and over longer distances than those in water (Young and Ritz 2005). Recent research has identified an insect-induced belowground indirect defence plant signal, (*E*)-β-caryophyllene, which strongly attracts an entomopathogenic nematode. Insect-damaged maize roots release the compound in response to insect herbivory, and this sesquiterpene attracts *Heterorhabditis megidis* entomopathogenic nematodes through soil (Rasmann et al. 2005). Further olfactometer experiments have revealed variable responses at the level of volatiles production in three plant species following elicitation by herbivores. The different volatile blends produced attracted the nematodes differentially, with some volatiles, namely (*E*)-β-caryophyllene, being more attractive than others. This suggests a degree of specificity in this below-ground tritrophic interaction (Rasmann and Turlings 2008).

Plant root recruitment of natural enemies of their parasitic nematode populations has thus far not been described, but such mechanisms are likely to exist, as communication in the rhizosphere that involves all key players has been reported (Johnson and Gregory 2006).

2.4 How Molecular Approaches Are Shaping Our Knowledge of Nematode Control in Natural Ecosystems

The drivers of biological control mechanisms in nature, their impact on selected populations, on the nematode community and on the soil community as a whole are still not clearly understood. To fully understand the ecology of nematode control mechanisms in natural systems, we must be able to address key questions: what is the identity of the nematodes and what is their fundamental niche; how are they distributed in soil and how diverse are their populations? Similar questions on the organisms they interact with need to be attended to. And when we get the necessary answers, we must direct our research effort to the functional aspects of the interactions: how are they processed; and what affects their outcome?

Conventional methods of nematode quantification and identification in soil are time-consuming and demand a high level of expertise, compromising the number of samples that can be processed. Even carefully-designed sampling methods will usually average the distribution of organisms, eliminate spatial structure or be biased for the particular sampling season and sampling time (Ettema and Wardle 2002). Extraction methods vary in their efficiency, influence the numbers of extracted nematodes and may preferentially extract certain groups or life-stages (McSorley and Frederick 2004). The identification of nematodes can itself be a herculean task. Their morphometrics are variable and key characteristics overlap for some species, with several specimens of different life-stages being required for identification to species level (Powers 2004).

If nematode populations are difficult to identify and quantify in soil, those of the microbial biota pose a larger problem still. Most of these organisms are unculturable and therefore cannot be counted in sequential dilution plates. The assessment of their community structure and dynamics was only made possible through the application of molecular profiling and biomass estimation techniques. Phospholipid fatty acid analyses (PLFA) have been elucidating how nematodes and the bacterial and fungal decomposer communities interact (Bardgett et al. 1996, 1999c, Denton et al. 1999, Laakso et al. 2000). These analyses have shown to be sensitive to microbial community changes induced in grassland and significantly upgraded other tools that measure microbial activity.

PLFA provide a fingerprint of the microbial community structure, being indicative of biomass content of fungi and various bacterial groups, through their phospholipid fatty acid signature (Bardgett et al. 1996). This has permitted the assessment of soil microbial activity in the fungal and bacterial decomposer channels separately and the calculation of the fungal:bacterial biomass ratio, which can be compared between samples (Bardgett et al. 1999c). However, although some bacteria can be classed into different groups through their fatty acid signature, all fungal biomass is measured through only one fatty acid, 18:2ω6 (Denton et al. 1999). PLFA are an extremely useful tool for measuring, and to an extent, describing the response of the soil bacterial community to changes in environmental conditions (O'Donnell et al. 2005), but do not give detailed indications of the identity or diversity of the microbial groups. Nevertheless, PLFA profiles of the microbial community coupled to nematode population studies in grassland soils have revealed the inter-connectedness of different nematode trophic groups through the food-web, and further implicated nematodes in nutrient mineralization and nutrient transfer between plants (Bardgett et al. 1999a).

Denaturing gradient gel electrophoresis (DGGE) produces a community fingerprint of large groups of organisms, providing a measure of their genetic diversity, and an indication of their abundance. This PCR-based method was initially adapted to assess bacterial communities in soil, through amplification and electrophoresis of amplified 16S rDNA fragments. This technique allows the separation of fragments of the same length but with different base-pair sequences in a denaturing gradient gel, as based on differential electrophoretic mobility of partially melted DNA molecules (Muyzer et al. 1993). Sequencing the obtained fragment bands can provide taxonomic information to complement the diversity and abundance profiling (De Mesel et al. 2004). PCR-based DGGE has been successfully applied to 18S rDNA templates extracted directly from soil to assess fungal communities, being indicative of the incidence and prevalence of specific fungi. But quantification of soil fungi by DGGE, like with other methods, is complicated by the inability to distinguish numbers of fungal spores from numbers of colony-forming mycelial fragments (van Elsas et al. 2000). Nevertheless, this technique allows the comparison of multiple fungal community profiles between different treatments and can be used to perform broad analyses of how the fungal community responds to changes in the rhizosphere. DGGE analyses have recently been applied to demonstrate that nematode populations induce changes in the fungal community structure in a plant-species specific way; these changes, however, did not seem to provide nematode control and therefore do not substantiate the existence of indirect mechanisms of plant defence (Wurst et al. 2009).

Using consensus primers designed for small subunit (SSU), or 18S, rDNA sequences, DGGE has also been applied to the study of nematode communities in soil with some success (Waite et al. 2003). The initial insufficient specificity of the primers for nematodes could be circumvented by extracting nematodes into a soil suspension and discarding other metazoans prior to DNA extraction, which was found to improve the accuracy of the method. However, due to PCR bias, nematode diversity as measured by MOTUs (molecular operational taxonomic unit) is still under-represented when challenged by conventional morphological analysis (Foucher et al. 2004). This is a common artefact of PCR-based molecular tools that depend on DNA content, body size, number of cells and number of copies of 18S rDNA of a mixed population of nematodes (Wu et al. 2009). As with DGGE analyses of other soil communities, a given population may be omitted if it represents under 1% of the biomass of the total community (Foucher et al. 2004; Muyzer et al. 1993). Further limitations of this method include the poor relatedness of obtained bands to MOTUs, and hence, to community diversity (De Mesel et al. 2004; Foucher et al. 2004) and the lack of functional meaning of the amplified fragments. Bacterial 16S rRNA, for example, can be as small as 0.05% of the total genome and its variability has little or no ecological and physiological meaning (Kowalchuk et al. 1997). Terminal Restriction Fragment Length Polymorphism (T-RFLP) of SSU rDNA is being proposed as an alternative molecular approach to obtain profiles of nematode communities in agricultural sites, which could be combined with nematode diversity indices (Donn et al. 2007).

Molecular barcodes, obtained through PCR following sequencing the SSU 18S rDNA of single nematodes, supply MOTU that represent a rapid assessment of nematode biodiversity in soils (Floyd et al. 2002). Molecular barcodes are being given a 'face': the obtained sequences can be blasted to known species sequences to provide species names; nematode molecular information is being compiled together with nematode images, specimen voucher lists and other material in online databases to aid nematode molecular diagnostics (Powers 2004). Barcodes can now be obtained not just from individual nematodes, but from bulk samples, with the difficult task of assigning MOTUs to the obtained sequences being made easy by available software. Whether or not identification to species level can be achieved, the use of molecular barcodes can give sufficient data on the diversity of nematodes (Blaxter et al. 2005). This method, however, is not as straightforward as DGGE, as it involves a number of steps, including DNA purification, cloning into recombinant plasmids, sequencing and bioinformatics tools that require more equipment and molecular expertise.

Knowledge gained from estimating the relative abundance and diversity of soil organisms, although highly valuable, can be of limited use in unravelling the intricate interactions between these organisms. But molecular approaches using sequencing of the SSU have gathered information that can be used to construct phylum-wide phylogenies that have brought novel interpretation of the evolution of parasitism in nematodes (Blaxter et al. 1998; Holterman et al. 2006). The ecological advantages and/or disadvantages of parthenogenesis in *Meloidogyne* have been the subject of much debate (reviewed in Trudgill and Blok 2001). Recently, the evolution of life history traits, including apomixis, of the Tylenchida plant parasites are being clarified through phylogenetic analyses (Holterman et al. 2009). Such findings can help elucidate the ecological role of nematodes and develop ecological theory on how it was achieved, but in order to understand finer, more subtle interactions that can, nevertheless, have large impacts on an ecosystem, other approaches are needed.

The sequencing of the *Caenorhabiditis elegans* nematode over 10 years ago (The *C. elegans* Sequencing Consortium 1998) was but a starting point in genomics research. The genome sequences of *Meloidogyne hapla* and *M. incognita* have recently been published (Opperman et al. 2008 and Abad et al. 2008, respectively), and provide exciting new opportunities for the investigation of plant-parasitic nematodes. The analyses of the information contained in these newly available genome sequences, compared to those of the genomes of the free-living nematodes *C. elegans* (The *C. elegans* Sequencing Consortium 1998), *C. briggsae* (Stein et al. 2003) and *Pristionchus pacificus* (Dieterich et al. 2008) and the draft genome of the human filarial parasite *Brugia malayi* (Ghedin et al. 2007), can yield exciting new research opportunities and guide the formulation of new hypotheses.

To be able to use the information present in genome sequences, however, scientists need to resort to functional genomics: studies of what genes are expressed and when, and of the gene products (Mitreva et al. 2005). For example, whilst genetic analyses have shown how root-knot nematodes have acquired plant-infection genes from rhizobia through horizontal gene transfer (Abad et al. 2008; Bauer and Mathesius 2004; Mathesius 2003; Opperman et al. 2008; Scholl et al. 2003), proteomic analysis is elucidating the mechanisms of the interaction between root-knot nematodes and rhizobia that alter the expression of stress and pathogenesis-related proteins by the plant host. The ecological consequences of such interactions are being further investigated in terms of the outcomes of such interactions for the host plants: could rhizobial associations 'defend' highly promiscuous exotic plants against root-knot nematodes? (Costa et al. 2008).

The authors of the *Meloidogyne* spp. genome sequences have indicated and began to investigate gene products that are putatively involved in the nematode-induced modification of plant cell walls to form giant (feeding) cells in the host (Abad et al. 2008; Opperman et al. 2008), which represents a possible future application of this work for bottom-up control of *Meloidogyne* sp. Eleven new putative parasitism genes expressed in the esophageal glands of *M. incognita* have been found, which will permit a better understanding of the evolution and biology of nematode-plant interactions and of plant parasitism in a wider scale. Specific innate immunity genes similar to those found in *C. elegans* were also found in the root-knot nematode, but in much smaller number; conversely, several candidate fucosyl-transferases can be expressed on the cuticle of *M. incognita* (Abad et al. 2008), and this could denote a much larger investment by root-knot nematodes on evading host recognition than on defence against natural enemy attack. Again, this new knowledge can be exploited on the development of natural control mechanisms towards biological (in this case top-down) control strategies. Once only female *Meloidogyne* spp. are parasitic on plant roots, the elucidation of the genetics of sex determination (Abad et al. 2008; Opperman et al. 2008), allied to ecological studies of the modulation of gene expression by environmental factors, could have great importance in re-thinking biological control strategies.

Studies on nematode control in natural ecosystems, and particularly top-down control, still depend more than would be desirable on the tentative interpretation of available ecological theory, namely that of insect control. Much data is still being gathered through population dynamics studies done through intensive sampling and conventional identification and enumeration of individual groups, and through mesocosm or pot experiments with the addition or subtraction of soil biota through physical techniques. Where molecular studies have been applied, the results have surprised us. Some aspects of soil ecology can only be understood through the investigation of genotypes, phenotypes and of their plasticity in response to biotic and abiotic factors.

The challenge for biological control scientists, as for current biology as a whole (Zheng and Dicke 2008), seems to remain one of integrating research schools that have traditionally been seen as separate, such as ecology and plant pathology, but also research disciplines that investigate different levels of organisation, from molecules to ecosystems.

References

Abad P, Gouzy J, Aury JM et al (2008) Genome sequence of the metazoan plant-parasitic nematode *Meloidogyne incognita*. Nat Biotechnol 26:909–915

Ahren D, Tunlid A (2003) Evolution of parasitism in nematode-trapping fungi. J Nematol 35:194–197

Allison SD (2006) Brown ground: a soil carbon analogue for the green world hypothesis? Am Nat 167:619–627

Anderson RM, May RM (1981) The population dynamics of micro-parasites and their invertebrate hosts. Philos Trans R Soc B 291:451–524

Avendano F, Pierce FJ, Melakeberhan H (2004) The relationship between soybean cyst nematode seasonal population dynamics and soil texture. Nematology 6:511–525

Bardgett RD, Wardle DA (2003) Herbivore-mediated linkages between aboveground and belowground communities. Ecology 84:2258–2268

Bardgett RD, Hobbs PJ, Frostegard A (1996) Changes in soil fungal: bacterial biomass ratios following reductions in the intensity of management of an upland grassland. Biol Fertil Soils 22:261–264

Bardgett RD, Cook R, Yeates GW et al (1999a) The influence of nematodes on below-ground processes in grassland ecosystems. Plant Soil 212:23–33

Bardgett RD, Denton CS, Cook R (1999b) Below-ground herbivory promotes soil nutrient transfer and root growth in grassland. Ecol Lett 2:357–360

Bardgett RD, Lovell RD, Hobbs PJ et al (1999c) Seasonal changes in soil microbial communities along a fertility gradient of temperate grasslands. Soil Biol Biochem 31:1021–1030

Barron GL (1977) The nematode destroying fungi. Canadian Publications Ltd., Ontario

Barron GL (2003) Predatory fungi, wood decay, and the carbon cycle. Biodivers 4:3–9

Bauer WD, Mathesius U (2004) Plant responses to bacterial quorum sensing signals. Curr Opin Plant Biol 7:429–433

Bell NL, Watson RN (2001) Population dynamics of *Paratylenchus nanus* in soil under pasture: 1. Aggregation and abiotic factors. Nematology 3:187–197

Berg G, Opelt K, Zachow C et al (2006) The rhizosphere effect on bacteria antagonistic towards the pathogenic fungus *Verticillium* differs depending on plant species and site. FEMS Microbiol Ecol 56:250–261

Bever JD (1994) Feedback between plants and their soil communities in an old field community. Ecology 75:1965–1977

Bever JD (2003) Soil community feedback and the coexistence of competitors: conceptual frameworks and empirical tests. New Phytol 157:465–473

Bezemer TM, Fountain M, Barea J et al (2010) Divergent composition but similar function of soil food webs beneath individual plants: plant species and community effects. Ecology. 91: 3027–3036

Blaxter ML, De Ley P, Garey JR et al (1998) A molecular evolutionary framework for the phylum *Nematoda*. Nature 392:71–75

Blaxter M, Mann J, Chapman T et al (2005) Defining operational taxonomic units using DNA barcode data. Philos Trans R Soc B 360:1935–1943

Bongers T (1990) The maturity index – an ecological measure of environmental disturbance based on nematode species composition. Oecologia 83:14–19

Bongers T, Bongers M (1998) Functional diversity of nematodes. Appl Soil Ecol 10:239–251

Brinkman EP, Duyts H, Van der Putten WH (2005a) Competition between endoparasitic nematodes and effect on biomass of *Ammophila arenaria* (marram grass) as affected by timing of inoculation and plant age. Nematology 7:169–178

Brinkman EP, Duyts H, Van der Putten WH (2005b) Consequences of variation in species diversity in a community of root-feeding herbivores for nematode dynamics and host plant biomass. Oikos 110:417–427

Brinkman EP, Troelstra SR, van der Putten WH (2005c) Soil feedback effects to the foredune grass *Ammophila arenaria* by endoparasitic root-feeding nematodes and whole soil communities. Soil Biol Biochem 37:2077–2087

Brinkman EP, Duyts H, van der Putten WH (2008) Interactions between root-feeding nematodes depend on plant species identity. Soil Biol Biochem 40:2186–2193

Christian CE (2001) Consequences of a biological invasion reveal the importance of mutualism for plant communities. Nature 413:635–639

Costa SR (2006) Interactions between nematodes and their microbial enemies in a sand dune ecosystem. PhD thesis biological sciences, Lancaster University, Lancaster

Costa SR, Kerry BR, Bardgett RD et al (2006) Exploitation of immunofluorescence for the quantification and characterization of small numbers of *Pasteuria* endospores. FEMS Microbiol Ecol 58:593–600

Costa SR, Freitas H, Mathesius U (2008) Interactions between nematodes and rhizobia: from proteomics to plant distribution. 5th international congress of nematology, Brisbane, 2008

De Deyn GB, Raaijmakers CE, van Ruijven J et al (2004) Plant species identity and diversity effects on different trophic levels of nematodes in the soil food web. Oikos 106:576–586

de Gives PM, Davies KG, Clark SJ et al (1999) Predatory behaviour of trapping fungi against *srf* mutants of *Caenorhabditis elegans* and different plant and animal parasitic nematodes. Parasitology 119:95–104

de la Pena E, Rodriguez-Echeverria S, van der Putten WH et al (2006) Mechanism of control of root-feeding nematodes by mycorrhizal fungi in the dune grass *Ammophila arenaria*. New Phytol 169:829–840

de la Pena E, Vandegehuchte M, Bonte D et al (2008) Analysis of the specificity of three root-feeders towards grasses in coastal dunes. Plant Soil 310:113–120

De Mesel I, Derycke S, Moens T et al (2004) Top-down impact of bacterivorous nematodes on the bacterial community structure: a microcosm study. Environ Microbiol 6:733–744

De Ruiter PC, Neutel AM, Moore JC (1995) Energetics, patterns of interaction strengths, and stability in real ecosystems. Science 269:1257–1260

Deacon JW, Saxena G (1997) Orientated zoospore attachment and cyst germination in *Catenaria anguillulae*, a facultative endoparasite of nematodes. Mycol Res 101:513–522

Denton CS, Bardgett RD, Cook R et al (1999) Low amounts of root herbivory positively influence the rhizosphere microbial community in a temperate grassland soil. Soil Biol Biochem 31:155–165

DeWalt SJ, Denslow JS, Ickes K (2004) Natural-enemy release facilitates habitat expansion of the invasive tropical shrub *Clidemia hirta*. Ecology 85:471–483

Dieterich C, Clifton SW, Schuster LN et al (2008) The *Pristionchus pacificus* genome provides a unique perspective on nematode lifestyle and parasitism. Nat Genet 40:1193–1198

Donn S, Daniell TJ, Griffiths BS et al (2007) T-RFLP approaches to nematode assemblage analysis. J Nematol 39:81

Dromph KM, Cook R, Ostle NJ et al (2006) Root parasite induced nitrogen transfer between plants is density dependent. Soil Biol Biochem 38:2495–2498

Duponnois R, Fargette M, Fould S et al (2000) Diversity of the bacterial hyperparasite *Pasteuria penetrans* in relation to root-knot nematodes (*Meloidogyne* spp.) control on *Acacia holosericea*. Nematology 2:435–442

Ebert D, Hamilton WD (1996) Sex against virulence: the coevolution of parasitic diseases. Trends Ecol Evol 11:A79–A82

Ehrlich PR, Raven PH (1964) Butterflies and plants: a study in coevolution. Evolution 18:586–608

Ettema CH (1998) Soil nematode diversity: species coexistence and ecosystem function. J Nematol 30:159–169

Ettema CH, Wardle DA (2002) Spatial soil ecology. Trends Ecol Evol 17:177–183

Ettema CH, Yeates GW (2003) Nested spatial biodiversity patterns of nematode genera in a New Zealand forest and pasture soil. Soil Biol Biochem 35:339–342

Ettema CH, Coleman DC, Vellidis G et al (1998) Spatiotemporal distributions of bacterivorous nematodes and soil resources in a restored riparian wetland. Ecology 79:2721–2734

Ettema CH, Rathbun SL, Coleman DC (2000) On spatiotemporal patchiness and the coexistence of five species of *Chronogaster* (Nematoda: Chronogasteridae) in a riparian wetland. Oecologia 125:444–452

FAOSTAT (2009) Food and agriculture organization of the United Nations http://faostat.fao.org. Accessed 21 Oct 2009

Farrell FC, Jaffee BA, Strong DR (2006) The nematode-trapping fungus *Arthrobotrys oligospora* in soil of the Bodega marine reserve: distribution and dependence on nematode-parasitized moth larvae. Soil Biol Biochem 38:1422–1429

Ferris H, Mullens TA, Foord KE (1990) Stability and characteristics of spatial description parameters for nematode populations. J Nematol 22:427–439

Floyd R, Abebe E, Papert A et al (2002) Molecular barcodes for soil nematode identification. Mol Ecol 11:839–850

Foucher A, Bongers T, Noble LR et al (2004) Assessment of nematode biodiversity using DGGE of 18 S rDNA following extraction of nematodes from soil. Soil Biol Biochem 36:2027–2032

Freckman DW, Mankau R (1986) Abundance, distribution, biomass and energetics of soil nematodes in a Northern Mojave Desert ecosystem. Pedobiologia 29:129–142

Freckman DW, Mankau R, Ferris H (1975) Nematode community structure in desert soils – nematode recovery. J Nematol 7:343–346

Ghedin E, Wang SL, Spiro D et al (2007) Draft genome of the filarial nematode parasite *Brugia malayi*. Science 317:1756–1760

Gommers FJ (1981) Biochemical interactions between nematodes and plants and their relevance to control. Helmin Abstr 50:9–24

Goralczyk K (1998) Nematodes in a coastal dune succession: indicators of soil properties? Appl Soil Ecol 9:465–469

Grayston SJ, Wang SQ, Campbell CD et al (1998) Selective influence of plant species on microbial diversity in the rhizosphere. Soil Biol Biochem 30:369–378

Gruntman M, Novoplansky A (2004) Physiologically mediated self/non-self discrimination in roots. Proc Natl Acad Sci USA 101:3863–3867

Haase S, Ruess L, Neumann G et al (2007) Low-level herbivory by root-knot nematodes (*Meloidogyne incognita*) modifies root hair morphology and rhizodeposition in host plants (*Hordeum vulgare*). Plant Soil 301:151–164

Hairston NG, Smith FE, Slobodkin LB (1960) Community structure, population control, and competition. Am Nat 94:421–425

Hastings A, Powell T (1991) Chaos in a 3-species food-chain. Ecology 72:896–903

Heemsbergen DA, Berg MP, Loreau M et al (2004) Biodiversity effects on soil processes explained by interspecific functional dissimilarity. Science 306:1019–1020

Hillocks RJ (2001) The implications for plant health of nematode-fungal interactions in the root zone. In: Jeger MJ, Spence NJ (eds) Biotic interactions in plant-pathogen associations. CAB International, Oxford, pp 269–283

Hodge A (2006) Plastic plants and patchy soils. J Exp Bot 57:401–411

Holterman M, van der Wurff A, van den Elsen S et al (2006) Phylum-wide analysis of SSU rDNA reveals deep phylogenetic relationships among nematodes and accelerated evolution toward crown clades. Mol Biol Evol 23:1792–1800

Holterman M, Karssen G, van den Elsen S et al (2009) Small subunit rDNA-based phylogeny of the Tylenchida sheds light on relationships among some high-impact plant-parasitic nematodes and the evolution of plant feeding. Phytopathology 99:227–235

Hutchinson GE (1959) Homage to Santa-Rosalia or why are there so many kinds of animals. Am Nat 93:145–159

Ingham RE, Trofymow JA, Ingham ER et al (1985) Interactions of bacteria, fungi, and their nematode grazers – effects on nutrient cycling and plant-growth. Ecol Mongr 55:119–140

Jaffee BA, Strong DR (2005) Strong bottom-up and weak top-down effects in soil: nematode-parasitized insects and nematode-trapping fungi. Soil Biol Biochem 37:1011–1021

Jaffee BA, Muldoon AE, Tedford EC (1992) Trap production by nematophagous fungi growing from parasitized nematodes. Phytopathology 82:615–620

Johnson SN, Gregory PJ (2006) Chemically-mediated host-plant location and selection by root-feeding insects. Physiol Entomol 31:1–13

Jones FGW, Jones MG (1964) Pests of field crops. Edward Arnold Ltd., London, p 406 p

Kawecki TJ (1998) Red queen meets Santa Rosalia: arms races and the evolution of host specialization in organisms with parasitic lifestyles. Am Nat 152:635–651

Kerry BR, Hominick WM (2002) Biological control. In: Lee DL (ed) The biology of nematodes. Taylor & Francis, London, pp 483–509

Klironomos JN (2002) Feedback with soil biota contributes to plant rarity and invasiveness in communities. Nature 417:67–70

Klironomos JN (2003) Variation in plant response to native and exotic arbuscular mycorrhizal fungi. Ecology 84:2292–2301

Korthals GW, Smilauer P, Van Dijk C et al (2001) Linking above- and below-ground biodiversity: abundance and trophic complexity in soil as a response to experimental plant communities on abandoned arable land. Funct Ecol 15:506–514

Kowalchuk GA, Stephen JR, DeBoer W et al (1997) Analysis of ammonia-oxidizing bacteria of the beta subdivision of the class proteobacteria in coastal sand dunes by denaturing gradient gel electrophoresis and sequencing of PCR-amplified 16S ribosomal DNA fragments. Appl Environ Microbiol 63:1489–1497

Kowalchuk GA, De Souza FA, Van Veen JA (2002) Community analysis of arbuscular mycorrhizal fungi associated with *Ammophila arenaria* in Dutch coastal sand dunes. Mol Ecol 11:571–581

Laakso J, Setala H, Palojarvi A (2000) Influence of decomposer food web structure and nitrogen availability on plant growth. Plant Soil 225:153–165

Lee DL (2002) Life cycles. In: Lee DL (ed) The biology of nematodes. Taylor & Francis, London

Lee BS, Lee HB, Choi SW et al (2005) Effective screening of antagonist for the biological control of soilborne infectious disease (damping-off). J Microbiol Biotechnol 15:701–709

Lewis T, Taylor LR (1967) Introduction to experimental ecology. Academic, London

Liang W, Steinberger Y (2001) Temporal changes in nematode community structure in a desert ecosystem. J Arid Environ 48:267–280

Luttbeg B, Rowe L, Mangel M (2003) Prey state and experimental design affect relative size of trait – and density-mediated indirect effects. Ecology 84:1140–1150

Mathesius U (2003) Conservation and divergence of signalling pathways between roots and soil microbes – the Rhizobium-legume symbiosis compared to the development of lateral roots, mycorrhizal interactions and nematode-induced galls. Plant Soil 255:105–119

Mauchline TH, Kerry BR, Hirsch PR (2004) The biocontrol fungus *Pochonia chlamydosporia* shows nematode host preference at the infraspecific level. Mycol Res 108:161–169

McSorley R, Duncan LW (2004) Population dynamics. Nematology: advances and perspectives, vol 1, Nematode morphology, physiology and ecology. CABI Publishing, Wallingford, pp 469–492

McSorley R, Frederick JJ (2004) Effect of extraction method on perceived composition of the soil nematode community. Appl Soil Ecol 27:55–63

Mitra RM, Shaw SL, Long SR (2004) Six nonnodulating plant mutants defective for Nod factor-induced transcriptional changes associated with the legume-rhizobia symbiosis. Proc Natl Acad Sci USA 101:10217–10222

Mitreva M, Blaxter ML, Bird DM et al (2005) Comparative genomics of nematodes. Trends Genet 21:573–581

Moore JC, Hunt HW (1988) Resource compartmentation and the stability of real ecosystems. Nature 333:261–263

Moore JC, Berlow EL, Coleman DC et al (2004) Detritus, trophic dynamics and biodiversity. Ecol Lett 7:584–600

Muyzer G, Dewaal EC, Uitterlinden AG (1993) Profiling of complex microbial populations by denaturing gradient gel-electrophoresis analysis of polymerase chain reaction-amplified genes coding for 16 s ribosomal-RNA. Appl Environ Microbiol 59:695–700

Neutel AM, Heesterbeek JAP, de Ruiter PC (2002) Stability in real food webs: weak links in long loops. Science 296:1120–1123

O'Donnell AG, Colvan SR, Malosso E et al (2005) Twenty years of molecular analysis of bacterial communities and what have we learned about function? In: Bardgett RD, Usher MB, Hopkins DW (eds) Biological diversity and function in soils. Cambridge University Press, Cambridge, pp 44–56

Opperman CH, Bird DM, Williamson VM et al (2008) Sequence and genetic map of *Meloidogyne hapla*: a compact nematode genome for plant parasitism. Proc Natl Acad Sci USA 105:14802–14807

Pen-Mouratov S, Steinberger Y (2005) Spatio-temporal dynamic heterogeneity of nematode abundance in a desert ecosystem. J Nematol 37:26–36

Piskiewicz AM, Duyts H, Berg MP et al (2007) Soil microorganisms control plant ectoparasitic nematodes in natural coastal foredunes. Oecologia 152:505–514

Piskiewicz AM, Duyts H, van der Putten WH (2008) Multiple species-specific controls of root-feeding nematodes in natural soils. Soil Biol Biochem 40:2729–2735

Piskiewicz AM, de Milliano MJK, Duyts H et al (2009a) Plant ectoparasitic nematodes prefer roots without their microbial enemies. Plant Soil 316:277–284

Piskiewicz AM, Duyts H, van der Putten WH (2009b) Soil microorganisms in coastal foredunes control the ectoparasitic root-feeding nematode *Tylenchorhynchus ventralis* by local interactions. Funct Ecol 23:621–626

Powers T (2004) Nematode molecular diagnostics: from bands to barcodes. Annu Rev Phytopathol 42:367–383

Price PW, Bouton CE, Gross P et al (1980) Interactions among 3 trophic levels – influence of plants on interactions between insect herbivores and natural enemies. Annu Rev Ecol Syst 11:41–65

Rantalainen ML, Haimi J, Setala H (2004) Testing the usefulness of habitat corridors in mitigating the negative effects of fragmentation: the soil faunal community as a model system. Appl Soil Ecol 25:267–274

Rantalainen ML, Haimi J, Fritze H et al (2006) Effects of small-scale habitat fragmentation, habitat corridors and mainland dispersal on soil decomposer organisms. Appl Soil Ecol 34:152–159

Rantalainen ML, Haimi J, Fritze H et al (2008) Soil decomposer community as a model system in studying the effects of habitat fragmentation and habitat corridors. Soil Biol Biochem 40:853–863

Rasmann S, Turlings TCJ (2008) First insights into specificity of belowground tritrophic interactions. Oikos 117:362–369

Rasmann S, Kollner TG, Degenhardt J et al (2005) Recruitment of entomopathogenic nematodes by insect-damaged maize roots. Nature 434:732–737

Reitz M, Rudolph K, Schroder I et al (2000) Lipopolysaccharides of *Rhizobium etli* strain G12 act in potato roots as an inducing agent of systemic resistance to infection by the cyst nematode *Globodera pallida*. Appl Environ Microbiol 66:3515–3518

Rodriguez-Echeverria S, Crisostomo J, Freitas H (2004) Arbuscular mycorrhizal fungi associated with *Ammophila arenaria* L. in European coastal sand dunes. In: Arianotsou M, Papanastasis D (eds) Proceedings of the 10th international conference on mediterranean climate ecosystems, Rhodes, IOS Press, Amsterdam, The Netherlands 2004, pp 1–7

Rodriguez-Kabana R (1991) Control biologico de nematodos parasitos de plantas. Nematropica 21:111–122

Schnee C, Kollner TG, Held M et al (2006) The products of a single maize sesquiterpene synthase form a volatile defense signal that attracts natural enemies of maize herbivores. Proc Natl Acad Sci USA 103:1129–1134

Scholl EH, Thorne JL, McCarter JP et al (2003) Horizontally transferred genes in plant-parasitic nematodes: a high-throughput genomic approach. Genome Biol 4:R19

Siddiqui ZA, Mahmood I (1996) Biological control of plant parasitic nematodes by fungi: a review. Biores Technol 58:229–239

Signor PW (1994) Biodiversity in geological time. Am Zool 34:23–32

Stein LD, Bao Z, Blasiar D et al (2003) The genome sequence of *Caenorhabditis briggsae*: a platform for comparative genomics. Public Libr Sci Biol 1:E45

Strauss SY (1991) Indirect effects in community ecology – their definition, study and importance. Trends Ecol Evol 6:206–210

Strong DR, Whipple AV, Child AL et al (1999) Model selection for a subterranean trophic cascade: root-feeding caterpillars and entomopathogenic nematodes. Ecology 80:2750–2761

Stubbs WJ, Wilson JB (2004) Evidence for limiting similarity in a sand dune community. J Ecol 92:557–567

Sturz AV, Kimpinski J (2004) Endoroot bacteria derived from marigolds (*Tagetes* spp.) can decrease soil population densities of root-lesion nematodes in the potato root zone. Plant Soil 262:241–249

The CeSC (1998) Genome sequence of the nematode *Caenorhabditis elegans*. A platform for investigating biology. Science 282:2012–2018

Trudgill DL, Blok VC (2001) Apomictic, polyphagous root-knot nematodes: exceptionally successful and damaging biotrophic root pathogens. Annu Rev Phytopathol 39:53–77

Uhlenbroek JH, Bijloo JD (1958) Investigations on nematicides. 1. Isolation and structure of a nematicidal principe occurring in Tagetes roots. Recl Trav Chim Pay B 77:1004–1009

van Den Berg W, Rossing WAH (2005) Generalized linear dynamics of a plant-parasitic nematode population and the economic evaluation of crop rotations. J Nematol 37:55–65

van der Heijden MGA, Klironomos JN, Ursic M et al (1998) Mycorrhizal fungal diversity determines plant biodiversity, ecosystem variability and productivity. Nature 396:69–72

van der Putten WH, Peters BAM (1997) How soil-borne pathogens may affect plant competition. Ecology 78:1785–1795

van der Putten WH, Van Dijk C, Peters BAM (1993) Plant-specific soil-borne diseases contribute to succession in foredune vegetation. Nature 362:53–55

van der Putten WH, Cook R, Costa S et al (2006) Nematode interactions in nature: models for sustainable control of nematode pests of crop plants? Adv Agron 89:227–260

van der Stoel CD, Duyts H, Van der Putten WH (2006) Population dynamics of a host-specific root-feeding cyst nematode and resource quantity in the root zone of a clonal grass. Oikos 112:651–659

van Elsas JD, Duarte GF, Keijzer-Wolters A et al (2000) Analysis of the dynamics of fungal communities in soil via fungal-specific PCR of soil DNA followed by denaturing gradient gel electrophoresis. J Microbiol Meth 43:133–151

van Valen L (1973) A new evolutionary law. Evol Theory 1:1–30

van Valen L (1975) Reply to Foin et al. Nature 257: 515–516

van Valen L (1976) The red queen. Am Nat 110:809–810

van Veen FJF, Morris RJ, Godfray HCJ (2006) Apparent competition, quantitative food webs, and the structure of phytophagous insect communities. Annu Rev Entomol 51:187–208

Viketoft M, Palmborg C, Sohlenius B et al (2005) Plant species effects on soil nematode communities in experimental grasslands. Appl Soil Ecol 30:90–103

Villenave C, Cadet P (1998) Interactions of *Helicotylenchus dihystera*, *Preatylenchus pseudopratensis*, and *Tylenchorhynchus gladiolatus* on two plants from te soudano-sahelian zone of West Africa. Nematropica 28:31–39

Waite IS, O'Donnell AG, Harrison A et al (2003) Design and evaluation of nematode 18 S rDNA primers for PCR and denaturing gradient gel electrophoresis (DGGE) of soil community DNA. Soil Biol Biochem 35:1165–1173

Wardle DA, Yeates GW, Williamson W et al (2003) The response of three trophic level soil food web to the identity and diversity of plant species and functional groups. Oikos 102:45–56

Wardle DA, Bardgett RD, Klironomos JN et al (2004) Ecological linkages between aboveground and belowground biota. Science 304:1629–1633

Watt M, Silk WK, Passioura JB (2006) Rates of root and organism growth, soil conditions, and temporal and spatial development of the rhizosphere. Ann Bot 97:839–855

Whipps JM, Davies KG (2000) Success in biological control of plant pathogens and nematodes by microorganisms. In: Gurr G, Wratten S (eds) Biological control: measures of success. Kluwer, Dordrecht, pp 231–269

White EM, Wilson JC, Clarke AR (2006) Biotic indirect effects: a neglected concept in invasion biology. Divers Distrib 12:443–455

Wooton JT (1994) The nature and consequences of indirect effects in ecological communities. Annu Rev Ecol Syst 25:443–466

Wu TH, Ayres E, Li G, Bardgett RD et al (2009) Molecular profiling of soil animal diversity in natural ecosystems: incongruence of molecular and morphological results. Soil Biol Biochem 41:849–857

Wurst S, van Beersum S, Wagenaar R et al (2009) Plant defence against nematodes is not mediated by changes in the soil microbial community. Funct Ecol 23:488–495

Yeates GW (1987) How plants affect nematodes. Adv Ecol Res 17:63–113

Yeates GW, Bongers T, de Goede RGM et al (1993) Feeding habits in soil nematode families and genera – an outline for soil ecologists. J Nematol 25:315–331

Yorozuya H (2006) Effects of parasitoids on a mycophagous drosophilid community in northern Japan and an evaluation of the disproportionate parasitism hypothesis. Entomol Sci 9:13–22

Young IM, Ritz K (2005) The habitat of soil microbes. In: Bardgett RD, Usher MB, Hopkins DW (eds) Biological diversity and function in soils. Cambridge University Press, Cambridge, pp 31–43

Zheng SJ, Dicke M (2008) Ecological genomics of plant-insect interactions: from gene to community. Plant Physiol 146:812–817

Chapter 3
Microbial Interactions with *Caenorhabditis elegans*: Lessons from a Model Organism

Maria J. Gravato-Nobre and Jonathan Hodgkin

Abstract In recent years, the study of invertebrate innate immune defense responses has been greatly expanded by the use of the powerful tractable model *Caenorhabditis elegans*. Because of the accessible mechanisms underpinning its innate immune system, the worm has become into a valuable model for identifying core strategies of microbial pathogenicity and host defense. *C. elegans*-microbial interaction studies have revealed a conservation of both pathogen virulence factors and metazoan immune repertoires. In *C. elegans* the signaling pathways involved in orchestrating immune responses are: three mitogen-activated protein kinases (p38, JNK and ERK), the unfolded protein response (serine threonine/kinase IRE-1 and PQN/ABU proteins), the transforming growth factor-β (TGF-β), the insulin-like receptor (DAF-2), the Wnt/Wingless β-catenin (BAR-1), and the component of programmed cell death BCL-2 homolog (CED-9). These pathways also serve major developmental, behavioral and metabolic functions.

Abbreviations

ABU	activated in blocked unfolded protein response
AGE-1	aging alteration
Apaf-1	apoptosis protease activating factor-1
ASK1	apoptosis signaling-regulating kinase-1
BAR-1	beta-catenin/armadillo related
CED	cell death abnormality
DAF	dauer formation abnormal
DBL-1	decapentaplegic/BMP-like
EGL	egg-laying defective

M.J. Gravato-Nobre (✉) and J. Hodgkin
Department of Biochemistry, University of Oxford, South Parks Road, Oxford OX1 3QU, UK
e-mail: Maria.Gravato-Nobre@bioch.ox.ac.uk

K. Davies and Y. Spiegel (eds.), *Biological Control of Plant-Parasitic Nematodes: Building Coherence between Microbial Ecology and Molecular Mechanisms*, Progress in Biological Control 11, DOI 10.1007/978-1-4020-9648-8_3,

ELT	erythroid-Like Transcription factor family
ERK	extracellular signal-regulated kinase
FOXO	forkhead box class O
IGF-1	insulin-like growth factor-1
INS-7	insulin-related
IRE-1	IRE1 kinase related
JNK	C-Jun amino-terminal kinase
KGB-1	kinase, GLH-binding
MAP	mitogen-activated protein
MEK-1	MAP kinase kinase
MKK	MAP kinase kinase
MKK7	MAP kinase kinase
MPK-1	MAP kinase
NIPI-3	no Induction of Peptide after *Drechmeria* Infection
NSY-1	neuronal symmetry
PMK-1	p38 MAP kinase family
SREC	scavenger receptors expressed by endothelial cells
SARM	sterile alpha and HEAT/Armadillo motif protein
SMA	mothers against decapentaplegic homolog
SUR-2	suppressor of activated let-60 Ras
TGF-β	transforming growth factor β
TIR-1	Toll and IL-1 receptor
TRF-1	tumor necrosis factor receptor associated factor (TRAF) homolog

3.1 Introduction

To survive multiple microbial encounters, invertebrates have to perceive insults by deleterious organisms and respond adequately by activating efficient immune defense responses. In the soil, nematodes are constantly threatened by large and diverse arrays of microbes and substantial evidence indicates that they have evolved immune defense systems to combat pathogens. As in all organisms, an efficient nematode defense response must have the capacity firstly to recognize potentially harmful pathogens; secondly activate immune defense networks; and thirdly suppress pathogen activity. These are intrinsic properties endowed by their "hard-wired" innate immune system. Despite the relative paucity of information regarding the ecology and the microbial flora in its natural environments, the free-living nematode *Caenorhabditis elegans* has been extensively used as an experimental platform to uncover signal transduction pathways involved in metazoan innate immune reactions, as well to reveal strategies deployed by pathogens to subvert such defenses (see reviews: Ausubel 2005; Gravato-Nobre and Hodgkin 2005; Kim and Ausubel 2005; Millet and Ewbank 2004; Nicholas and Hodgkin 2004b; Schulenburg and Ewbank 2007).

To combat infection, *C. elegans* uses an array of preexisting and inducible immune strategies, which in part, are shared with higher organisms. Such defense responses result in the production of many immune effectors with antibacterial and fungal activities. Potent antimicrobial peptides, lysozymes, lipases and C-type lectins are amongst the most likely candidate effectors. In this review we consider some of the major functional and ecological differences between free-living and plant-parasitic nematodes. We discuss some presumptive *C. elegans*-microbial associations likely encountered in the wild by highlighting the most prominent molecular players of such interactions. We provide an overview of the mechanisms underlying *C. elegans* immune signaling responses.

3.2 The *C. elegans* Paradigm for Host-Microbial Associations

Why has *C. elegans* been extensively used to investigating host-microbial associations? One answer lies in the powerful resources available for forward and reverse genetic analysis in this organism. Mutagenesis to generate *C. elegans* strains resistant or hypersensitive to pathogens can be easily accomplished (Darby et al. 2007; Gravato-Nobre and Hodgkin 2005; Marroquin et al. 2000; Pradel and Ewbank 2004; Tan and Ausubel 2000; Yook and Hodgkin 2007). *C. elegans* is a convenient, simple, ethically acceptable, and inexpensive model host. With its 3-day generation time at 20ºC, and 1 mm adult length, the worm can easily be grown in microtiter plates and is small enough for automated liquid sorting. This feature makes it amenable for whole-organism *in vivo* screening of libraries of antimicrobial compounds (Breger et al. 2007; Moy et al. 2006). The small cell number (959 somatic nuclei) and the transparency of its body also make *C. elegans* attractive for studying microbial infections. By using transgenic lines bearing fluorescent reporters, *C. elegans* is well suited for *in vivo* imaging of genes specifically modulated upon infection. Moreover, appropriately tagged pathogens can easily be followed in vivo and used to dissect biological aspects of their host pathogenicity (Sifri et al. 2003; Tenor and Aballay 2008).

In laboratory culture, *C. elegans* is grown on bacterial lawns. Such property is advantageous for two main reasons: firstly, because it provides a convenient route of microbial infection analysis and, secondly, bacterial feeding can be used for the application of RNA interference (RNAi)-based gene silencing. Genome-scale screens using feeding RNAi are now commonplace (Boutros and Ahringer 2008). RNAi has been efficiently used to inhibit gene function and to identify new resistance and susceptibility genes that cannot easily be targeted by using mutagenesis-based approaches. Homologues of genes implicated in defense in other organisms can rapidly be tested in *C. elegans* by this means (Alper et al. 2007, 2008). Using a combination of genetic strategies, several groups have now been able to establish the roles of transducers that mediate interactions between *C. elegans* and pathogenic microorganisms. One theme emerging from such approaches is that the innate immune mechanisms of *C. elegans* and that of the flies, mammals and plants, share striking similarities (Alper et al. 2007; Kurz and Ewbank 2003; O'Rourke et al. 2006; Tan and Ausubel 2000).

3.3 Nematode Feeding Styles Diversity: of Muscles, and Needles

Nematodes are among the most abundant and successful of all the metazoa on Earth (Kimpinski and Sturz 2003). Their success may be due to the common features of a simple body plan, highly resistant collagenous cuticle and ready adaptation to unfavorable environmental conditions. Despite these similarities, the phylum Nematoda embraces a considerable diversity of body specializations, life-styles and ecological niches. First of all, fundamental differences exist within nematodes in respect to their feeding styles and requirements. While there are vast numbers of free-living nematodes (FLN) that feed on bacteria, fungi and other small organisms, there are also many nematodes that parasitize plants and animals. Plant-parasitic nematodes (PPN) have profound impact on plant health and productivity. They represent at least three distinct clades in the nematode tree of life, reflective of their long and independent evolutionary histories (Baldwin et al. 2004). However, they all share a functional needle-like protrusible stylet, in some cases a tooth. This spear is used to puncture holes and to withdraw the contents of host plant cells. This morphological specialization is a key adaptation to plant parasitism. The majority of PPN have an obligatory requirement for their cognate host plants for development and reproduction. Micro-herbivores such as *C. elegans*, on the other hand, are common filter-feeders that do not feed on living plant material. Instead of a stylet, they employ a pharynx with considerable muscular pumping power. Thus, bacteria that are devoured through the buccal capsule are broken down by the mechanical action of the pharyngeal grinder (Avery 1993; Avery and Horvitz 1989; Avery and Shtonda 2003). This mechanism, which physically destroys microbial cells, plays a crucial role in preventing live pathogens from reaching the worm's intestine. Animals in which the pharyngeal muscle is compromised, such as the *C. elegans phm-2* mutant, are remarkably sensitive to pathogenic bacteria (Kim et al. 2002; Labrousse et al. 2000; Tan 2002).

3.4 Nematode Life Styles and Ecological Niches Diversity

Another conspicuous difference between FLN and PPN is the duration of their life cycle: while *C. elegans* take approximately 3 days to complete a generation, PPN exhibit a much longer reproductive cycle, taking at least 2–4 weeks. Moreover, soil nematodes can also inhabit very different ecological niches. For example, although *C. elegans* has been recognized as a probable human commensal that spread around the world in association with agriculture (Thomas 2008) it is primarily found in ephemeral habitats like fruits rotting below trees, and decaying organic matter (Barriere and Felix 2007) at the soil-air interface. The distribution of PPN, on the other hand, is greatest around the roots of host plants, which they follow sometimes to considerable depths (from 30 to 200 cm or more).

3.5 Nematode-Microbial Associations' Diversity: Enemies of Their Own?

In their natural habitat, all soil nematodes are surrounded by a plethora of microorganisms, some of which may present potential threats. Due to their different lifestyles, microbial and plant feeders are bound to experience distinct immune challenges, and are likely to harbor their own specialized lineages of pathogens. In the soil, ingestion of microorganisms by the microbial feeders is probably the main route of nematode infection. In contrast, due to their obligatory plant feeding requirements, PPN cannot feed on microbes even while searching for their host in the soil. Some similarities do exist: the second-stage larvae of the endoparasite *Meloidogyne* spp. can be compared to the well-characterized dauer larvae of *C. elegans*, because both of these juvenile stages are non-feeding, developmentally arrested, slow aging and long-lived (Opperman et al. 2008). Nevertheless, because of morphological adaptations, and because endoparasitic nematodes spend most their lives embedded in the host tissues protected from external challenges, the impact of edible microorganisms on these PPN is likely to be reduced. Hence, major differences between saprophytic and stylet-bearing nematodes are that while the first group may carry viable microorganisms in their digestive tract, PPN do not. Given these discrepancies it is surmised that each of these two groups will likely be targeted by distinct sets of microbial pathogens. A full list of such associations is beyond the scope of this review. However, some examples should be noted. One significant association is when endosymbionts engage in elaborate relationships with nematodes. Intimate interactions have been found between bacteria and nematodes, and demonstrate that microorganisms can live within the host's differentiated cells. One case is the obligate intracellular association between rickettsia-like alphaproteobacteria and the soybean and potato cyst nematodes *Heterodera* spp and *Globodera rostochiensis*, respectively (Endo 1979; Walsh et al. 1983). Although it is unclear whether these organisms are commensals or deleterious to their hosts, the stable interactions they establish suggest that specialized events must have led to the uptake of the prokaryote by the eukaryotic host cells. Examples of other soil nematode endosymbionts come from the entomopathogenic members of the Enterobacteriacae. *Xenorhabdus* and *Photorhabdus* have a mutualistic association with soil-dwelling steinernematid and heterorhabditid nematodes, respectively (David 2008). The symbiotic association of infective juvenile stage *Heterorhabditis bacteriophora* by *P. luminescens* requires a set of elaborate and specific events whereby the symbiont cells can only adhere to and invade certain host cells at specific life-stages of the nematode development. The targets for this invasion are the rectal gland and the pharyngeal valve cells (Ciche et al. 2008). The mutualistic nature of *P. luminescens* and *X. nematophila* does not seem to extend to other nematodes. In fact, *P. luminescens* was shown to be pathogenic to *C. elegans* (Couillault and Ewbank 2002; Sicard et al. 2007). A heat-stable diffusible factor is responsible for the killing ability of *X. nematophila*, while it does not seem to contribute for the *P. luminescens* lethality effect. *X. nematophila* colonizes the head of *C. elegans* displaying a biofilm-like structure that resembles the one formed by *Yersinia* sp (Couillault and Ewbank 2002). A natural endosymbiont of *C. elegans* has yet to be found. The second example relates to virus associations. The vector relationship between plant-parasitic nematodes and plant viruses has been long known. Nematode vectors demonstrate great efficiency and specificity in those viruses they transmit; of the several hundred PPN species no more than few species in the genera *Xiphinema*, *Longidorus*, *Trichodorus* and *Paratrychodorus* are able to transmit viruses. Virus particles are retained on the cuticular lining of the oesophagus and can be maintained after extended periods of starvation. During molting, and as they shed their cuticles, juveniles also lose their ability to transmit viruses. It has been generally assumed that nematodes are only passive virus vectors. This is because virions do not replicate within their hosts and do not seem to affect nematode reproduction (Das and Raski 1969). However, it is possible that these nematode-virus associations do inflict some subtle fitness penalties on the nematodes themselves. The possible consequences of such PPN-virus relationships deserve to be revisited.

Natural viruses for the nematode *C. elegans* have been discovered recently (Félix et al. 2011). In fact, this organism was not considered an appropriate model organism for studying virus–host interactions. Previous efforts to define virus-host responses in *C. elegans* were pursued by several independent groups who, using artificial techniques of introducing viruses into animals, where able to demonstrate that viruses that infect mammalian cells can indeed infect, replicate, and assemble within *C. elegans* cells (Liu et al. 2006; Lu et al. 2005; Schott et al. 2005; Wilkins et al. 2005). Wilkins et al. and Schott et al. were able to show that cultured primary cells from *C. elegans* embryos could be infected with vesicular stomatitis virus (VSV), a negative-strand RNA virus that infects both insects and mammals. In a different set of experiments, Lu et al. examined the replication of the Flock house virus (FHV), a positive-strand RNA virus that has a large spectrum of hosts, including plants, yeast, insects, and mammalian cells. In their studies, FHV RNA was produced in a *C. elegans* transgene which harbored the viral DNA driven by a heat-shock promoter. A *C. elegans* mutation in *rde-1* (rendering the worms RNAi resistant) allowed the FHV RNA to be replicated non-efficiently, suggesting that RNAi may contribute to natural antiviral defense in *C. elegans*.

3.6 *C. elegans* as a Host for Many Pathogenic Bacteria and Fungi

3.6.1 Oral Infection and Colonization of the Gut

As mentioned above, *C. elegans* serves as a facile experimental model host for studying the invasion and colonization by bacterial and fungal pathogens. Pathogens can easily be tested by replacing the normal laboratory food source (*E. coli*), with the pathogen of interest. In this way, many human pathogens have been introduced to worms and shown to be capable of causing visible signs of infection leading to morbidity and mortality. These include the Gram-negative bacteria *Pseudomonas aeruginosa* (Darby et al. 1999; Mahajan-Miklos et al. 1999; Tan et al. 1999a, b), *Salmonella typhimurium* and other *S. enterica serovars* (Labrousse et al. 2000) *Serratia marcescens* (Mallo et al. 2002), *Yersinia pestis* (Styer et al. 2005) and *Burkholderia* (O'Quinn et al. 2001; Kothe et al. 2003), the Gram-positive bacteria *Enterococcus faecalis* (Garsin et al. 2003; Sifri et al. 2002), *Staphylococcus aureus* and *S. epidermis* (Begun et al. 2005, 2007) and the fungi *Cryptococcus neoformans* (Mylonakis et al. 2002), and *Candida albicans* (Breger et al. 2007).

Basically, at least two distinct mechanisms exist by which pathogens can cause infectious disease in *C. elegans*: through the direct pathogen colonization and proliferation of the mucosa (intestinal lumen or cuticular/glycocalyx surfaces) and through toxin-mediated molecules. The majorities of the pathogens that have been tested in *C. elegans* fall into the first group and accumulate within the intestinal tract, after oral uptake. The pathological symptoms they induce can be quantified as decreased lifespan of the nematodes. One distinction between the mammalian and the *C. elegans* infection system is that while the intracellular life style of some mammalian pathogens is well-documented, this pattern of infectious diseases has not been reported until recently in *C. elegans*. Of great interest is the recent report by Troemel et al. that describes a microsporidian fungus able to form colonies and multiply inside the intestinal cells of the nematode *C. elegans* (Troemel et al. 2008).

Some bacteria are able not only to colonize but also to persist in the nematode intestine. Striking examples of bacterial persistence in the gut are those that involve *S. typhimurium*, *S. marcescens* and *E. faecalis* associations (Aballay et al. 2000; Garsin et al. 2001; Kurz et al. 2003). Naïve worms exposed to *S. typhimurium* for a few hours and then transferred to lawns of *E. coli*, died with similar kinetics to those observed in worms that were continuously exposed to the bacterial pathogen (Aballay et al. 2000). In contrast, other pathogens like *S. aureus* do not persistently colonize the digestive tract of *C. elegans* and can be cleared from the nematode gut once the worms are transferred to a new food source (Sifri et al. 2003). Similarly, after transfer to fresh liquid medium, *C. elegans* exposed to *Cryptococcus neoformans* lawns seem able to clear the fungal infection by defecation of the cryptococcal cells (Breger et al. 2007; Mylonakis et al. 2002).

Other bacteria that are not pathogenic to mammals have also been tested in *C. elegans*, some of which reduce the worm lifespan. These include the phytopathogens *Agrobacterium tumefaciens*, *Erwinia carotovora* and *E. chrysanthemi* (Couillault and Ewbank 2002). There are likely to be many more phytopathogens with such an effect, as few have been tested thus far.

3.6.2 Surface Infections

Despite the fact that most of the *C. elegans*-pathogen interactions involve oral ingestion, some microorganisms can cause extra-intestinal infections. They do so by two distinct mechanisms: (1) Ectoparasitism, whereby pathogens adhering to the worm surface, and without physically penetrating the epidermis, are able to establish a harmful interaction; (2) Endoparasitism in which the infectious agents can only cause disease if they gain access to the worm body's cavity, by breaching its cuticle.

The surface of worm can support a diverse array of microorganisms which in turn are able to establish specific associations and cause disease.

One such groups, is the mammalian pathogens which exhibit a distinctive interaction, adhering to the surface of the worm and forming biofilms. This is the case of the plague bacterium *Yersinia pestis*, and the closely-related species *Y. pseudotuberculosis*. These bacteria attach to the cuticle in the head region and form an obstructive matrix that accumulates over the time and prevents normal feeding (Darby et al. 2002; Joshua et al. 2003). Biofilms are polysaccharide rich and their formation requires the hmsHFRS operon in both *Yersinia* species. Disruptions in this locus abrogate the ability of the pathogens to form obstructive biofilms and renders *Yersinia avirulent*. Some *C. elegans* mutants with altered surface characteristics such as *srf-2, srf-3* and *srf-5* are resistant to *Yersinia* colonization by preventing biofilm attachment (Cipollo et al. 2004; Hoflich et al. 2004; Joshua et al. 2003). The loss of *srf-3*-encoded nucleotide sugar transporter of the Golgi leads to altered surface glyconjugates and enable the mutant nematodes to grow unaffected in the presence of the pathogenic bacteria. Interestingly, these *C. elegans* mutants are similarly resistant to the bacterial pathogen *M. nematophilum* (Hodgkin et al. 2000; Hoflich et al. 2004). Cross-genus infection assays have shown that glycoconjugates present on the surface of the worm are critical for the attachment by these two bacterial pathogens. To understand the mechanisms of biofilm formation Darby et al. performed a genetic screen for *C. elegans* mutants with a biofilm absent on the head (Bah) phenotype. ENU mutagenesis identified *bus-4, bus-12*, and *bus-17* and three novel genes: *bah-1-bah-3* mutants displayed distinct cuticle/surface properties but were not resistant to *M. nematophilum*. These mutations suggest that there are unique components in the surface of *C. elegans* that mediate *Yersinia* biofilm formation (Darby et al. 2007).

In flea host, biofilm formation by *Yersinia* is important, but virulence does not involve biofilm formation. Biofilm-independent virulence can also be investigated in *C. elegans* and may lead to the identification of virulence factors that are important for mammalian pathogenesis (Styer et al. 2005). Styer and coworkers described a biofilm-deficient mutant of *Y. pestis* (without the hmsHFRS genes) that causes a persistent and lethal intestinal infection in *C. elegans* by a mechanism different from blocking food intake. They were not only able to identify several genes that had been previously associated with virulence in other bacterial pathogens, such as ompT, y3857 and yapH, but also novel genes that had not been previously implicated in *Y. pestis* pathogenicity in mammals. *Yersinia pestis* KIM5 exposed worms also exhibited a swollen tail which resembled that seen in animals that are infected with a specific *C. elegans* pathogen, *Microbacterium nematophilum*.

Microbes that are non-pathogenic to mammals, and may be natural nematode pathogens, have also been studied with *C. elegans*. These include the fungi *Drechmeria coniospora* (Jansson 1994; Pujol et al. 2008a, b), *Monacrosporium haptotylum* (Fekete et al. 2008), *Streptoverticillium albireticuli* (Park et al. 2002),

Duddingtonia flagrans (Mendoza De Gives et al. 1999) and Gram positive bacteria *Leucobacter chromiireducens* subsp. *solipictus* (Muir and Tan 2008) and *Microbacterium nematophilum* (Hodgkin et al. 2000). In common they are extra-intestinal and their pathological effects are due to their interaction with the nematode surfaces. Some like *Monacrosporium haptotylum* and *Duddingtonia flagrans*, develop traps which adhere and penetrate to the whole cuticle. Trapped nematodes then become paralyzed, and as hyphae penetrate and invade the internal body, rapid tissue degradation occurs which eventually lead to nematode death. Endoparasitic fungi such *S. albireticuli*, on the other hand, do not form trapping organs, but instead use their spores to infect and invade the nematode host (Park et al. 2002). Belonging to the same group is the fungal pathogen *D. coniospora*, which establishes a rather more localized association with *C. elegans*. Conidia from this obligatory endoparasitic pathogen attach to the surface in the region of the mouth and vulva. Subsequently, specialized appressoria are formed and penetrate the cuticle, allowing hyphae to spread throughout the whole body of the nematode, leading to its death (Couillault et al. 2004; Jansson 1994).

At present, two Microbacteriaceae have been described to cause infection in *C. elegans*. These are *L. chromiireducens* (Muir and Tan 2008) and *M. nematophilum* (Hodgkin et al. 2000) which cause localized infections in the uterus and the rectal region, respectively. However, whereas *L. chromiireducens* cause lethality in *C. elegans*, *M. nematophilum* establishes a non-lethal interaction but the worms show severe constipation and a decrease in the overall fecundity (Hodgkin et al. 2000).

The ability of *C. elegans* to survive infection by *M. nematophilum* does not extend to all the members of the genus *Caenorhabditis*. In fact, a number of *Caenorhabditis* species succumb to this bacterial infection during early larval development (Akimkina et al. 2006). Especially susceptible are hatchlings of *C. plicata* and *C. drosophilae*, which rarely or never mature to adulthood in the presence of *M. nematophilum*. While a natural isolate of *M. nematophilum* has not yet been found in the wild, it is interesting that a related Microbacterium species, the endophytic bacterium *M. esteraromaticum* has been shown to play a role in the suppression of the root-lesion nematode, *Pratylenchus pentrans* through the attenuation of nematode proliferation in the soil (Sturz and Kimpinski 2004).

Infection by *M. nematophilum* proved to be a highly productive tool for identifying surface-determining genes. Extensive screening for Bus mutants, bacterially unswollen upon exposure to the bacterial pathogen, identified 19 loci and also yielded alleles of *srf-2, srf-3*, and *srf-5* (Gravato-Nobre et al. 2005; Wilkins et al. 2005). These genetic screens added to the methodological repertoire a simple assay for cuticle fragility, providing confirmation that cuticles are compromised in many of the Bus mutants. In essence, these studies show that bacterial adhesion and glycosylation and/or other post-translational modifications pathways that take place at the worm's surface are tightly linked. The importance of such events is reflected in the loss-of-function mutations that include: the essential glycosyl-transferase, *bus-8* (Partridge et al. 2008), the acyltransferase *bus-1* (Gravato-Nobre and Hodgkin 2008), the galactosyltransferases *bus-17* (Yook and Hodgkin 2007), *bus-2* and *bus-4* (Gravato Nobre et al. 2011), the nucleotide sugar transporter of unknown specificity *bus-12*, and the novel protein, *bus-19* (Yook and Hodgkin 2007). Further evidence that the glycosylation events that take place at the level of the epidermis provide important general substrates for bacterial adhesion come also from data showing that *srf-3, bus-2, bus-4* and *bus-12* mutants exhibit resistance to both *M. nematophilum* and *Yersinia* (Darby et al. 2007; Gravato-Nobre et al. 2005) as a result of non-adhesion of bacteria to target tissues. All these glycosylation genes express in the nematode seam cells, specialized cells that secrete components of the cuticle (Gravato Nobre et al. 2011). The end-product glycoconjugates that are responsible for these surface phenotypes have yet to be identified. The screens for *M. nematophilum* resistance have also yielded mutants with disruptions in four known loci which had been previously identified in unrelated contexts. These corresponded to the Hox gene *egl-5*, the suppressor of activated *let-60* ras, *sur-2*, the β-subunit of phospholipase C (PLCβ) *egl-8*, and the cyclic nucleotide-gated (CNG) channel subunit, *tax-4*. Mutations in these genes affect posterior cell fate (Chisholm 1991), vulval development (Singh and Han 1995), locomotion and egg laying (Miller et al. 1999), and thermosensation/chemosensation and social aggregation behaviors (Komatsu et al. 1996), in *C. elegans* respectively.

3.6.3 Virulence Factor-Mediated Infections

A prominent aspect of *C. elegans* pathogenicity studies has been the demonstration that many virulence factors required for full pathogenicity in mammalian systems have also been shown to play important roles in the infectious process in the worm. Many mutations in microorganisms that reduce pathogenesis in mammalian systems also result in diminished killing of *C. elegans*. Conversely, when the worm was used to screen for less virulent pathogen mutants, many such mutants exhibited also reduced virulence in mammalian models of infection (Mylonakis and Aballay 2005; Sifri et al. 2005). This has been shown in *P. aeruginosa*, *B. pseudomallei*, *S.enterica*, *S. marcescens*, *Y. pseudotuberculosis*, *Staphylococcus aureus*, *S. pneumoniae* and *C. neoformans* (Alegado and Tan 2008; Garsin et al. 2001; Joshua et al. 2003; Kurz et al. 2003; Mylonakis et al. 2002; Sifri et al. 2002; Tan et al. 1999a, b; Tenor et al. 2004). Overall, several of the virulence-related genes that are required for mammalian pathogenesis have proved to be essential for pathogenicity in *C. elegans*. The gacA-gacS and LasR quorum sensing systems in *P. aeruginosa* and the PhoP/Q master regulators in *Salmonella* are among the virulence genes studied (Alegado and Tan 2008; Tan et al. 1999b).

3.6.4 Toxin-Mediated Infections

Pathogens can also cause morbidity and mortality in *C. elegans* by implementing significantly different pathogenic processes which involve toxin production, rather than adhesion and colonization to mucosa. Some examples will be provided below.

When grown in high osmolarity rich medium, *P. aeruginosa* strain PA14 produces low molecular weight diffusible toxins of the pyocyanin-phenazine class that rapidly kill the worms (Mahajan-Miklos et al. 1999). This "fast killing" is not associated with the proliferation of the bacteria in the intestine, but is rather the result of the production of reactive oxygen species that are directly toxic to the worm. Unlike PA14, the *P. aeruginosa*, PA01 strain, is able to cause rapid irreversible nematode paralysis by producing of hydrogen cyanide (Gallagher and Manoil 2001). The production of such toxic compound is regulated by the quorum-sensing activators LasR and RhIH and the gacS regulator gene. In a genetic screen for *C. elegans* mutants resistant to *P. aeruginosa* PA01-induced paralysis, Darby and colleagues isolated two alleles of the *egl-9* gene (Darby et al. 1999). *egl-9* mutants were resistant to killing by live bacteria and responded to cyanide-mediated killing in a dose dependent manner (Gallagher and Manoil 2001). EGL-9 encodes a dioxygenase that functions as negative regulator of a hypoxia-response pathway (Epstein et al. 2001).

Some Gram-positive bacteria, including *E. faecium* and a range of streptococcal species, can kill worms through the production of hydrogen peroxide (Jansen et al. 2002; Moy et al. 2004). In some cases however, worm killing appears as the result of a combinatorial action of intestinal colonization and the production of several diffusible toxins. This seems apply to the *B. pseudomallei* and *B. cepacia*-mediated killing (Gan et al. 2002).

3.6.5 Pore-Forming Toxins-Mediated Infections

Pore-forming toxins represent the largest class of bacterial protein toxins. The best studied of these are the crystal toxins Cry and Cyt toxins of *B. thuringiensis*. *B. thuringiensis* (Bt) is a Gram-positive, sporulating pathogen that can be found in the soil. Cry5B toxicity is directed against the intestinal epithelium of the midgut and leads to vacuole, pore formation, pitting and eventual degradation of whole intestine (Marroquin et al. 2000). Upon exposure to this toxin, the worms turn pale, their intestinal tissue is destroyed and they rapidly die. This pathology can be recapitulated by expressing Cry5B in *E. coli*, demonstrating that the damage is toxin-specific and not a general response to pathogen. The mechanism by which *C. elegans* is sensitive to Bt toxin CRY5B has been elucidated with the isolation of five *C. elegans* bre mutants (*Bacillus* toxin resistant), *bre-1-bre-5*. While *bre-2-bre-5* encode four glycosyltransferases that appear to function in a single pathway required for the intestinal uptake of Cry5B, *bre-1* encodes a protein with significant homology to a GDP-mannose 4,6-dehydratase, which catalyzes the first step in the biosynthesis of GDP-fucose from GDP-mannose (Barrows et al. 2007; Griffitts et al. 2001, 2003). Further analysis showed that the genetic mechanism for Bt toxin resistance in *C. elegans* entails the loss of certain glycolipid carbohydrates in the intestine (Griffitts et al. 2005).

3.7 *C. elegans* Innate Immune System

3.7.1 Physical/Chemical Barriers and Evasion Behavior

Metazoans are endowed with an innate immune defense that provides protection against infection and consists of five interconnected components: (1) physical barriers to prevent microbial invasion (2) constitutive chemical shields to inhibit microbial growth; (3) recognition systems to identify the entry of foreign microorganisms; (4) inducible antimicrobial responses triggered by the recognition system; (5) cellular recruitment processes to amplify and enhance defense (Akira et al. 2006).

Currently, it seems that the innate immune response of nematodes relies on four of these components. Like all metazoans *C. elegans* is equipped with a panoply of defense mechanisms, both constitutive and inducible. However, an obvious cellular defense response consisting of phagocytosis and/or encapsulation of invading microorganisms remains to be demonstrated in the worm. Although the body cavity of the worm is filled with pseudocoelomic fluid that contains coelomocytes, which are candidate macrophages, a role in microorganism engulfment and disposal has not been established. Likewise, *C. elegans* lacks several characteristics of animal humoral immune systems such as a phenoloxidase pathway. In *Drosophila* this defense cascade is believed to target Gram-positive bacterial challenges and involves the release of cytotoxic reactive oxygen species.

Primary defense in nematodes is provided by the multi-layered cuticle, which offers a superb physical barrier against external aggressions. Secondly, they are equipped not only with a muscular grinder that breaks down bacteria but also with an intestine which generates an environment hostile to microbial colonization. A complete transcript inventory of the *C. elegans* intestine reveals an arsenal of secretory proteins with roles in bacterial digestion (lysozymes, saposins, lipases, lectins and proteases), detoxification and stress responses (thaumatin-like, ABC transporters) (McGhee et al. 2007). Furthermore, the worm has a sophisticated chemosensory system, which enables it to sense different bacteria and to learn how to discriminate between innocuous and pathogenic microbes (reviews by Schulenburg and Boehnisch 2008; Zhang 2008). For an organism that lives in decaying matter and feeds on microorganisms, an efficient pathogen avoidance behavior appears to be one of the best strategies to escape infectious diseases. In the case of *S. marcescens* Db10, for example the nematode avoids this pathogenic bacterium by detecting the natural secreted product serrawettin W2 (Pradel et al. 2007). Physical evasion of *S. marcescens* and serrawettin requires the function of the only *C. elegans* Toll-like receptor gene, *tol-1* (Pradel et al. 2007; Pujol et al. 2001). This is substantiated by data showing that partial loss-of-function mutants are defective in avoiding the bacterial lawn of Db10. Another example comes from work with *M. nematophilum*, where it has been shown that *C. elegans* avoids the smell of this infectious bacterial pathogen through *tax-4* and *tax-2* cGMP-gated channels (Yook and Hodgkin 2007). Although these genes are required in the chemosensory neurons for pathogen and hyperoxia avoidance (Chang et al. 2006), they are also likely to play a role in coordinating the secretion of the components of the cuticle that mediate adhesion by *M. nematophilum* to the rectum of infected animals.

3.7.2 Basal/Constitutive Response

One question in the field of *C. elegans* immunity research is whether pathogen avoidance behavior is part of a general stress response and whether such responses can be legitimately considered as basal/constitutive nematode defenses. Attention has focused on a major regulator of a basal or constitutively expressed response to pathogens, the FOXO/forkhead transcription factor DAF-16.

In *C. elegans*, the DAF-2/DAF-16, the insulin-like signaling pathway regulates dauer formation, longevity, and the responses to environmental stressors and pathogens. It consists of the transmembrane tyrosine kinase insulin-like receptor DAF-2 and its downstream transducer DAF-16. The activation of this signaling cascade leads to the cytoplasmic retention of DAF-16, whereas its downregulation induces translocation of DAF-16 to the nucleus. In such circumstances, nuclear DAF-16 can activate the transcription of two classes of proteins: (1) stress resistance proteins, which include those involved in detoxification (e.g. metallothioneins), oxidative stress (superoxidase dismutase, glutathione-S-transferase, catalase) and heat shock responses; (2) the antimicrobial immune effectors (lysozymes, LYS-7, LYS-8, saposins, SSP-1, SSP-9, SSP-12, and thaumatins among others) (Lee et al. 2003; Murphy et al. 2003). DAF-16 can act as a transcriptional repressor or activator of gene expression (Shivers et al. 2008). Given the role of DAF-16 in the activation of a number of immune effectors it is striking that there seem to be little overlap of between those genes positively regulated by this transcription factor and those that are induced upon pathogen infection (Troemel et al. 2006). In fact, current data suggests that DAF-16 can actually repress the expression of many pathogen-induced genes (Shapira et al. 2006; Troemel et al. 2006). This paradoxical situation has not been resolved.

DAF-16 is required in the intestine to extend the life-span of *daf-2* mutants (Libina et al. 2003). Long-lived worms with mutations in *daf-2* show increased resistance to many pathogens (Garsin et al. 2003; Troemel et al. 2006) and have also enhanced evasion behavior (Hasshoff et al. 2007). It is therefore possible that one of the roles of the DAF-16 immune effector targets is to contribute to longevity of the worm. Reflecting also the overlap in the mechanisms by which DAF-16 mediates pathogen resistance and responses to stress, Chavez et al. observed that genes required for oxidative stress under the control of DAF-16 (including the superoxidase dismutase *sod-3*, and the catalases *ctl-1* and *ctl-2*) are also required for the enhanced pathogen resistance afforded by *daf-2* mutants. The authors hypothesized that by overproducing oxidative stress enzymes, *daf-2* mutants may increase their fitness towards controlling the damage occurring during the pathogen interaction (Chavez et al. 2007).

Several lines of evidence support that the nervous system and the innate immune response are tightly interconnected. However, how neuronal mechanisms can regulate innate immunity is still unclear. Recent advance in this field came from work by two independent laboratories and led to the identification of specific neuroendocrine signals that can modulate immune responses in *C. elegans*. First, the Tan laboratory has shown that an increase in neurosecretion involving dense core vesicles (DCV) can reduce the resistance of *C. elegans* to the pathogenic bacteria to *P. aeruginosa* PA14 infection, by modulating innate immunity genes in the worm. Typically, enhanced resistance was observed in mutants with deficient DCV exocytosis such as *unc-31* and *unc-64* or in animals defective in the insulin-like neuropeptide *ins-7*. All these mutants seemed to be able to clear the bacterial infection in the intestine to a far greater extent than wild-type worms and in addition they expressed higher levels of known immune effectors. This work has also established a pathogen specific role for PA14 infection in the worm, involving neuronal secretion of the neuropeptide *ins-7*, which acts in the intestine to activate the DAF-2/DAF-16 pathway; other bacterial pathogens fail to induce this pathway. This work lent support to the view that a PA14-specific innate immune response and aging can be regulated by genetically distinct mechanisms in *C. elegans* (Evans et al. 2008; Kawli and Tan 2008).

Reinforcement of the idea that specific genes and neurons are responsible for effective innate immune responses, separable from canonical neuronal/behavior phenotypes, comes also from the Aballay laboratory (Styer et al. 2008). Styer and colleagues have shown that *npr-1*, a G protein-coupled receptor (GPCR) related to mammalian neuropeptide Y receptors, the cGMP-gated-ion channel *tax-2* and *tax-4*, and the soluble guanylate cyclase GCY-35 act to suppress *C. elegans* innate immune responses to multiple pathogens. *npr-1* mutants are behaviorally abnormal and appear immuno-compromised upon pathogen exposure but their enhanced susceptibility seems to involve more than defective pathogen avoidance behavior. In *C. elegans*, NPR-1 is involved in a neuronal circuit that integrates behavioral responses to oxygen, food, and the presence of other animals. Restoration of NPR-1 activity in the sensory neurons AQR, PQR and URX, significantly increases *npr-1* survival on *P. aeruginosa*. It is plausible that neuroendocrine signals such as INS-7 sent by the NPR-1 neurons, reach non-neuronal immunocompetent tissues through the pseudocoelomic fluid which baths the worm body (Styer et al. 2008).

3.7.3 Signaling Pathways

Following exposure to pathogen microorganisms, the epithelial layer of mucosal surfaces is confronted with the task of recognizing pathogen-associated molecular patterns. The mechanisms by which *C. elegans* recognizes pathogens have yet to be elucidated. Nevertheless, the genetic pathways operating in the worm's response to microbial challenges have been extensively characterized in recent years (see reviews Ewbank 2006; Gravato-Nobre and Hodgkin 2005; Kim and Ausubel 2005; Schulenburg and Boehnisch 2008). It has emerged that most of the signal transduction components activated in *C. elegans* upon pathogen recognition have also been implicated in innate responses in plants, mammalians and the flies. We will focus on the recent findings in the identification of such signaling components.

C. elegans immune responses defense rely on a diverse array of signaling cascades: three mitogen-activated protein kinase (MAPK) pathways (p38 MAPK, c-Jun N-terminal kinase, JNK; extracellular signal-regulated kinase, ERK), the unfolded protein response, the transforming growth factor-β (TGF-β) pathway, the insulin-like receptor (ILR), the Wnt/Wingless β-catenin (BAR-1) and the programmed cell death (PCD) pathway (Table 3.1). A recurrent theme is that these conserved signaling pathways involved in innate defenses involve transducers that play essential roles in various developmental processes in the worm and have been extensively studied in these contexts.

Table 3.1 Genes involved in the activation of *C. elegans* innate immune responses

Pathway	Putative human orthologues	References
Toll		Liberati et al. (2004), Tenor and Aballay (2008a, b)
TOL-1	Toll-like receptor	
TRF-1	TNF receptor-associated factor 1	
P38 MAPK		Aballay et al. (2003), Couillault et al. (2004), Huffman et al. (2004a), Kim et al. (2002, 2004, Liberati et al. (2004), Pujol et al. 2008a, b, Sifri et al. (2003), Troemel et al.(2006)
NIPI-3	Tribbles homolog 1 (TRIB1/hTribbles)	
TIR-1	SARM	
RAB-1	Ras-related GTPase Rab-1A	
R53.1	ATP synthase subunit	
NSY-1	ASK1 MAPKKK	
SEK-1	MKK3, MKK6 MAPKK	
PMK-1	P38 MAPK	
UPR		Boutros and Ahringer (2008), Haskins et al. (2008)
IRE-1	ERN1	
XBP-1	HAC1-like transcription factor	
ABU-1	SREC	
ABU-11	SREC	
JNK MAPK		Huffman et al. (2004a), Kim et al. (2004)
MEK-1	MKK7 MAPKK	
KGB-1	JNK MAPK	
VHP-1	MKP7 MAPK phosphatase	

(continued)

Table 3.1 (continued)

Pathway	Putative human orthologues	References
ERK MAPK		Nicholas and Hodgkin (2004a)
LIN-45	B-Raf serine/threonine protein kinase	
MEK-2	ERK MAPKK2	
MPK-1	ERK MAPK	
KSR-1	Kinase suppressor of Ras	
SUR-2	MED23 mediator subunit	
TGF-β		Mallo et al. (2002)
DBL-1	TGF-β ligand	
SMA-6	Type I TGF-β receptor	
SMA-2	Smad protein	
SMA-3	Smad protein	
SMA-4	Smad protein	
Wnt/Wingless		Irazoqui et al. (2008)
BAR-1	B-catenin	
EGL-5	Homeobox	
Insulin-like receptor (ILR)		Chavez et al. (2007), Evans et al. (2008), Kawli and Tan (2008), Kerry et al. (2006), Singh and Aballay (2006a), Troemel et al. (2006)
INS-7	Insulin-like peptide	
DAF-2	Insulin/IGF-1 receptor	
AGE-1	Phosphatidylinositol 3-kinase	
DAF-16	FOXO family transcription factor	
Programmed cell death (PCD)		Aballay and Ausubel (2001), Aballay et al. (2003)
CED-9	Bcl-2W	
CED-4	Apaf-1, caspase activator	
CED-3	Caspase	
EGL-1	Protein with BH3 domain	
GATA transcription factor		Kerry et al. (2006), Pujol et al. (2008b), Shapira et al. (2006)
ELT-2	GATA transcription factor	
ELT-3	GATA transcription factor	
Heat shock factor		Singh and Aballay (2006b)
HSF-1	Heat shock factor	

In mammals, a family of conserved transmembrane Toll-like receptors (TLRs), function directly as pattern recognition receptors (PRRs) for microbe-associated molecules. TLRs are characterized by an extracellular leucine-rich repeat (LRR) domain and an intracellular TIR (the Toll-interleukin 1, IL-1 receptor) domain. *C. elegans* genome has a single TIR domain adaptor protein, which is homologous to the mammalian SARM. Reduction of function of *tir-1* by RNAi results not only in enhanced susceptibility to *Drechmeria coniospora* (Couillault et al. 2004) but also to *P. aeruginosa* (Liberati et al. 2004). In an immune context in *C. elegans*, TIR-1 acts as an upstream component of the conserved p38 MAPK pathway, in a TOL-1–independent fashion (Liberati et al. 2004; Pujol et al. 2008b); TIR-1 activity is necessary for the induction of the anti-fungal genes NLP-29 and NLP-31 (Couillault et al. 2004). It has now emerged that the putative receptor upstream of *tir-1* encodes a protein kinase with a catalytic domain that is most similar to the human Tribbles homolog 1 (Pujol et al. 2008a). The kinase NIPI-3 (no induction of peptide infection) was found in an elegant screen for mutants that were unable to activate *nlp-29*:GFP expression after *Drechmeria* infection. Remarkably, the isolation of the *nipi-3* mutant has also revealed that in *C. elegans*, separable signaling cascades control gene expression after infection and wounding. While *nipi-3* serves for the specific activation of the AMPs *nlp-29* or *nlp-31* in the fungal infection context, the signaling process activated by this protein kinase is dispensable for the response to epidermal injury by needle pricking (Pujol et al. 2008a).

The Toll pathway is an evolutionary conserved signaling cascade that plays a key role in *Drosophila* and mammalian immune responses. In insects, the activation of Toll causes the nuclear import of NF-κB family transcription factors and massive expression of antifungal peptides. Although several Toll pathway orthologues are present in the *C. elegans* genome, there is no obvious NF-κB homolog in the worm. A recent analysis of a *tol-1* mutant allele, in which the cytoplasmic domain of TOL-1 had been removed, has revealed a novel role for this gene in protecting the worm pharyngeal tissues against *Salmonella enterica* but apparently not other bacterial pathogens. TOL-1 is required for expression of the effectors defensin-like molecule ABF-2 and the heat-shock protein HSP-16.41 (Tenor and Aballay 2008).

The MAPK signaling cascades are likely to be the most ancient components of immune defense and are found from plants to invertebrates and mammals. In *C. elegans*, besides its role in immune signaling, the p38 MAPK PMK-1, also mediates resistance to osmotic stress (Solomon et al. 2004) and contributes to protection against pore-forming toxins (Huffman et al. 2004a, b). Using microarrays to identify differentially regulated genes after exposure to Cry5B, Huffman and co-workers were able to show that the signaling cascade implicated in the response to the toxin involved both PMK-1 and KGB-1. The combined action of these two transducers resulted in the increased expression of *ttm-1* and *ttm-2* (toxin-regulated targets of MAPK) effectors, which function to increase resistance to pore-forming toxin. KGB-1 is a component of a JNK-like MAPK pathway involved in heavy metal stress responses (Mizuno et al. 2004). Subsequently, a comparative study involving genes induced by Cry5B and *P. aeruginosa* infection has revealed that there is a substantial overlap between genes activated in response to PFT and

P. aeruginosa insults (Troemel et al. 2006). This study lends support to the view that in *C. elegans* part of the transcriptional response to *P. aeruginosa* infection is likely to be due to an intestinal damage.

The PMK-1/p38 MAPK cascade does not appear to play a role in the *C. elegans-M. nematophilum* association. However, the extracellular signal-regulated ERK protein kinase MAPK pathway has been shown to mediate both tail swelling and a protective response against this Gram-positive bacterial pathogen (Nicholas and Hodgkin 2004a).

The unfolded protein response (UPR) is a highly conserved cellular response that balances ER secretory load with protein folding capacity. Recent work in *C. elegans* has shown that the UPR branches involving IRE-1/XBP-1 (Bischof et al. 2008) or a network of PQN/ABU proteins contribute to proper defense against toxins (Haskins et al. 2008).

For the cellular response to the PFT Cry5B, worms appear to rely partly on a UPR pathway that involves the recruitment of the *ire-1*, *xbp-1* or the *atf-6* transducers. Mutations *ire-1-xbp-1* or the *atf-6* rendered worms hypersensitive to PFT (Breger et al. 2007). While this pathway protects the intestinal cells of the worm against PFT, its role for *C. elegans* defense against other toxic insults seems dispensable. Intestinal cells challenged by PFT require p38MAPK, and the inactivation of the p38MAPK *pmk-1* resulted in an inability to turn on UPR, pathway thus suggesting that p38MAPK – *ire-1-xbp-1* form a module that transduce a defense signal against the pathogen toxin. It became also apparent that the UPR-mediated activation in response to PFTs is separable from the classical UPR-pathway which mediates unfolded proteins.

Another report relating an UPR to *C. elegans* immune responses come from work by Haskins et al. (2008). The authors have shown that the apoptotic corpse receptor CED-1 and a network of UPR proteins are part of a pathway that prevents *S. enterica* invasion of the pharyngeal tissues. Transcriptional profiling analysis has revealed a new role for this receptor, through the up-regulation of a family of genes, referred to as abu genes (activated in a blocked unfolded protein response) and which encode proteins with prion-like glutamine/asparagine (Q/N)-rich domains. Such proteins are known to be mediators of the UPR. Interestingly, loss-of-function analysis demonstrated that CED-1 is required for immunity to *S. enterica* but not as a component of the classical apoptotic pathway that involves the CED-9/CED-4/CED-3 axis. CED-1 is known as a scavenger receptor that binds to dying cells, however its contribution to control innate immunity by promoting an UPR cascade has yet to be fully understood. In addition, this pathogen-induced UPR that mediates the *pqn/abu* gene products appeared to be independent of the canonical UPR that involves *xbp-1* transcription factor (Haskins et al. 2008).

In *C. elegans* the evolutionarily conserved transcriptional cofactor bar-1/β-catenin has been involved in a number of distinct developmental processes that are part of signalling pathways downstream of Wnt secreted glycoprotein ligands. The Ausubel group has now shown that this pathway plays also a role in the *C. elegans* immune response to *S. aureus*. Mutations in *bar-1* and its downstream transducer HOX gene *egl-5* render the worms hypersensitive to this bacterial pathogen.

Epistasis analysis has demonstrated that *bar-1* and *egl-5* function in parallel to the immune signalling cascades *daf-2/daf-16* insulin and *tir-1/pmk-1* p38 MAPK (Irazoqui et al. 2008).

3.7.4 Regulation

The up-regulation of candidate antimicrobials that occurs following infection appears primarily regulated at the transcriptional level. Besides DAF-16, additional regulators of the immune response include the heat shock factor HSF-1 (Singh and Aballay 2006a, b) the GATA transcription factors ELT-2 (Kerry et al. 2006; Shapira et al. 2006) and ELT-3 (Pujol et al. 2008b). While ELT-2 is required for all intestinal expression and may be part of a general defense pathway that regulates intestinal innate immunity genes (Kerry et al. 2006), ELT-3 appears to function in the epidermis to protect the worm against fungal infection as well as to environmental damage and stress (Pujol et al. 2008b). The HSF-1 pathway regulates immunity independently of p38 MAPK and is needed for the effects of DAF-2 and DAF-16 in multi-pathogen defenses (Singh and Aballay 2006a, b).

3.8 Monitoring the Worm Innate Immune Response Following Different Modes of Pathogen Infections

A hallmark of the innate immune response in *C. elegans*, as in other invertebrates, is the challenge-induced synthesis of a battery of antimicrobial peptides and proteins, which are expressed in tissues in contact with invading microorganisms. *C. elegans* encodes a wide diversity of candidate immune effectors, which include: lysozymes, caenopores or saposin-like proteins (ssp), antimicrobial caenacins (cnc) and neuropeptide-like proteins (nlp), thaumatins (thm), PR-1 plant antimicrobial homologues, lipases, proteins with Metridin-like ShK toxin domain SHKT, defensin like ABF peptides (abf), C-type lectins (clec) and CUB-like gene family members (see reviews by Schulenburg et al. 2008; Nicholas and Hodgkin 2004b). Microarray experiments designed to identify pathogen-induced genes have revealed many effectors which may recognize or eliminate pathogens. Among them are the C-type lectin domain containing proteins (CTLD) which are part of a large family of mostly secreted proteins whose exact role is unknown; the membrane anchored members may serve as adhesion as well as pathogen recognition receptors (Cambi et al. 2005). In *C. elegans*, the role of CTLD proteins in immunity is indicated by their up-regulation upon pathogen exposure (Alper et al. 2007; Mallo et al. 2002; O'Rourke et al. 2006; Troemel et al. 2006; Wong et al. 2007). Microarray experiments have also identified molecules that are broadly induced in response to immune challenges (Shivers et al. 2008). These include: lysozymes, which cleave cell walls of bacteria and have been shown to have antimicrobial activity; ShK-like toxins, proteins that share similarity to a toxin produced by sea anemones; saposins, plant secondary metabolites that contain antimicrobial activity, among others. These biological read-outs are now proving to be convenient markers to monitor signal transduction activation, and there is evidence that they contribute to resistance (Alper et al. 2007; Irazoqui et al. 2008; O'Rourke et al. 2006; Wong et al. 2007).

3.9 Concluding Remarks

As a free-living soil nematode, *C. elegans* spends its entire life in a habitat enriched with microorganisms. We know little about the diversity of the microbes that infect or attack *C. elegans* in the wild, but we know even less about virulence mechanisms that natural pathogens employ or how natural pathogens interact with the nematode immune system in their natural habitats. Nevertheless, their evolutionary success indicates that nematodes must possess strategies to fight microbial infections. In recent years much information has accumulated concerning the molecular basis of the worm's basal and inducible defense responses against its pathogens. Expression profiling analyses using *C. elegans* challenged by a number of pathogens have pinpointed PMK-1/p38 MAPK and DAF-16 as key signal transducers of the worm defense response. These studies reveal that stress response, life span, nematode development and innate immunity can be mediated by signaling pathways used in different ways. It is emerging also that *C. elegans* is able to distinguish between the effects of different pathogens, and between pathogens and toxic/stress insults. Overall, *C. elegans* antimicrobial responses represent major changes in gene expression that not only result in the production of antimicrobial effectors, but also modulate other defense mechanisms, such as those that deal with physical and toxic insults. Major challenges are to understand the relative contribution of these effectors to the total host response, to test their specificity towards different pathogens and to determine potential synergisms between them.

The extensive experimental resources of *C. elegans* have provided powerful methods for exploring pathogen virulence and metazoan innate immune pathways. Simple models such as *C. elegans* and *Drosophila* can be used effectively to identify universal pathogen virulence factors and host immune defenses. On the other hand, the diversity of nematodes makes it likely that each group examined will have unique features in its interactions with microbes. The recent sequencing of *Meloidogyne incognita* and *M. hapla* genomes (Abad et al. 2008; Opperman et al. 2008) reveals conservation of pathways for the key processes implicated in immunity and development. However, some immune effectors such as lysozymes, C-type lectins and chitinases were much less abundant in *M. incognita* than in *C. elegans*. Notably absent from *M. incognita* genome were immune effectors such as the antibacterial *abf* and *spp* gene families, and the antifungal *nlp*, *cnc* gene families.

Future research will be needed to establish the general properties of immune response in nematodes and to define both universal and species-specific elements of the innate immunity in *C. elegans*.

References

Abad P, Gouzy J, Aury JM et al (2008) Genome sequence of the metazoan plant-parasitic nematode *Meloidogyne incognita*. Nat Biotechnol 26:909–915

Aballay A, Ausubel FM (2001) Programmed cell death mediated by ced-3 and ced-4 protects *Caenorhabditis elegans* from *Salmonella typhimurium*-mediated killing. Proc Natl Acad Sci USA 98:2735–2739

Aballay A, Yorgey P, Ausubel FM (2000) *Salmonella typhimurium* proliferates and establishes a persistent infection in the intestine of *Caenorhabditis elegans*. Curr Biol 10:1539–1542

Aballay A, Drenkard E, Hilbun LR et al (2003) *Caenorhabditis elegans* innate immune response triggered by *Salmonella enterica* requires intact LPS and is mediated by a MAPK signaling pathway. Curr Biol 13:47–52

Akimkina T, Yook K, Curnock S et al (2006) Genome characterization, analysis of virulence and transformation of *Microbacterium nematophilum*, a coryneform pathogen of the nematode *Caenorhabditis elegans*. FEMS Microbiol Lett 264:145–151

Akira S, Uematsu S, Takeuchi O (2006) Pathogen recognition and innate immunity. Cell 124:783–801

Alegado RA, Tan MW (2008) Resistance to antimicrobial peptides contributes to persistence of *Salmonella typhimurium* in the *C. elegans* intestine. Cell Microbiol 10:1259–1273

Alper SS, McBride JB, Lackford JH et al (2007) Specificity and complexity of the *Caenorhabditis elegans* innate immune response. Mol Cell Biol 27:5544–5553

Alper SR, Laws B, Lackford WA et al (2008) Identification of innate immunity genes and pathways using a comparative genomics approach. Proc Natl Acad Sci USA 105:7016–7021

Ausubel FM (2005) Are innate immune signaling pathways in plants and animals conserved? Nat Immunol 6:973–979

Avery L (1993) The genetics of feeding in *Caenorhabditis elegans*. Genetics 133:897–917

Avery L, Horvitz HR (1989) Pharyngeal pumping continues after laser killing of the pharyngeal nervous system of *C. elegans*. Neuron 3:473–485

Avery L, Shtonda BB (2003) Food transport in the *C. elegans* pharynx. J Exp Biol 206:2441–2457

Baldwin JG, Nadler SA, Adams BJ (2004) Evolution of plant parasitism among nematodes. Annu Rev Phytopathol 42:83–105

Barriere A, Felix MA (2007) Temporal dynamics and linkage disequilibrium in natural *Caenorhabditis elegans* populations. Genetics 176:999–1011

Barrows BD, Haslam SM, Bischof LJ et al (2007) Resistance to *Bacillus thuringiensis* toxin in *Caenorhabditis elegans* from loss of fucose. J Biol Chem 282:3302–3311

Begun J, Sifri CD, Goldman S et al (2005) *Staphylococcus aureus* virulence factors identified by using a high-throughput *Caenorhabditis elegans*-killing model. Infect Immun 73:872–877

Begun J, Gaiani JM, Rohde H et al (2007) Staphylococcal biofilm exopolysaccharide protects against *Caenorhabditis elegans* immune defenses. PLoS Pathog 3:e57

Bischof LJ, Kao CY, Los FC et al (2008) Activation of the unfolded protein response is required for defenses against bacterial pore-forming toxin in vivo. PLoS Pathog 4:e1000176

Boutros M, Ahringer J (2008) The art and design of genetic screens: RNA interference. Nat Rev Genet 9:554–566

Breger JB, Fuchs B, Aperis G et al (2007) Antifungal chemical compounds identified using a *C. elegans* pathogenicity assay. PLoS Pathog 3:e18

Cambi A, Koopman M, Figdor CG (2005) How C-type lectins detect pathogens. Cell Microbiol 7:481–488

Chang AJ, Chronis N, Karow DS et al (2006) A distributed chemosensory circuit for oxygen preference in *C. elegans*. PLoS Biol 4:e274

Chavez VA, Mohri-Shiomi A, Maadani LA et al (2007) Oxidative stress enzymes are required for DAF-16-mediated immunity due to generation of reactive oxygen species by *Caenorhabditis elegans*. Genetics 176:1567–1577

Chisholm A (1991) Control of cell fate in the tail region of *C. elegans* by the gene egl-5. Development 111:921–932

Ciche TA, Kim KS, Kaufmann-Daszczuk B et al (2008) Cell invasion and matricide during *Photorhabdus luminescens* transmission by *Heterorhabditis bacteriophora* nematodes. Appl Environ Microbiol 74:2275–2287

Cipollo JF, Awad AM, Costello CE et al (2004) srf-3, a mutant of *Caenorhabditis elegans*, resistant to bacterial infection and to biofilm binding, is deficient in glycoconjugates. J Biol Chem 279:52893–52903

Couillault C, Ewbank JJ (2002) Diverse bacteria are pathogens of *Caenorhabditis elegans*. Infect Immun 70:4705–4707

Couillault C, Pujol N, Reboul J et al (2004) TLR-independent control of innate immunity in *Caenorhabditis elegans* by the TIR domain adaptor protein TIR-1, an ortholog of human SARM. Nat Immunol 5:488–494

Darby C, Cosma CL, Thomas JH et al (1999) Lethal paralysis of *Caenorhabditis elegans* by *Pseudomonas aeruginosa*. Proc Natl Acad Sci USA 96:15202–15207

Darby C, Hsu JW, Ghori N et al (2002) *Caenorhabditis elegans*: plague bacteria biofilm blocks food intake. Nature 417:243–244

Darby C, Chakraborti A, Politz SM et al (2007) *Caenorhabditis elegans* mutants resistant to attachment of *Yersinia* biofilms. Genetics 176:221–230

Das S, Raski D (1969) Effect of grapevine fanleaf virus on the reproduction and survival of its nematode vector *Xiphinema index* Thorne and Allen. J Nematol 1:107–110

David JC (2008) *Photorhabdus*: a model for the analysis of pathogenicity and mutualism. Cell Microbiol 10:2159–2167

Endo BY (1979) The ultrastructure and distribution of an intracellular bacterium-like microorganism in tissues of larvae of the soybean cyst nematode, *Heterodera glycines*. J Ultrastruct Res 67:1–14

Epstein AC, Gleadle JM, McNeill LA et al (2001) *C. elegans* EGL-9 and mammalian homologs define a family of dioxygenases that regulate HIF by prolyl hydroxylation. Cell 107:43–54

Evans EA, Chen WC, Tan MW (2008) The DAF-2 insulin-like signaling pathway independently regulates aging and immunity in *C. elegans*. Aging Cell 7:879–893

Ewbank JJ (2006) Signaling in the immune response. WormBook:1–12. http://www.wormbook.org/chapters/www_signalingimmuneresponse/signalingimmuneresponse.html

Fekete CM, Tholander B, Rajashekar D et al (2008) Paralysis of nematodes: shifts in the transcriptome of the nematode-trapping fungus *Monacrosporium haptotylum* during infection of *Caenorhabditis elegans*. Environ Microbiol 10:364–375

Félix M-A, Ashe A, Piffaretti J et al (2011) Natural and experimental infection of *Caenorhabditis* nematodes by novel viruses related to nodaviruses. PLoS Biol 9(1):e1000586

Gallagher LA, Manoil C (2001) *Pseudomonas aeruginosa* PAO1 kills *Caenorhabditis elegans* by cyanide poisoning. J Bacteriol 183:6207–6214

Gan YH, Chua KL, Chua HH et al (2002) Characterization of *Burkholderia pseudomallei* infection and identification of novel virulence factors using a *Caenorhabditis elegans* host system. Mol Microbiol 44:1185–1197

Garsin DA, Sifri CD, Mylonakis E et al (2001) A simple model host for identifying Gram-positive virulence factors. Proc Natl Acad Sci USA 98:10892–10897

Garsin DA, Villanueva JM, Begun J et al (2003) Long-lived *C. elegans* daf-2 mutants are resistant to bacterial pathogens. Science 300:1921

Gravato-Nobre MJ, Hodgkin J (2005) *Caenorhabditis elegans* as a model for innate immunity to pathogens. Cell Microbiol 7:741–751

Gravato-Nobre MJ, Hodgkin J (2008) The acyltransferase gene *bus-1* exhibits conserved and specific expression in nematode rectal cells and reveals pathogen-induced cell swelling. Dev Dyn 237:3762–3776

Gravato-Nobre MJ, Nicholas HR, Nijland R et al (2005) Multiple genes affect sensitivity of *Caenorhabditis elegans* to the bacterial pathogen *Microbacterium nematophilum*. Genetics 171:1033–1045

Gravato-Nobre MJ, Stroud D, O'Rourke D et al (2011) Glycosylation genes expressed in Seam cells determine complex surface properties and bacterial adhesion to the cuticle of *Caenorhabditis elegans*. Genetics 187:141–155

Griffitts JS, Whitacre JL, Stevens DE et al (2001) Bt toxin resistance from loss of a putative carbohydrate-modifying enzyme. Science 293:860–864

Griffitts JS, Huffman DL, Whitacre JL et al (2003) Resistance to a bacterial toxin is mediated by removal of a conserved glycosylation pathway required for toxin-host interactions. J Biol Chem 278:45594–45602

Griffitts JS, Haslam SM, Yang T et al (2005) Glycolipids as receptors for *Bacillus thuringiensis* crystal toxin. Science 307:922–925

Haskins KA, Russell JF, Gaddis N et al (2008) Unfolded protein response genes regulated by CED-1 are required for *Caenorhabditis elegans* innate immunity. Dev Cell 15:87–97

Hasshoff MC, Bohnisch D, Tonn B et al (2007) The role of *Caenorhabditis elegans* insulin-like signaling in the behavioral avoidance of pathogenic *Bacillus thuringiensis*. FASEB J 21:1801–1812

Hodgkin JP, Kuwabara E, Corneliussen B (2000) A novel bacterial pathogen, *Microbacterium nematophilum*, induces morphological change in the nematode *C. elegans*. Curr Biol 10:1615–1618

Hoflich JP, Berninsone C, Gobel MJ et al (2004) Loss of srf-3-encoded nucleotide sugar transporter activity in *Caenorhabditis elegans* alters surface antigenicity and prevents bacterial adherence. J Biol Chem 279:30440–30448

Huffman DL, Abrami L, Sasik R et al (2004a) Mitogen-activated protein kinase pathways defend against bacterial pore-forming toxins. Proc Natl Acad Sci USA 101:10995–11000

Huffman DL, Bischof LJ, Griffitts JS et al (2004b) Pore worms: using *Caenorhabditis elegans* to study how bacterial toxins interact with their target host. Int J Med Microbiol 293:599–607

Irazoqui JE, Ng A, Xavier RJ et al (2008) Role for beta-catenin and HOX transcription factors in *Caenorhabditis elegans* and mammalian host epithelial-pathogen interactions. Proc Natl Acad Sci USA 105:17469–17474

Jansen WT, Bolm MR, Balling GS et al (2002) Hydrogen peroxide-mediated killing of *Caenorhabditis elegans* by *Streptococcus pyogenes*. Infect Immun 70:5202–5207

Jansson H-B (1994) Adhesion of conidia of *Drechmeria coniospora* to *Caenorhabditis elegans* wild type and mutants. J Nematol 26:430–435

Joshua GW, Karlyshev AV, Smith MP et al (2003) A *Caenorhabditis elegans* model of *Yersinia* infection: biofilm formation on a biotic surface. Microbiology 149:3221–3229

Kawli T, Tan MW (2008) Neuroendocrine signals modulate the innate immunity of *Caenorhabditis elegans* through insulin signaling. Nat Immunol 9:1415–1424

Kerry S, TeKippe M, Gaddis NC et al (2006) GATA transcription factor required for immunity to bacterial and fungal pathogens. PLoS ONE 1:e77

Kim DH, Ausubel FM (2005) Evolutionary perspectives on innate immunity from the study of *Caenorhabditis elegans*. Curr Opin Immunol 17:4–10

Kim DH, Feinbaum R, Alloing G et al (2002) A conserved p38 MAP kinase pathway in *Caenorhabditis elegans* innate immunity. Science 297:623–626

Kim DH, Liberati NT, Mizuno T et al (2004) Integration of *Caenorhabditis elegans* MAPK pathways mediating immunity and stress resistance by MEK-1 MAPK kinase and VHP-1 MAPK phosphatase. Proc Natl Acad Sci USA 101:10990–10994

Kimpinski J, Sturz AV (2003) Managing crop root zone ecosystems for prevention of harmful and encouragement of beneficial nematodes. Soil Tillage Res 72:213–221

Komatsu H, Mori I, Rhee JS et al (1996) Mutations in a cyclic nucleotide-gated channel lead to abnormal thermosensation and chemosensation in *C. elegans*. Neuron 17:707–718

Kothe M, Antl M, Huber B et al (2003) Killing of *Caenorhabditis elegans* by *Burkholderia cepacia* is controlled by the cep quorum-sensing system. Cell Microbiol 5:343–351

Kurz CL, Ewbank JJ (2003) *Caenorhabditis elegans*: an emerging genetic model for the study of innate immunity. Nat Rev Genet 4:380–390

Kurz CL, Chauvet S, Andres E et al (2003) Virulence factors of the human opportunistic pathogen *Serratia marcescens* identified by in vivo screening. EMBO J 22:1451–1460

Labrousse AS, Chauvet C, Couillault CL et al (2000) *Caenorhabditis elegans* is a model host for *Salmonella typhimurium*. Curr Biol 10:1543–1545

Lee SS, Kennedy S, Tolonen AC et al (2003) DAF-16 target genes that control *C. elegans* life-span and metabolism. Science 300:644–647

Liberati NT, Fitzgerald KA, Kim DH et al (2004) Requirement for a conserved Toll/interleukin-1 resistance domain protein in the *Caenorhabditis elegans* immune response. Proc Natl Acad Sci USA 101:6593–6598

Libina N, Berman JR, Kenyon C (2003) Tissue-specific activities of *C. elegans* DAF-16 in the regulation of lifespan. Cell 115:489–502

Liu WH, Lin YL, Wang JP et al (2006) Restriction of vaccinia virus replication by a ced-3 and ced-4-dependent pathway in *Caenorhabditis elegans*. Proc Natl Acad Sci USA 103:4174–4179

Lu R, Maduro M, Li F et al (2005) Animal virus replication and RNAi-mediated antiviral silencing in *Caenorhabditis elegans*. Nature 436:1040–1043

Mahajan-Miklos S, Tan MW, Rahme LG et al (1999) Molecular mechanisms of bacterial virulence elucidated using a *Pseudomonas aeruginosa-Caenorhabditis elegans* pathogenesis model. Cell 96:47–56

Mallo GV, Kurz CL, Couillault C et al (2002) Inducible antibacterial defense system in *C. elegans*. Curr Biol 12:1209–1214

Marroquin LD, Elyassnia D, Griffitts JS et al (2000) *Bacillus thuringiensis* (Bt) toxin susceptibility and isolation of resistance mutants in the nematode *Caenorhabditis elegans*. Genetics 155:1693–1699

McGhee JD, Sleumer MC, Bilenky M et al (2007) The ELT-2 GATA-factor and the global regulation of transcription in the *C. elegans* intestine. Dev Biol 302:627–645

Mendoza De Gives PM, Davies KG, Clark SJ et al (1999) Predatory behavior of trapping fungi against srf mutants of *Caenorhabditis elegans* and different plant and animal parasitic nematodes. Parasitology 119(Pt 1):95–104

Miller KG, Emerson MD, Rand JB (1999) Goalpha and diacylglycerol kinase negatively regulate the Gqalpha pathway in *C. elegans*. Neuron 24:323–333

Millet AC, Ewbank JJ (2004) Immunity in *Caenorhabditis elegans*. Curr Opin Immunol 16:4–9

Mizuno T, Hisamoto N, Terada T et al (2004) The *Caenorhabditis elegans* MAPK phosphatase VHP-1 mediates a novel JNK-like signaling pathway in stress response. EMBO J 23:2226–2234

Moy TI, Mylonakis E, Calderwood SB et al (2004) Cytotoxicity of hydrogen peroxide produced by *Enterococcus faecium*. Infect Immun 72:4512–4520

Moy TI, Ball AR, Anklesaria Z et al (2006) Identification of novel antimicrobials using a live-animal infection model. Proc Natl Acad Sci USA 103:10414–10419

Muir RE, Tan MW (2008) Virulence of *Leucobacter chromiireducens* subsp. *solipictus* to *Caenorhabditis elegans*: characterization of a novel host-pathogen interaction. Appl Environ Microbiol 74:4185–4198

Murphy CT, McCarroll SA, Bargmann CI et al (2003) Genes that act downstream of DAF-16 to influence the lifespan of *Caenorhabditis elegans*. Nature 424:277–283

Mylonakis E, Aballay A (2005) Worms and flies as genetically tractable animal models to study host-pathogen interactions. Infect Immun 73:3833–3841

Mylonakis E, Ausubel FM, Perfect JR et al (2002) Killing of *Caenorhabditis elegans* by *Cryptococcus neoformans* as a model of yeast pathogenesis. Proc Natl Acad Sci USA 99:15675–15680

Nicholas HR, Hodgkin J (2004a) The ERK MAP kinase cascade mediates tail swelling and a protective response to rectal infection in *C. elegans*. Curr Biol 14:1256–1261

Nicholas HR, Hodgkin J (2004b) Responses to infection and possible recognition strategies in the innate immune system of *Caenorhabditis elegans*. Mol Immunol 41:479–493

O'Quinn AL, Wiegand EM, Jeddeloh JA (2001) *Burkholderia pseudomallei* kills the nematode *Caenorhabditis elegans* using an endotoxin-mediated paralysis. Cell Microbiol 3:381–393

O'Rourke D, Baban D, Demidova M et al (2006) Genomic clusters, putative pathogen recognition molecules, and antimicrobial genes are induced by infection of *C. elegans* with *M. nematophilum*. Genome Res 16:1005–1016

Opperman CH, Bird DM, Williamson VM et al (2008) Sequence and genetic map of *Meloidogyne hapla*: a compact nematode genome for plant parasitism. Proc Natl Acad Sci USA 105:14802–14807

Park JO, El-Tarabily KA, Ghisalberti EL et al (2002) Pathogenesis of *Streptoverticillium albireticuli* on *Caenorhabditis elegans* and its antagonism to soil-borne fungal pathogens. Lett Appl Microbiol 35:361–365

Partridge FA, Tearle AW, Gravato-Nobre MJ et al (2008) The *C. elegans* glycosyltransferase BUS-8 has two distinct and essential roles in epidermal morphogenesis. Dev Biol 317:549–559

Pradel E, Ewbank JJ (2004) Genetic models in pathogenesis. Annu Rev Genet 38:347–363

Pradel E, Zhang Y, Pujol N et al (2007) Detection and avoidance of a natural product from the pathogenic bacterium *Serratia marcescens* by *Caenorhabditis elegans*. Proc Natl Acad Sci USA 104:2295–2300

Pujol N, Link EM, Liu LX et al (2001) A reverse genetic analysis of components of the Toll signaling pathway in *Caenorhabditis elegans*. Curr Biol 11:809–821

Pujol N, Cypowyj S, Ziegler K et al (2008a) Distinct innate immune responses to infection and wounding in the *C. elegans* epidermis. Curr Biol 18:481–489

Pujol N, Zugasti O, Wong D et al (2008b) Anti-fungal innate immunity in *C. elegans* is enhanced by evolutionary diversification of antimicrobial peptides. PLoS Pathog 4:e1000105

Schott DH, Cureton DK, Whelan SP et al (2005) An antiviral role for the RNA interference machinery in *Caenorhabditis elegans*. Proc Natl Acad Sci USA 102:18420–18424

Schulenburg H, Boehnisch C (2008) Diversification and adaptive sequence evolution of *Caenorhabditis lysozymes* (Nematoda: Rhabditidae). BMC Evol Biol 8:114

Schulenburg H, Ewbank JJ (2007) The genetics of pathogen avoidance in *Caenorhabditis elegans*. Mol Microbiol 66:563–570

Schulenburg H, Hoeppner MP, Weiner J et al (2008) Specificity of the innate immune system and diversity of C-type lectin domain (CTLD) proteins in the nematode *Caenorhabditis elegans*. Immunobiology 213:237–250

Shapira M, Hamlin BJ, Rong J et al (2006) A conserved role for a GATA transcription factor in regulating epithelial innate immune responses. Proc Natl Acad Sci USA 103:14086–14091

Shivers RP, Youngman MJ, Kim DH (2008) Transcriptional responses to pathogens in *Caenorhabditis elegans*. Curr Opin Microbiol 11:251–256

Sicard M, Hering S, Schulte R et al (2007) The effect of *Photorhabdus luminescens* (Enterobacteriaceae) on the survival, development, reproduction and behavior of *Caenorhabditis elegans* (Nematoda: Rhabditidae). Environ Microbiol 9:12–25

Sifri CD, Mylonakis E, Singh KV et al (2002) Virulence effect of *Enterococcus faecalis* protease genes and the quorum-sensing locus fsr in *Caenorhabditis elegans* and mice. Infect Immun 70:5647–5650

Sifri CD, Begun J, Ausubel FM et al (2003) *Caenorhabditis elegans* as a model host for *Staphylococcus aureus* pathogenesis. Infect Immun 71:2208–2217

Sifri CD, Begun J, Ausubel FM (2005) The worm has turned–microbial virulence modeled in *Caenorhabditis elegans*. Trends Microbiol 13:119–127

Singh V, Aballay A (2006a) Heat-shock transcription factor (HSF)-1 pathway required for *Caenorhabditis elegans* immunity. Proc Natl Acad Sci USA 103:13092–13097

Singh V, Aballay A (2006b) Heat shock and genetic activation of HSF-1 enhance immunity to bacteria. Cell Cycle 5:2443–2446

Singh N, Han M (1995) sur-2, a novel gene, functions late in the let-60 ras-mediated signaling pathway during *Caenorhabditis elegans* vulval induction. Genes Dev 9:2251–2265

Solomon A, Bandhakavi S, Jabbar S et al (2004) *Caenorhabditis elegans* OSR-1 regulates behavioral and physiological responses to hyperosmotic environments. Genetics 167:161–170

Sturz A, Kimpinski J (2004) Endoroot bacteria derived from marigolds (*Tagetes* spp.) can decrease soil population densities of root-lesion nematodes in the potato root zone. Plant Soil 262: 241–249

Styer KL, Hopkins GW, Bartra SS et al (2005) *Yersinia pestis* kills *Caenorhabditis elegans* by a biofilm-independent process that involves novel virulence factors. EMBO Rep 6:992–997

Styer KL, Singh V, Macosko E et al (2008) Innate immunity in *Caenorhabditis elegans* is regulated by neurons expressing NPR-1/GPCR. Science 322:460–464

Tan MW (2002) Identification of host and pathogen factors involved in virulence using *Caenorhabditis elegans*. Meth Enzymol 358:13–28

Tan MW, Ausubel FM (2000) *Caenorhabditis elegans*: a model genetic host to study *Pseudomonas aeruginosa* pathogenesis. Curr Opin Microbiol 3:29–34

Tan MW, Mahajan-Miklos S, Ausubel FM (1999a) Killing of *Caenorhabditis elegans* by *Pseudomonas aeruginosa* used to model mammalian bacterial pathogenesis. Proc Natl Acad Sci USA 96:715–720

Tan MW, Rahme LG, Sternberg JA et al (1999b) *Pseudomonas aeruginosa* killing of *Caenorhabditis elegans* used to identify *P. aeruginosa* virulence factors. Proc Natl Acad Sci USA 96:2408–2413

Tenor JL, Aballay A (2008) A conserved Toll-like receptor is required for *Caenorhabditis elegans* innate immunity. EMBO Rep 9:103–109

Tenor JL, McCormick BA, Ausubel FM et al (2004) *Caenorhabditis elegans*-based screen identifies *Salmonella virulence* factors required for conserved host-pathogen interactions. Curr Biol 14:1018–1024

Thomas JH (2008) Genome evolution in *Caenorhabditis*. Brief Funct Genomic Proteomic 7:211–216

Troemel ER, Chu SW, Reinke V et al (2006) p38 MAPK regulates expression of immune response genes and contributes to longevity in *C. elegans*. PLoS Genet 2:e183

Troemel ER, Félix M-A, Whiteman NK et al (2008) Microsporidia are natural intracellular parasites of the nematode *C. elegans*. PLoS Biol 6:2736–2752

Walsh JA, Lee DL, Shepherd AM (1983) The distribution and effect of intracellular Rickettsia-like micro-organisms infecting adult males of the potato cyst-nematode *Globodera rostochiensis*. Nematologica 29:227–239

Wilkins C, Dishongh R, Moore SC et al (2005) RNA interference is an antiviral defence mechanism in *Caenorhabditis elegans*. Nature 436:1044–1047

Wong D, Bazopoulou D, Pujol N et al (2007) Genome-wide investigation reveals pathogen-specific and shared signatures in the response of *Caenorhabditis elegans* to infection. Genome Biol 8:R194

Yook K, Hodgkin J (2007) Mos1 mutagenesis reveals a diversity of mechanisms affecting response of *Caenorhabditis elegans* to the bacterial pathogen *Microbacterium nematophilum*. Genetics 175:681–697

Zhang Y (2008) Neuronal mechanisms of *Caenorhabditis elegans* and pathogenic bacteria interactions. Curr Opin Microbiol 11:257–261

Chapter 4
Exploiting Genomics to Understand the Interactions Between Root-Knot Nematodes and *Pasteuria penetrans*

Jenn E. Schaff, Tim H. Mauchline, Charles H. Opperman, and Keith G. Davies

Abstract *Caenorhabditis elegans* was the first multicellular organism to have its genome sequenced and has proved useful in the investigations of innate immunity, the generic science that underpins the biology of host-pathogen interactions. This chapter explores the sequencing of plant-parasitic nematodes and microbial genomes and shows how this knowledge can help in understanding the biology of the interaction between *Meloidogyne* spp. and the bacterial nematode parasite *Pasteuria penetrans*. Three examples examine how genomic information can help in developing new approaches to the problems associated with using *Pasteuria* as a biological control agent: initially one focuses on the transportome and how genomics might help to understand the fastidious nature of *Pasteuria* growth in the nematode; secondly, comparative genomics is used to explore the phosphorylation pathway important in initiating sporulation; and, thirdly, comparative genomics is exploited to understand endospore attachment to the nematode cuticle where, in comparisons with other animal parasitic *Bacillus* spp., collagen-like fibres have been implicated. Finally the chapter suggests that genomics paves the way for the development of designer control agents but such an approach would not be without its critics.

4.1 Introduction

The focus of this chapter is to review the knowledge of genomics and how this might help in developing new approaches to the biological control of nematode pests, in particular focusing on the *Pasteuria* group of bacteria that has potential to be devel-

J.E. Schaff and C.H. Opperman
Plant Pathology Department, North Carolina State University, Raleigh, NC 27695, USA
e-mail: jeschaff@ncsu.edu

T.H. Mauchline and K.G. Davies (✉)
Department of Plant Pathology and Microbiology, Rothamsted Research, Harpenden, Hertfordshire AL5 2JQ, UK
e-mail: keith.davies@rothamsted.ac.uk

K. Davies and Y. Spiegel (eds.), *Biological Control of Plant-Parasitic Nematodes: Building Coherence between Microbial Ecology and Molecular Mechanisms*, Progress in Biological Control 11, DOI 10.1007/978-1-4020-9648-8_4,

oped into biological control agents (Stirling 1991). The first multicellular animal to be sequenced was the free-living nematode *Caenorhabditis elegans* (The *C. elegans* Sequencing Consortium 1998) and this has built a technological platform (Chap. 3) from which to investigate innate immunity, the generic science that underpins the biology of host-pathogen interactions. The price of nucleic acid sequencing has approximately halved every 2 years since 1965, a phenomenon known as Hodgkin's Law (Dawkins 2009), and cost reduction has facilitated the sequencing of increasing numbers of genomes. The last 3 years represent a milestone for parasitic nematode genomics; in addition to the release of a draft genome sequence of the human parasite nematode *Brugia malayi* (Ghedin et al. 2007), the first two complete genomes of plant-parasitic nematodes have been obtained, both from the root-knot nematode genus: *Meloidogyne incognita* and *M. hapla* (Abad et al. 2008; Opperman et al. 2008). The *M. hapla* genome sequencing project was coordinated by the Center for Biology of Nematode Parasitism in Raleigh NCSU (USA). Two distinct gene prediction algorithms, FgeneH and GlimmerHMM, were independently trained on hand-curated *M. hapla* gene models based on full length ESTs. In combination with the PASA algorythym these were used for *ab initio* gene predictions. The *M. incognita* genome was sequenced in France under the initiative of the Nematology group at INRA Sophia Antipolis in close collaboration with the Génoscope at Evry (the French centre for sequencing), and the Bioinformatic platform at INRA Toulouse. The ensemble of predicted and automatically annotated protein coding genes using the Eugene gene predictor trained with *M. incognita* ESTs was manually hand annotated by a consortium of 27 laboratories.

Currently, another 21 nematode genomes are being sequenced four of which are plant-parasitic nematodes and with the increasing power of the next generation of sequencing technologies it is the intention to sequence a total of 959 across the phylum Nematoda (http://www.nematodes.org/nematodegenomes/index.php/959_Nematode_Genomes). Following the sequencing of *C. elegans* it was suggested that it would provide a good reference from which to compare hypothesised biochemical and developmental pathways in plant-parasitic nematodes (Bird et al. 1999). However, now over a decade later, the data accruing from genome sequencing projects have grown immeasurably and, although it is possible to compare catalogues of genes at the whole genome level, it is also possible to construct subtle hypotheses around groups of genes with possible roles in host-parasite interactions and from this start to predict the likely effects on parasite biology and vice-a-versa. This next step in exploiting sequence data requires an understanding of genomic and operonic structures before going on to explore the relationship between sequence and worm biology.

4.2 Genomics

The genome sequence of *C. elegans* has become an indispensible resource for studying host-parasite interactions (Chap. 3 above). Genome sequences are also available for several species of pathogenic *Bacillus* spp. of which *Pasteuria* is a close relative

(Charles et al. 2005). The interaction between *Pasteuria penetrans* and root-knot nematodes provides a model by which host-parasitic interactions can be investigated. The recent publication of the root-knot nematode genome sequences together with the partial genome sequence of *Pasteuria* will help generate insights into *Pasteuria*-nematode biology which in turn will inform strategies for the control of these important pests. However, before such interactions can be analysed in detail a broad knowledge of genomic and operonic structures is required.

4.2.1 Genome Structures and Operonic Structures of Root-Knot Nematodes

Substantial differences between organization of the *M. hapla* and *M. incognita* genomes have been shown. *M. hapla* VW9 is diploid with 16 chromosomes. The assembled scaffolds results in a 10.4X coverage of the approximately ~54 Mb and spans >99% of the genome. The genome possesses a relatively small percentage of moderately repetitive DNA (~12%) that is made up primarily of low complexity sequence. Therefore, this genome represents not only the smallest nematode genome yet completed, but also the smallest metazoan genome characterized so far.

The *M. incognita* genome seems to be more complex. The assembled sequence reads gave a total coverage of 86 Mb, which is almost twice the estimated size of 47–51 Mb per haploid genome of this species (Pableo and Triantaphyllou 1989; Leroy et al. 2003). Therefore, we suspect that *M. incognita* is a fixed heterozygous organism. Indeed, an all-against-all comparison of super-contig sequences revealed that most of the genome is present as homologous but diverged segment pairs that might represent former alleles. The average sequence divergence between the aligning regions is 7–8%, which is among the highest observed until now for a sequenced heterozygous organism. All together, these observations are consistent with the strictly parthenogenetic lifestyle of *M. incognita*, in which the absence of meiotic recombination may allow alleles to diverge considerably, as hypothesized for bdelloid rotifers (Welch et al. 2004). Repetitive/transposable elements (TEs) are quite abundant in *M. incognita* genome (36%) compared with the *M. hapla* and *C. elegans* genomes. In both *M. hapla* and *M. incognita* genomes, no DNA attributable to bacterial endosymbiont genome was identified.

The gene content also differs greatly between these two RKN genomes. *M. incognita* is estimated to possess 19,212 genes, while *M. hapla* appears to carry 14,420 genes. This substantial difference in gene number could be explained by the *M. incognita* genome organization depicted above. It is tempting to correlate the larger gene set in *M. incognita* with its strictly mitotic parthenogenetic reproductive mode, which allows maintenance of functional divergent ancient alleles and duplications. Gene density in *M. incognita* is very similar to that in *C. elegans*, while *M. hapla* shows substantially greater gene density. This likely reflects the small genome size of *M. hapla*.

The overall G+C content of *M. incognita* (31.4%) is comparable to that of *B. malayi* (30.5%) and lower than that of the free-living nematodes *C. elegans* (35.4%) or *C. briggsae* (37.4%), whilst *M. hapla* exhibits a significantly lower G+C content of 27.4%. Additionally, in the two RKN genomes, satellite DNA families were found, as were the rRNA sequences (16S-5.8S-28S) organised in clusters.

One striking characteristic of nematode genomes is the presence of operons (Guiliano and Blaxter 2006). In *C. elegans*, operons are defined as sets of adjacent co-transcribed genes that are *trans*-spliced to SL (spliced-leader) and SL-like exons. In *M. incognita*, potential operonic structures were thus searched considering gene pairs with an intergenic distance less than 1,000 bp. This led to the identification of 1,585 candidate operons containing a total of 3,966 genes with 516 of those having a *C. elegans* orthologue. Only nine of them are fully conserved between these two species. In the *M. hapla* genome, a comparison of all 1:1 orthologs from *M. hapla* to *C. elegans* results in the identification of 140 operons from *C. elegans* that are at least partially conserved in *M. hapla*. Hence, operons are a dynamic structure of nematode genome architecture, but the conservation of synteny is not present between these species.

4.2.2 Gene Families in Root-Knot Nematodes

Although the genomes of *M. hapla* and *M. incognita* differ in numerous important ways, they also share a number of common characteristics. For example, both species carry substantially fewer G-protein coupled receptors (147 and 108 in *M. hapla* and *M. incognita*, respectively) than does the free-living nematode *C. elegans* (1,280 genes).

Collagens are ubiquitous structural proteins that play an essential role as shown by the range of defects identified by mutational analysis in *C. elegans*. The cuticle collagens are an abundant gene family in *C. elegans*, with over 180 members grouped into six sub-families according to homology relationships (Page and Johnstone 2007). The RKN genomes revealed a great reduction with the presence of 122 genes in *M. incognita* and 81 genes in *M. hapla*.

The superfamily of nuclear receptors (NR) is of widespread relevance to almost all aspects of physiology since they are involved in the regulation of gene expression. The evolutionary history of nematode NRs is known to be quite complex. Many NRs that are of significant physiological importance in other animals are not found in *C. elegans*, but are present in *B. malayi*. In *M. hapla*, only 25% of the *C. elegans* genes were identified. In *M. incognita*, the situation is more complex. Among the 92 predicted NRs identified, clear orthologs to some known nematode NRs were found, mainly from the *B. malayi* source. In addition, the presence of a great number of supplementary NRs was observed in *M. incognita*. These findings implicate multiple duplication events started before and after the split between the *B. malayi, C. elegans* and *M. incognita* lineages.

Taken together, data from this preliminary comparative analysis of some important traits in nematode physiology highlight the fact that the model species *C. elegans* is not reflective of genomic diversity displayed in the phylum Nematoda, especially when parasitic species are considered. This is not to say that *C. elegans* is not a extremely valuable resource to understand other nematode species, but rather that different species of nematodes show amazing adaptation to their specific niches.

4.2.3 Developmental Pathways Conserved in Nematodes

The pathway of genes responsible for sex determination in *C. elegans* has been studied in detail and is linked to the dosage compensation pathway (Zarkower 2006). In spite of their different mode of reproduction, *M. incognita* and *M. hapla* homologues of at least one member of each stage of the sex determination cascade were identified, including genes from the dosage compensation pathway, from the sex determination pathway itself and from numerous downstream genes such as a gene that represses male promoting genes and which controls male differentiation and behaviour. However, genes upstream in the pathway were not detected, suggesting a divergence between the RKN species and *C. elegans* in signals that trigger these sex determination pathways.

Since RNAi can be induced in different RKN species, we therefore expected to find components of the RNAi pathway in the *M. incognita* and *M. hapla* genomes. Although many of the components of the RNAi are indeed found in these two genomes, the red4 gene was not found, similar to the animal-parasitic nematode *Haemonchus contortus* (Zawadzki et al. 2006). In addition, as reported for *B. malayi* and *H. contortus*, homologues of *sid-1, sid-2, rsd-2* and *rsd-6* involved in systemic RNAi and dsRNA spreading to surrounding cells, were not found either. Novel or poorly conserved factors of spreading could explain the systemic RNAi reported in *M. incognita*. In order to identify pathways unique to nematode development and parasitism that can serve as new targets for nematicides, the RNAi experiment repository in Wormbase was examined to identify potentially lethal phenotypes; these were then used to search for orthologous genes in the *M. incognita* database and were retrieved. Among them more than 340 *M. incognita* genes were identified as potential nematode targets for anti-parasitic drug design.

Globally, one of the most important conclusions of this genomic analysis is the striking reduction of the *M. hapla* genome size coupled with significant gene loss when compared to the model species *C. elegans*. Consistent with the genome size reduction observed in *Brugia malayi*, this gene loss seems to be a clear attribute of the parasitic life style.

However, this is not the case for the *M. incognita* genome, where evolution in the absence of sex towards effective haploidy through the Meselson effect led to the maintenance of functional divergent ancient alleles, which probably accounts for the larger number of genes present. For genes involved in the host-parasite interface, such genetic plasticity could explain the extremely wide host-range and geographic distribution of this nematode, contributing to its successful establishment as a polyphagous plant-parasite. Conversely, *M. hapla* can be considered as an evolutionary basal species, and deeper comparative analysis of these two genomes will likely shed light on the evolution of *Meloidogyne* spp. and point to both the basal gene complement and genes involved in host range.

4.3 Inter- and Intra-Specific Variation Amongst *Meloidogyne* spp.

It was proposed by Haldane (1949) that disease is a driver for the generation of genetic diversity and the Red Queen Hypothesis suggests that sexual reproduction is maintained because it enables species to respond to their changing biotic environment (Van Valen 1973; Otto and Nuismer 2004), which includes pathogenic organisms. Over long periods of time interactions between hosts and pathogens leads to co-evolutionary developments and for plant-parasitic nematodes this can be between either a nematode and a plant or a nematode and a pathogen. As different root-knot nematodes have adopted different reproductive strategies (Evans 1998) studies of inter- and intra-specific variation will produce insights into their evolutionary development as driven by host-parasite interactions.

4.3.1 Variation of Subspecies of Meloidogyne hapla

Variability in the genome is not only limited to that between *Meloidogyne* species; genotypic differences exist between RKN subspecies, or races, and are corroborated by differences in behavioural characteristics, such as the ability to infect certain species of plants. Hartman and Sasser (1985) capitalized on this by designing a diagnostic assay based on host-specificity to distinguish between subspecies of both *M. incognita* and *M. arenaria*. Characteristics that differ between races of *M. hapla* include gall size, tendency to aggregate (personal communication, VM Williamson, UC Davis), and the ability in some subspecies to overcome resistance in alfalfa (Griffin and McKenry 1989), *Solanum* (Van der Beek et al. 1998) and common bean cultivars (Chen and Roberts 2003). Liu and Williamson (2006) obtained two geographic isolates of *M. hapla* and developed them into inbred lines, VW8 and VW9, the same race that was used to generate the *M. hapla* sequence (Opperman et al. 2008). As VW8 and VW9 differ in the characteristics mentioned above, Liu et al. (2007) took advantage of the availability of both outcrossing and parthenogenetic reproduction of *M. hapla* and used AFLP techniques to develop a genetic map and place markers near these variable traits. The genetic map merged with the genomic sequence provides a powerful tool to look at variation between races, and will potentially reveal the causes for differences in some behaviour, including clues into host-specificity.

Variation between VW8 and VW9 was further explored by skimming the genomic sequence of VW8. A 2x coverage (totalling 125 Mb) of the *M. hapla* genome was generated by sequencing DNA extracted from 10 pooled progeny of a VW8xVW9 cross. Resulting sequences were mapped back onto 83% of the contigs that make up the VW9 reference sequence (VM Williamson and JE Schaff, manuscript in preparation). Further exploration between these two genomes will hopefully provide clues as to why there are differences in behaviour (specifically, infection capability) as well as insight into the structure of genomes and why some regions are more prone to polymorphisms than others.

4.3.2 *The Hypotheses for RKN Evolution*

As presented above, RKN have evolved very diverse reproductive strategies and undergone extensive cytogenetic differentiation. In parallel, a general pattern of relationship between host specificity and reproduction mode has been proposed, although some exceptions may occur (Jepson 1987). Indeed, most amphimictic RKN species are host-specific (e.g. *M. megatyla* and *M. pini*, which are both restricted to *Pinus* spp.), while parthenogenetic species in general have a wider host range. In particular, the major mitotic species *M. arenaria*, *M. incognita* and *M. javanica* exhibit extreme polyphagy, with a potential host range that encompasses the majority of the estimated 250,000 flowering plants (Trudgill and Blok 2001). Whether this ability to exploit of a very large range of plant genotypes by parthenogenetic RKN compared to their amphimictic relatives is a direct consequence of their mode of reproduction remains unanswered. In addition, adaptation to the selective pressure of plant resistance genes is well-documented in natural RKN populations (Castagnone-Sereno 2002). Such genes trigger a localized hypersensitive reaction (HR) of plant cells at the infection site, which prevents installation and further development of avirulent nematodes (Williamson 1999). Since virulent populations, i.e. able to develop on resistant plants without eliciting the HR, have been reported from both meiotic or mitotic RKN species, response to plant resistance seems to be a unifying example of the capacity of stable genetic variation and adaptative evolution of these parasites, and will be developed in more details below looking at the interactions between root-knot nematodes and the obligate bacterial parasite *Pasteuria penetrans*.

4.4 *Pasteuria* – Root-Knot Interactions

The use of natural enemies for the control of plant-parasitic nematodes has a long history (see Chap. 1) but research on soils being suppressive to nematode pests has only been the focus of intensive study over the last 30–40 years. The motivation for research on microbial enemies of nematodes that produce suppressive nematode soils and the use of this knowledge to develop pest management strategies is the subject of this book and details of the different organisms are covered in other chapters.

The focus of this chapter is to review the knowledge of genomics and how this might help in developing new approaches to the biological control of nematode pests, in particular focusing on the *Pasteuria* group of bacteria. The most intensively studied bacterium of the *Pasteuria* group is *P. penetrans*, the parasite of root-knot nematodes and, although there is increasing interest in other species of *Pasteuria*, such as *P. nishizawae* and *P. usage,* which are parasites of soybean cyst nematode (*Heterodera glycines*; Noel et al. 2005) and sting nematode (*Belonolaimus longicaudatus*; Giblin-Davis et al. 2003) respectively, the following sections will focus on *Pasteuria penetrans*.

Pasteuria penetrans is a member of the endospore forming group of Gram-positive bacteria and the life-cycle is initiated when infective second-stage juvenile nematodes, migrating through the soil towards plant roots, come into contact with endospores that lie dormant in the soil. These endospores are the resting stages of the bacterium and can remain viable for many years (Giannakou et al. 1997). Although there appears to be a preference for endospores to adhere to the head region of the second-stage juvenile, in a compatible interaction endospores can adhere to any part of the second-stage juvenile cuticle (Davies unpublished data). In a compatible interaction, the numbers of endospores that adhere to a juvenile can range from one to around 20 in field soils; however, in endospore attachment assays in which spores and juveniles are centrifuged together, over 100 spores per second-stage juvenile have been observed in some tests. The motility of second-stage juveniles encumbered with >15 spores is affected and this reduces their ability to migrate to and invade plant roots (Davies et al. 1988, 1991). Germination of the endospore takes place once the spore encumbered juvenile enters the plant root and establishes a feeding site and prior to the moult to a third-stage juvenile. In some other nematode species, such as *Heterodera avenae,* the endospore can germinate before the juvenile has entered the root and this reduces the ability of the second-stage juvenile to find and invading its host (Davies et al. 1990). Recent studies of developing females infected with *P. penetrans* using electron microscopic techniques have shown that following germination rhizoid structures grow out from the site of infection into the pseudoceolomic cavity and granular masses of rod shaped bacilli have been observed (Davies 2009). Similar rod shaped bacilli have also been seen growing in in vitro cultures of *P. penetrans* (Hewlett et al. 2004). It is likely that these rod shaped bacilli, in the right growth conditions, can multiply exponentially and move as some of these rod shaped bacteria observed in the pseudocoelom appear to have a single polar flagellum (Davies 2009; Davies et al. 2010). Although it has suggested that these rod shaped bacteria are *Pasteuria* (Davies 2009; Davies et al. 2011), this has not been shown unequivocally. There are reports in the literature that helper bacteria may be involved in the growth of *P. penetrans* (Duponnois et al. 1999; Gerber and White 2001; Hewlett et al. 2004); hence, because of the obligate nature of the *Pasteuria* life-cycle, and the difficulty of generating a population of the bacterium from a single individual to undertake experiments that fulfil Koch's postulates, the life-cycle proposed by Davies (2009) will remain open to question. Root-knot nematode females infected with *P. penetrans* produce few, if any, progeny as their reproductive system quickly degenerates (Davies et al. 2008). Sporogenesis begins when unidentified triggers, perhaps when certain key nutrients are limiting, lead to the initiation development of microcolonies. These consist of clumps of dichotomously branching mycelia-like structures, which subsequently fragment into quartets and doublets. This process continues until single, separate sporangia are produced each containing a single endospore. An individual female that is infected can contain over 2×10^6 endospores and infected females after 6–8 weeks, often become larger than uninfected females (Davies et al. 1988). The endospores are released back into the soil when infected nematodes and plant roots decay.

4.4.1 State of the Pasteuria Sequencing

As discussed above, the last decade has seen the sequencing of eukaryotic and prokaryotic organisms and with the development of more efficient sequencing technologies the sequencing of organisms has become increasingly routine. Comparing the genomes of different organisms can often lead to insights into their evolutionary history and help to answer questions regarding how organisms with similar developmental processes and genetics have very different life forms, and conversely, how very similar life forms can have very dissimilar developmental processes and genetics (Frutos et al. 2006, Cañestro et al. 2007). Computer software is being developed to make the comparisons between nematode species (Harris et al. 2003) and bacterial species (Field et al. 2005) easy and accessible. With the sequencing of plant-parasitic nematodes, it could be of huge interest to also have sequenced a microbial parasite with biological control potential, as this provides an opportunity to gain an understanding of a host – parasite interactions from a very different perspective than the more usual nematode-plant interaction studies. The study of root-knot nematode – *Pasteuria* interactions may therefore produce unique insights that can lead to the development of novel control strategies. A survey sequence of the genome (Bird et al. 2003) has led to around 4,000 nucleotide sequences being available through GenBank and EMBL. Although at present there is no completed *Pasteuria* spp. genome, recent sequencing strategies using 454 based technologies have increased the coverage of the genome (Table 4.1). Even without having a completely sequenced genome it is possible to start making comparisons between closely related species and gain an understanding into key biological processes, examples of which will be given below. As *P. penetrans* has potential for being developed into a biological control agent, understanding of this particular host-parasite interaction is essential if it is to be developed into a commercial control agent. There are two aspects that are currently prohibiting its commercial development: (1) the inability to mass culture the bacterium in vitro and (2) its restricted host range. Focusing on these two fundamental problems, it is therefore possible to gain insights from genomic comparisons of the host-parasite interactions which may help development of the bacterium as a biological control agent.

Table 4.1 Comparison of *Pasteuria* genome sequencing undertaken using a Sanger and 454 sequencing platforms

	Sanger sequencing	454 sequencing
Base pairs (Mbp)	2.5	8.6
Number of contigs	1,500	5,964 (782>2 kb)
Largest contig (Kbp)	2.5	54.7

4.4.2 Comparative Genomics and In Vitro Culture

Until very recently the mass production of *P. penetrans* for the control of plant-parasitic nematodes has relied on in vivo culturing methods. The majority of these methods are adaptations of the method developed by Stirling and Wachtel (1980). This method requires females infected with *P. penetrans* spores being collected and a suspension of endospores prepared by homogenising these females in water. Infective root-knot nematode juveniles are then encumbered with 5–10 endospores by exposing juvenile to mature spores and they are then placed around the roots of a tomato plant. After 6–8 weeks the nematode infested roots containing *Pasteuria*-infected nematodes are washed free of soil, air dried and the roots milled to produce an inoculum for application to soil. Such milled tomato root powder can contain as many as 1.3×10^9 endospores per gram of root powder but the number of spores in each batch is highly variable (Pembroke and Gowen, personal communication). Although this production method is good enough to produce sufficient spores for the use of small scale growers, large growers will require levels of mass production that would be better suited to an in vitro culturing method. Attempts to grow *Pasteuria* in vitro (Williams et al. 1989; Bishop and Ellar 1991) have produced very limited success. Bishop and Ellar (1991) produced two media, one of which would sustain vegetative growth and another led to the production of endospores, but because at no point did the bacteria grow exponentially they were never able to produce enough for commercial application. More recently Pasteuria Bioscience LLC, Florida, has developed media in which it is possible to grow vegetative stages of *Pasteuria* (Hewlett et al. 2004) and these have been successfully deployed for the control of sting nematode on golf courses. A major issue in developing in vitro production methods is the ability to stimulate exponential growth of vegetative stages and then induce these growth forms to sporulate. Both Bishop and Ellar and Pasteuria Bioscience LLC realised that nutrition allows vegetative cells to grow exponentially and the switch from vegetative growth to sporulation; these are two key aspects of being able to develop in vitro culturing methods. The following two sections offer the types of insights that can be developed from having genomic information.

4.4.3 Transportome and Vegetative Growth

The transportome is the range of genes that an organism possesses that encode protein molecules that contribute to transport of molecules across biological membranes. The array of outer membrane bound transporters available to the bacterial cell governs the uptake of solutes and signalling molecules into the cell above ambient concentrations as well as controlling the secretion of proteins and excretion out of the cell. The outer membrane of bacterial cells is the first port of call for detection of environmental cues and it is therefore logical to assume that differences in the diversity, abundance and expression of the *Pasteuria* transportome compared to other bacteria is likely to be provide insights into the nutritional requirements of this obligate parasite.

The current nucleotide sequence information for *Pasteuria* transporters is poor and although a search in PubMed for these terms retrieves 103 nucleotide sequences, the majority of these are for ATP binding cassette (ABC) transporters that map to just a few contigs of the partially sequenced *Pasteuria* genome, and show almost 100% identity to each other. As such it is likely that these are duplicate entries for the same transport system. There are unique entries amounting to approximately two different solute binding proteins, five ATP binding cassettes and 10 integral membrane permeases. The remaining four entries are for a major facilitator superfamily (MFS) transporter as well as for three unclassified transporters with homology to *Clostridium* spp. This ensemble is inadequate to perform a transportome analysis of *P. penetrans*, but what we can do before a complete genome of this organism is obtained, is to analyse the transportome profile of related bacteria and see if there are any differences between pathogenic and non-pathogenic bacteria.

Phylogenetic analyses of *P. penetrans* using a range of 33 housekeeping genes was recently performed by Charles et al. (2005), and they reported with a high degree of confidence that *P. penetrans* clusters tightly in the *Bacillus-Clostridium* clade. Additionally, it was found that *P. penetrans* is more closely related to the non-pathogenic *Bacillus subtilis* and *B. halodurans* as opposed to virulent *Bacillus* species such as *B. anthracis*, *B. thuringiensis* and *B. cereus*. However, based on the composition of the exosporium of these bacteria, *Pasteuria* spp. can be considered more closely related to the pathogenic *Bacillus* species as they are also parasitic organisms and they also have the ability to synthesize exosporium bound collagen-like repeats, which are considered to be essential for host attachment in *Pasteuria* spp. as well as *B. thuringiensis*, *B. cereus* and *B. anthracis* (Todd et al. 2003). With these factors in mind, it is difficult to predict how the genetic composition and transportome of *Pasteuria* spp. is likely to be arranged. Nevertheless, it is of great interest to explore and further understand the genomes of the sequenced non-pathogenic and pathogenic *Bacillus* spp. to aid future comparative genomic studies with *Pasteuria* spp., which will be very important for the understanding of the biology of this bacterium.

4.4.3.1 Comparative Genomics: *Bacillus* spp. and Other Soil Bacteria

The sequenced *Bacillus* spp., like *Pasteuria* spp., are spore forming bacteria of the Firmicutes division. Broadly speaking the more pathogenic species have larger chromosomes than the less pathogenic species (around 5.2 Mb as opposed to 4.2 Mb). Additionally, the more pathogenic species such as *B. cereus*, *B. anthracis* and *B. thuringiensis* also house plasmids encoding toxin genes, bolstering their genome size further. It is difficult to speculate on the size of the *P. penetrans* genome as although *P. penetrans* is not thought to produce nematicidal toxins, it is nonetheless a nematode parasite. As such we can make a broad prediction that its genome size should be between 4.2 and 5 Mb.

A survey was conducted to compare the genome sizes of 244 sequenced soil bacteria; the information to perform this survey was obtained from the transport DB website (www.membranetransport.org). The soil bacteria were divided into their respective Divisions (the number of bacteria representing each Division is shown in brackets in Table 4.2) and the mean characteristics of each Division as well as the mean of the non-pathogenic and pathogenic Bacilli were compared with each other. This shows that the non-pathogenic *Bacillus* spp. have a larger genome size than the 'average' soil organism and is ranked joint fourth with the Actinobacteria out of the 12 groupings (Table 4.2). Additionally, it was found that the pathogenic *Bacillus* spp. grouping had the largest genome size (Table 4.2). This can be partly explained by the fact that the pathogenic *Bacillus* spp. are often known to harbour large toxin encoding plasmids that significantly bolster the genome size.

4.4.3.2 Transportome of Sequenced *Bacillus* Species

The survey was expanded to compare the transportomes of the non-pathogenic and pathogenic *Bacillus* spp. with each other as well as with other common soil bacteria. It was found that of all the groupings the pathogenic *Bacillus* spp. have the highest average complement of transporter encoding genes in their genomes, and that they also have the highest average density of transporters when compared with the 'average' soil bacterium as well as other individual Divisions of bacteria (Table 4.2). Additionally, the non-pathogenic *Bacillus* spp. are ranked second of 12 groupings for both total number of transporters as well as density of transporters (Table 4.2).

The reasons for these trends are likely to be linked to the pathogenic nature of the virulent species requiring them to possess specific transporters in order to cause disease. However, the explanation of why the non-pathogenic *Bacillus* spp. also possesses a relatively high number and density of transporters is less clear, as high numbers and densities of transporter genes are normally associated with organisms that occupy several distinct niches such as some examples of the Rhizobiaceae, or as mentioned above disease causing organisms. However, *B. subtilis* and *B. halodurans* that comprise the non-pathogenic *Bacillus* spp. group are ubiquitous soil saprophytes and so one possible explanation for the plethora of transporters observed in the non-pathogenic grouping is that they have a solute scavenging style of nutrient uptake. Additionally, *B. halodurans* C-125 is an alkaliphilic bacterium and so is likely to require specific transporters for homeostasis and *B. subtilis* is a competent rhizosphere coloniser, often used as a fungicide on plants, roots due to its rapid root system colonisation. As such this bacterium is also likely to possess specific transporters to occupy this niche successfully.

The ABC and MFS super families of transporters are the two most abundant classes of transporters found in nature. ABC transporters are primary active transporters,

Table 4.2 Showing the Divisions of bacteria surveyed (number of bacteria surveyed for each division/grouping in brackets). Non-pathogenic bacilli – *Bacillus subtilis*; *B. halodurans*; *B. licheniformis*, Pathogenic bacilli – *B. anthracis*, *B.cereus*; *B. thuringiensis*. The 7 sequenced strains of *Bacillus anthracis* were considered as one organism, and their details were averaged so as not to skew the details of the pathogenic bacilli. For columns other than the first the bracketed numbers show the rank of each group for each feature

Group	Genome size (Mb)	Total transporter proteins	Transporters/Mb	ABC	MFS	APC	DMT	Sugar PTS
Actinobacteria (18)	4.2 (4)	205.1 (5)	47.9 (9)	71.1 (5)	45.2 (3)	9.9 (3)	5.7 (10)	2.7 (6)
All (244)	3.26 (8)	187.3 (7)	53.03 (5)	63.1 (6)	26.8 (7)	5.8 (6)	9 (6)	6.1 (5)
Bacteroidetes (5)	4.4 (3)	178.4 (9)	44.5 (10)	39.2 (11)	21.4 (9)	1.4 (9)	6.6 (8)	0.4 (11)
Cyanobacteria (13)	3.19 (9)	118.7 (11)	37.9 (13)	53.8 (10)	7.3 (12)	1 (11)	3.5 (12)	0 (12)
Firmicutes (68)	2.62 (11)	178.9 (8)	61.2 (3)	59.8 (7)	22.9 (8)	9.5 (4)	7.2 (7)	12.5 (2)
Non-pathogenic Bacilli (3)	4.2 (4)	311 (2)	73.92 (2)	92 (2)	52.3 (2)	16 (2)	13.7 (2)	19 (1)
Pathogenic Bacilli (3)	5.25 (1)	411.3 (1)	78.5 (1)	115 (1)	70.76 (1)	25 (1)	30.05 (1)	12.42 (3)
Proteobacteria Alpha (48)	3.6 (6)	204.9 (6)	51.7 (6)	85.5 (3)	27.1 (6)	2.1 (8)	13.6 (3)	1.5 (10)
Proteobacteria Beta (14)	4.5 (2)	235.1 (3)	50.3 (8)	71.9 (4)	41 (4)	3.9 (7)	13.6 (3)	2.2 (8)
Proteobacteria Delta (7)	2.8 (10)	151 (10)	38.18 (12)	54.6 (9)	11.1 (10)	1.1 (10)	6 (9)	1.6 (9)
Proteobacteria Epsilon (8)	1.5 (13)	98.3 (12)	51.5 (7)	27.9 (13)	9.25 (11)	0.5 (13)	4.4 (11)	0 (12)
Proteobacteria Gamma (57)	3.55 (7)	212.6 (4)	54.8 (4)	57 (8)	33.6 (5)	7.1 (5)	10.5 (5)	8.1 (4)
Spirochaetes (6)	2 (12)	78.3 (13)	41.5 (11)	30.8 (12)	2.5 (13)	0.8 (12)	1.7 (13)	2.3 (7)

ABC – *ATP* binding cassette transporter, *MFS* Major Facilitator Superfamily, *APC* Amino acid/polyamine/organocation transporters, *DMT* Drug metabolite transporter, Sugar *PTS* sugar specific phospho-transferase systems

and in bacteria often rely on a solute binding protein (SBP) for transport of solutes. The MFS on the other hand are secondary transporters and no SBP is associated with these transporters. Both superfamilies function to import and export a staggeringly large array of molecules. Compounds transported by these systems include simple sugars, oligosaccharides, inositols, drugs, amino acids, nucleosides, organophosphate esters, Krebs cycle metabolites, and a large variety of organic and inorganic anions and cations (Pao et al. 1998; Higgins 1992). When examining their presence in soil organisms it was found that these systems were most abundant in the pathogenic *Bacillus* spp. grouping, with the non-pathogenic *Bacillus* species being ranked second of 12 groupings (Table 4.2). It is likely that these transporters are important for the pathogenic as well as the non-pathogenic *Bacillus* species during their saprophytic growth phase in the soil environment and that additional transporters are available to the pathogenic species for overcoming host defences or causing disease as well as for nutrient utilisation of additional nutrient sources once inside the host organism. For example, an ABC transporter in *B. anthracis* was found to transport bicarbonate, and this was shown to be imperative for initiation of virulence in the human host (Wilson et al. 2008).

Further analysis of the transportome database revealed that the sequenced pathogenic and non–pathogenic *Bacillus* spp. have an increase of three other classes of transporter when compared with most groupings of bacteria examined in this survey. These are the amino acid/polyamine/organocation (APC) superfamily, the Drug Metabolite Transporter (DMT) superfamily and the sugar-specific phosphotransferase family (PTS).The APC and DMTs are both secondary active transporters, with the APC systems functioning as solute:cation symporters and solute:solute antiporters (Jack et al. 2000). There are several DMT families, and the largest of these is the drug metabolite exporter (DME) family, and as its name suggests transporters belonging to this family are largely concerned with cellular export of molecules (Jack et al. 2001). Finally, the sugar PTS, are a type of group translocation system, which couple translocation of a substrate to its chemical modification, resulting in release of a modified substrate at the opposite side of the membrane (Law et al. 2008). More specifically, the sugar-PTS are important for carbohydrate import as well as chemotaxis toward PTS substrates (Postma et al. 1993). It is likely that the APC and sugar-PTS transporters function largely for active solute uptake and some examples of the APC permeases demonstrate exchange transport (Saier 2000). Although the DMTs are likely to also be involved in solute importation, the DME family as mentioned above is implicated in the export of molecules.

Currently, the role of the transporters described in the biology of this group of bacteria is not fully understood, though they are likely to be implicated primarily in nutrient acquisition. If the density and range of transporters found in the sequenced *Bacillus* species is found to be indicative of the range of transporters present in *Pasteuria* species, then it is likely that that *Pasteuria* species would require some of these transporters for nutrient uptake once the bacterium has penetrated the nematode host as *Pasteuria* species are unable to reproduce outside of their host organism. Additionally, a significant proportion of the *Bacillus* species ABC, MFS as well as DMT transporters are putative drug efflux systems. If this is also the case in

Pasteuria species then it is possible that orthologues of some of these transporters could be used for the bacterium to secrete anti-microbial molecules to stave off any potential secondary invading microbes. Such a strategy would secure the nematode host nutrient source for the bacterium's own ends as well as to provide protection against host-defence compounds or to mediate the secretion of host-specific toxins. It will be useful to use the transportome information highlighted here as a start point to perform comparative genomics with the *P. penetrans* sequence when it becomes available. This will hopefully provide information of the roles of transporters in the life-cycle of this organism and lead to clues into the nature of an obligate life-style that may enable improvements in the methods of in vitro culturing that are currently available.

4.4.4 Getting Vegetative Cells to Sporulate

The initiation of sporulation in *B. subtilis* has been studied extensively and it has been shown to be controlled by a phosphorelay pathway (Burbulys et al. 1991). In this pathway a phosphoryl group is transferred to the regulator Spo0F through a group of five kinases that are under environmental regulation. This phosphoryl group is then transferred to the phosphotransferase Spo0B, which in turn passes it onto the regulator/transcription factor Spo0A. Phosphorylation of Spo0A enhances the activation and repression of approximately 500 stationary phase and sporulation genes (Fawcett et al. 2000). Like all known regulators Spo0F requires a divalent metal ion to be present in the conserved aspartic acid pocket in order for phosphorylation to occur (Grimshaw et al. 1998) and Mg^{2+} has been shown to be important (Zapf et al. 1996). More recently it has been suggested that metal cations other than Mg^{2+} may play a role in the structure and function of Spo0F and its involvement in the initiation of sporulation (Mukhopadhyay et al. 2004). Investigations of the effects of the divalent cations Ca^{2+}, Cu^{2+}, Mg^{2+}, and Mn^{2+} on the structure and function Spo0F of *B. subtilis* showed that they bound to the aspartic acid pocket and that, while Mg^{2+} supports phosphotransfer from the kinase KinA to Spo0F, the copper cation Cu^{2+} inhibited their phosphotransfer (Kojetin et al. 2005).

Searches of the *Pasteuria* survey sequence (using BlastP) revealed a large number of genes (~6%) that had a high degree of similarity to genes involved in sporulation (Bird et al. 2003) and this included Spo0F. Alignment of Spo0F between *B. subtilis, B. anthracis* and *B. thuringiensis* and *P. penetrans* showed that key amino acids that form the aspartic acid pocket are conserved across these groups. From the results discussed above it was hypothesised that the presence of Cu^{2+}, at non-lethal concentrations in the sporulation media for *B. subtilis* and the related bacterium *P. penetrans*, might inhibit endospore formation while continuing to permit vegetative growth. Indeed, subsequent experiments revealed that the absence of Cu^{2+} in the media showed an increased number of sporulating cells (Kojetin et al. 2005). This result suggests that reduced availability of Cu^{2+} could be used to induce vegetative cells to enter sporulation.

4.4.5 Endospore – Cuticle Interactions

The infection of second-stage root-knot juveniles by *P. penetrans* endospores is initiated when viable endospores adhere to the cuticle surface of migrating nematodes as they move through the soil. Attachment can therefore be seen as the key to the commencement of infection. Studies (Stirling 1985; Davies et al. 1988, 1990; Channer and Gowen 1992; Sharma and Davies 1996; Espanol et al. 1997; Mendoza de Gives et al. 1999; Wishart et al. 2004) have shown that different populations of endospores do not adhere to all cuticles of all populations of nematodes and inter- and intra-attachment specificity is usual. Indeed, it has been shown that cuticle heterogeneity as exhibited by endospore attachment is not linked in any simple way to the phylogeny of the nematode (Davies et al. 2001) and, in addition, in standard attachment assays differences can also be found between different stages of the same nematode population (Davies and Williamson 2006). Interestingly, inter- and intra-specific functional variation as measured by *Pasteuria* spore attachment assays has shown an equal amount of variation even between amphimictic and parthenogenetically reproducing species of root-knot nematodes (Davies et al. 2008). If biological control is to work in a predictive manner and the correct spore populations applied to control susceptible nematode populations it will be important to understand the nature and mechanism of spore/cuticle attachment compatibility.

Henriques and Moran (2007) recently reviewed the structure and function of bacterial endospores. The endospore coat is the outermost layer of the spore; however in some bacterial species the spore is surrounded by an additional layer called the exosporium. *Pasteuria penetrans* is a species that possesses an exosporium, which provides it with resistance to chemical and enzymic treatments and is likely to give the spore its adhesive properties (Kozuka and Tochikubo 1985; Takumi et al. 1979). Fibrils are known to be important in the attachment of many bacteria to host surfaces and their decoration with sugars has been observed to confer host specificity (Benzi and Schmidt 2002; Power and Jennings 2003; Takeuchi et al. 2003). The exposure of endospores to HCl removes its central body to reveal a structure containing fibrils (Persidis et al. 1991) and scanning electron microscope studies on intact endospores have revealed that the parasporal fibres are positioned in such a way around the central body of the endospore to produce a *skirt*-like structure in which the under-surface of the endospore is in intimate contact with the nematode cuticle. Electron micrographs of the endospore reveal that the surface of the *skirt*-like structure, made up of parasporal fibres, are covered with other fine fibres both on the upper and lower surface and that the fibres on the concave surface of the endospore are more densely distributed than on the upper surface and it has been proposed that these fibres are involved in attachment of the mature endospore to the nematode cuticle (Davies 2009).

The structure of the exosporium in other closely related bacteria, *B. cereus, B. thuringiensis* and *B. anthracis* is species and strain specific (Plomp et al. 2005a, b) and it is clear that they also have an outer surface covered with a hair-like nap (Wehrli et al. 1980; DesRosier and Lara 1981) similar to *P. penetrans*. In *B. anthracis* the hair-like nap appears to be formed by a single collagen-like protein BclA in which the length of the filaments is related to the number of G-X-Y repeats (Sylvestre et al. 2002, 2003, 2005; Boydston et al. 2005). Homologous genes to *bclA* have been identified in other *Bacillus* spp. and they reside in a rhamnose cluster operon that contains around 30 genes within which are a number of glycosyl-transferases that form an exosporium island (Charlon et al. 1999; Steichen et al. 2003; Todd et al. 2003). In the initial genomic survey using Sanger sequencing of *P. penetrans* (Bird et al. 2003) four genes, with e-values $<e^{-14}$, were recognised using BlastP against *B. anthracis, B. cereus* and *B. thuringiensis*, within the rhamnose cluster including a collagen-like sequence that was phylogenetically more closely related to the bacterial collagens (Davies and Opperman 2006); subsequent 454 sequencing increased this to 12 genes and also included collagen-like sequences (Fig. 4.1). Several collagen-like sequences were identified each containing 28, 36 and 87, collagen-like G-X-Y repeats respectively and from which it was possible to predict that the *P. penetrans* hair-like nap would be made-up of filaments with lengths ranging from 56 to over 200 nm in length (Davies and Opperman 2006). Transmission electron microscope studies of endospores of *P. penetrans* have so far not provided evidence of fibres with a length significantly greater than 100 nm, but exosporial filaments ranging in length from 20 to over 100 nm have been identified (van de Meene, Rowe and Davies unpublished data). Conclusive evidence showing that these fibres on the surface of the endospore are collagen-like will need further investigation. However, results from a series of experiments in which endospores were either incubated in collagenase (Davies and Danks 1993) or were pretreated with fibronectin (Davies and Redden 1997; Mohan et al. 2001) suggested that the fibres are collagen-like, as these treatments reduced the ability of endospores to adhere to the nematode cuticle.

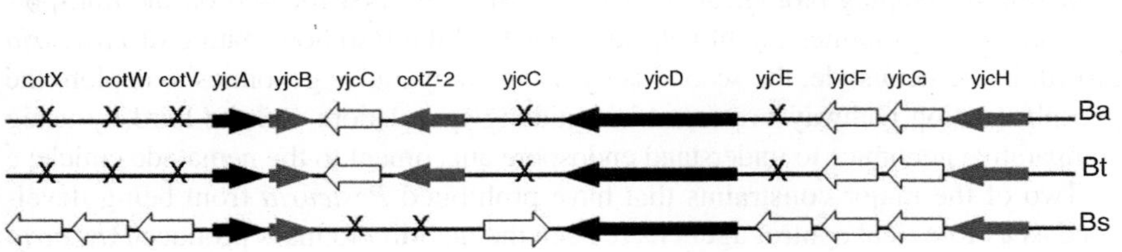

Fig. 4.1 Rhamnose cluster operon with alignments of genes for *Bacillus anthracis* (Ba), *B. thuringiensis* (Bt) and *B. subtilis* (Bs) BlastP hits with E-values$<e^{-14}$ from Sanger sequencing (*black arrows*) and 454 sequencing (*grey arrows*); X gene missing (Adapted from Todd et al. 2003)

The nature of the receptor on the nematode cuticle has not as yet been determined, however it is thought to involve some form of carbohydrate – lectin interaction (Davies and Danks 1993). Mucins are a family of polypeptides associated with both the innate and adapted immune systems and can be secreted or membrane bound to form a protective barrier that covers epithelial surfaces (Strous and Dekker 1992; Magalhães et al. 2010). The surface coat of *C. elegans* is a thin layer that is secreted onto the cuticle surface known to contain both mucin-like proteins amongst other glycosylated protein secretions (Hemmer et al. 1991; Gems and Maizels 1996). Mucin-like proteins are rich in serine and threonine and are highly glycosalated and it has been suggested that they play a role in immune defence (Hall and Altun 2008). Mucin-like proteins identified in *C. elegans* appear to have orthologues in *Meloidogyne* spp. (Davies 2009). RNAi experiments knocking down these mucin-like proteins have been shown to lead to changes in the recognition of the adult cuticle surface of *C. elegans* (Davies et al. 2009). It can therefore be hypothesised that similar proteins may be involved, directly or indirectly, in the endospore attachment process.

4.5 General Conclusions and Future Prospects

Two root-knot nematode genomes are now completed, the potato-cyst nematode (*Globodera pallida*) will be finished shortly, and a draft sequence of soya bean cyst (*Heterodera glycines*) is also available. The number plant-parasitic nematode genomes available will increase dramatically over the next few years as sequencing cost continue to fall. Concomitantly with this, the number of microbial genomes will also increase and these will include bacteria and fungi that are pathogens of nematodes. Therefore the tools will become increasingly available to dissect out and understand the mechanisms by which compatible and non-compatible nematode-microbial interactions occur.

In this chapter we have attempted to explore how genomics can be exploited and used to shed light into the interactions between *P. penetrans* and root-knot nematodes. After giving a current update on where the sequencing stands on plant-parasitic nematodes and *Pasteuria,* three examples were given of how this information could be used to help in developing biological control strategies: the first focused on the transportome and how genomics might help to understand the fastidious nature of *Pasteuria* growth in the nematode; the second is by using comparative genomics to explore the phosphorylation pathway important in initiating sporulation; and the third by using comparative genomics to understand endospore attachment to the nematode cuticle.

Two of the major constraints that have prohibited *Pasteuria* from being developed as a biological control agent have been the inability to mass produce *Pasteuria in vitro* and its host specificity. The first of these constraints, the inability to culture *Pasteuria* in vitro, has been circumvented by an empirical approach that has lead to the mass production of *Pasteuria* in a proprietary fermentation system, and their first product Econem™ (Pasteuria Bioscience LLC; www.pasteuriabio.com) is currently available for the control of *Belonolaimus* spp. However, the fact that endospores of *Pasteuria* do not attach to all populations of a given species suggests that host-specificity will remain a problem. Understanding the attachment process therefore still remains a key constraint for the use of *Pasteuria* as a biological control agent and genomic technologies will help to understand the molecular genetics of this interaction.

Genomic technologies have recently developed the first artificially constructed genome (Gibson et al. 2010) and can arguably claim to herald the beginning of the epoch of synthetic biology. This technology therefore brings into question traditional inundative approaches to biological control. Traditional approaches have relied on identifying pathogens of pests, screening them for efficacy, mass culturing them, and then releasing them back to control the pest. To date, biological control of nematodes has only met with limited success but synthetic biology, with the construction of de novo microorganisms with exactly the desired functionality required, offers the prospect of designer biological control agents. Although such approaches are in the medium to long term they will not be without their critics.

Acknowledgments KGD and THM would like to thank The Biotechnological and Biological Sciences Research Council of the United Kingdom. This work was also supported in part by the North Carolina Biotechnology Centre and the North Carolina Agricultural Research Service.

References

Abad P, Gouzy J, Auryet J-M et al (2008) Genome sequence of the metazoan plan-parasitic nematode *Meloidogyne incognita*. Nat Biotechnol 8:909–915

Benzi Z, Schmidt MA (2002) Never say never again: protein glycosylation in pathogenic bacteria. Mol Microbiol 45:267–276

Bird DMcK, Opperman CH, Jones SJ (1999) The *Caenorhabditis elegans* genome: a guide in the post genomics age. Annu Rev Phytopathol 37:347–365

Bird DM, Opperman CH, Davies KG (2003) Interactions between bacteria and plant-parasitic nematodes: now and then. Int J Parasitol 33:1269–1276

Bishop AH, Ellar DJ (1991) Attempts to culture *Pasteuria penetrans in vitro*. Biocontrol Sci Technol 1:101–114

Boydston JA, Chen P, Steichen CT et al (2005) Orientation within the exosporium and structural stability of the collagen-like glycoprotein BclA of *Bacillus anthracis*. J Bacteriol 187:5310–5317

Burbulys D, Trach KA, Hoch JA (1991) Initiation of sporulation in *B.subtilis* is controlled by a multicomponent phosphorelay. Cell 64:545–552

Cañestro C, Yokoi H, Postlethwait JH (2007) Evolutionary developmental biology and genomics. Nat Rev Genet 8:932–942

Castagnone-Sereno P (2002) Genetic variability of nematodes: a threat to the durability of plant resistance genes? Euphytica 124:193–199

Channer AG, Gowen SR (1992) Selection for increased host resistance and increased pathogen specificity in the *Meloidogyne-Pasteuria penetrans* interaction. Fundam Appl Nematol 15:331–339

Charles L, Carbone I, Davies KG et al (2005) Phylogenetic analysis of *Pasteuria penetrans* by useof multiple genetic loci. J Bacteriol 187:5700–5708

Charlon S, Moir AJG, Baillie L et al (1999) Characterisation of the exosporium of *Bacillus cereus*. J Appl Microbiol 87:241–245

Chen P, Roberts PA (2003) Virulence in *Meloidogyne hapla* differentiated by resistance in common bean (*Phaseolus vulgaris*). Nematology 5:39–47

Davies KG (2009) Understanding the interaction between an obligate hyperparasitic bacterium, *Pasteuria* obligate plant-parasiuriatic nematode host, *Meloidogyne* spp. Adv Parasitol 68:211–245

Davies KG, Danks C (1993) Carbohydrate/protein interactions between the cuticle of infective juveniles of *Meloidogyne incognita* and spores of the obligate hyperparasite *Pasteuria penetrans*. Nematologica 39:54–64

Davies KG, Opperman CH (2006) A potential role for collagen in the attachment of *Pasteuria penetrans* to nematode cuticle. In: Raaijmakers JM, Sikora RA (eds) Multitrophic interactions in the soil and integrated control. IOBC/WPRS Bull 29:11–15, General Secretariat IOBC/WPRS Dijon France

Davies KG, Redden M (1997) Diversity and partial characterisation of putative virulence determinants in *Pasteuria penetrans*, the hyperparasite of root-knot nematodes. J Appl Microbiol 83:227–235

Davies KG, Williamson VM (2006) Host-specificity exhibited by populations of endospores of *Pasteuria penetrans* to the juvenile and male cuticles of *Meloidogyne hapla*. Nematology 8:475–476

Davies KG, Kerry BR, Flynn CA (1988) Observations on the pathogenicity of *Pasteuria penetrans*, a parasite of root-knot nematodes. Annu Appl Biol 112:1491–1501

Davies KG, Flynn CA, Laird V et al (1990) The life-cycle, population dynamics and host specificity of a parasite of *Heterodera avenae,* similar to *Pasteuria penetrans*. Rev Nématol 13:303–309

Davies KG, Laird V, Kerry BR (1991) The motility, development and infection of *Meloidogyne incognita* encumbered with spores of the obligate hyperparasite *Pasteuria penetrans*. Rev Nématol 14:611–618

Davies KG, Fargette M, Balla G (2001) Cuticle heterogeneity as exhibited by *Pasteuria* spore attachment is not linked to the phylogeny of parthenogenetic root- knot nematodes (*Meloidogyne* spp.). Parasitology 122:111–120

Davies KG, Rowe J, Williamson VM (2008) Cuticle variation amongst amphimictic and parthenogenetic populations of nematode (*Meloidogyne* spp.) as exhibited by a bacterial parasite (*Pasteuria penetrans*). Int J Parasitol 38:851–859

Davies KG, Gravato-Nobre M, Hodgkin, J (2009) Knock-down of mucin-like genes using RNAi changes lectin staining in adult hermaphrodites of Caenorhabditis elegans. International Worm Meeting, Wormbase (abs), Los Angeles

Davies KG, Rowe J, Manzanilla-Lopez R et al (2011) Re-evaluation of the life-cycle of the nematode parasitic bacterium *Pasteuria penetrans* in root-knot nematodes *Meloidogyne* spp. Nematology (in press)

Dawkins R (2009) The greatest show on earth: the evidence for evolution. Transworld (United Kingdom and Commonwealth), New York

DesRosier JP, Lara JC (1981) Isolation and properties of pili from spores of *Bacillus cereus*. J Bacteriol 145:613–619

Duponnois R, Ba AM, Mateille T (1999) Beneficial effects of *Enterobacter cloacae* and *Pseudomonas mendocina* for the biocontrol of *Meloidogyne incognita* with the endospore-forming bacterium *Pasteuria penetrans*. Nematology 1:95–101

Espanol M, Verdejo-Lucas S, Davies KG et al (1997) Compatibility between *Pasteuria penetrans* and *Meloidogyne* populations from Spain. Biocontrol Sci Technol 7:219–230

Evans AAF (1998) Reproductive mechanisms. In: Perry RN, Wright DJ (eds) The physiology and biochemistry of free-living and plant-parasitic nematodes. CABI Publishing, Wallingford

Fawcett P, Eichenberger P, Losick R et al (2000) The transcriptional profile of early to middle sporulation in *Bacillus subtilis*. Proc Natl Acad Sci USA 97:8063–8068

Field D, Feil EJ, Wilson GA (2005) Databases and software for the comparison of prokaryotic genomes. Microbiology 151:2125–2132

Frutos R, Viari A, Ferraz C et al (2006) Comparative genomic analysis of three strains of *Ehrlichia ruminantium* reveals an active process of genome size plasticity. J Bacteriol 188:2533–2542

Gems DH, Maizels RM (1996) An abundantly expressed mucin-like protein from *Toxocara canis* infective larvae: the precursor of the larval surface coat glycoproteins. Proc Natl Acad Sci USA 93:1665–1670

Gerber JF, White JH (2001) Materials and methods for the efficient production of *Pasteuria*. International patent application WO 01/11017 A2

Ghedin E, Wang S, Spiro D et al (2007) Draft genome of the filarial nematode parasite *Brugia malayi*. Science 317:1756–1760

Giannakou IO, Pembroke B, Gowen SR et al (1997) Effects of long term storage and above normal temperatures on spore adhesion of *Pasteuria penetrans* and infection of root-knot nematode *Meloidogyne javanica*. Nematologica 43:185–192

Giblin-Davis RM, Williams DS, Bekal S et al (2003) *Candidatus* Pasteuria usgae' sp. nov., an obligate endoparasite of the phytoparasitic nematode, *Belonolaimus longicaudatus*. Int J Syst Evol Microbiol 53:197–200

Gibson DG, Glass JI, Lartigue C et al (2010) Creation of a bacterial cell controlled by a chemically synthesized genome. Science 329:52–56

Griffin GD, McKenry MV (1989) Susceptibility of Nevada synthetic XX germplasm to a California race of *Meloidogyne Hapla*. J Nematol 21:292–293

Grimshaw CE, Huang S, Hanstein CG (1998) Synergistic kinetic interactions between components of the phosphorelay controlling sporulation in *Bacillus subtilis*. Biochemistry 37:1365–1375

Guiliano DB, Blaxter ML (2006) Operon conservation and the evolution of trans-splicing in the phylum Nematoda. PLoS Genet 2(11): e198

Haldane JBS (1949) Disease and evolution. La Ricerca Sci Suppl 19:1–11

Hall DH, Altun ZF (2008) *C. elegans* Atlas. Cold Spring Harbor Laboratory Press, New York

Harris TW, Lee R, Schwarz R et al (2003) Wormbase: a cross-species database for comparative genomics. Nucleic Acids Res 31:133–137

Hartman KM, Sasser JN (1985) Identification of *Meloidogyne* species on the basis of differential host test and perineal-pattern morphology. In: Barker KR, Carter CC, Sasser JN (eds) An advanced treatise on *Meloidogyne*, Vol II. North Carolina State University, Raleigh, pp 69–77

Hemmer RM, Donkin SG, Chin KJ et al (1991) Altered expression of an L1-specific, O-linked cuticle surface glycoprotein in mutants of the nematode *Caenorhabditis elegans*. J Cell Biol 115:1237–1247

Henriques AP, Moran CP Jr (2007) Structure, assembly, function of the spore surface layers. Annu Rev Microbiol 61:555–588

Hewlett TE, Gerber JF, Smith KS (2004) In vitro culture of *Pasteuria penetrans*. In: Cook RC, Hunt DJ (eds) Nematology monographs and perspectives, vol 2, Proceedings of the fourth international congress of nematology. Brill, Leiden, pp 175–185

Higgins CF (1992) ABC transporters: from microorganisms to man. Annu Rev Cell Biol 8:67–113

Jack DL, Paulsen IT, Saier MH (2000) The amino acid/polyamine/organocation (APC) superfamily of transporters specific for amino acids, polyamines and organocations. Microbiology 146:1797–1814

Jack DL, Yang NM, Saier MH Jr (2001) The drug/metabolite transporter superfamily. Eur J Biochem 268:3620–3639

Jepson SB (1987) Identification of root-knot nematodes *(Meloidogyne species)*. CAB International, Wallingford

Kojetin DJ, Thompson RJ, Benson LM et al (2005) Structural analysis of divalent metals binding to the *Bacillus subtilis* response regulator Spo0F: the possibility for *in vitro* metalloregulation in the initiation of sporulation. Biometals 18:449–466

Kozuka S, Tochikubo K (1985) Properties and origin of filamentous appendages on spores of *Bacillus cereus*. Microbiol Immunol 29:21–37

Law CJ, Maloney PC, Wang DN (2008) Ins and outs of major facilitator superfamily antiporters. Annu Rev Microbiol 62:289–305

Leroy S, Duperray C, Morand S (2003) Flow cytometry for parasite nematode genome size measurement. Mol Biochem Parasitol 128:91–93

Liu QL, Williamson VM (2006) Host-specific pathogenicity and genome differences between inbred strains of *Meloidogyne hapla*. J Nematol 38:158–164

Liu QL, Thomas VP, Williamson VM (2007) Meiotic parthenogenesis in a root-knot nematode results in rapid genomic homozygosity. Genetics 76:1483–1490

Magalhães A, Ismail MN, Reis AC (2010) Sweet receptors mediate the adhesion of the gastric pathogen *Helicobacter pylori*: glycoproteomic stratergies. Expert Rev Proteomics 7:307–310

MendozadeGives P, Davies KG, Morgan M (1999) Attachment tests of *Pasteuria penetrans* to the cuticle of plant- and animal-parasitic nematodes, free-living nematodes and *srf* mutants of *Caenorhabditis elegans*. J Helminthol 73:67–71

Mohan S, Fould S, Davies KG (2001) The interaction between the gelatine-binding domain of fibronectin and the attachment of *Pasteuria penetrans* endospores to nematode cuticle. Parasitology 123:271–276

Mukhopadhyay D, Sen U, Zapf J et al (2004) Metals in the sporulation phosphorelay: manganese binding by the response regulator Spo0F. Acta Cryst D60:638–645

Noel GR, Atibalentja N, Domier LL (2005) Emended description of *Pasteuria nishizawae*. Int J Syst Evol Microbiol 55:1681–1685

Opperman CH, Bird DM, Williamson VM et al (2008) Sequence and genetic map of *Meloidogyne hapla*: a compact nematode genome for plant parasitism. Proc Natl Acad Sci USA 105:14802–14807

Otto SP, Nuismer SL (2004) Species interactions and the evolution of sex. Science 304:1018–1020

Pableo EC, Triantaphyllou AC (1989) DNA complexity of the root-knot nematode (*Meloidogyne* spp.) genome. J Nematol 21:260–263

Page AP, Johnstone IL (2007) The cuticle. In *WormBook*, ed. The *C. elegans* research community, WormBook, doi/10.1895/wormbook.1.138.1, http://www.wormbook.org

Pao SS, Saier PIT, Jr MH (1998) Major facilitator superfamily. Microbiol Mol Biol Rev 62:1–34

Persidis A, Lay JG, Manousis T (1991) Characterisation of potential adhesions of the bacterium *Pasteuria penetrans*, and of putative receptors on the cuticle of *Meloidogyne incognita*, a nematode host. J Cell Sci 100:613–622

Plomp M, Leighton TJ, Wheeler KE et al (2005a) The high-resolution architecture and structural dynamics of *Bacillus* spores. Biophys J 88:603–608

Plomp M, Leighton TJ, Wheeler KE (2005b) Architecture and high-resolution structure of *Bacillus thuringiensis* and *Bacillus cereus* spore coat surfaces. Langmuir 21:7892–7898

Postma PW, Lengeler JW, Jacobson GR (1993) Phosphoenolpyruvate :carbohydrate phosphotransferase systems of bacteria. Microbiol Rev 57:543–594

Power PM, Jennings MP (2003) The genetics of glycosylation in Gram-negative bacteria. FEMS Microbiol Lett 218:211–222

Saier MH Jr (2000) Families of transmembrane transporters selective for amino acids and their derivatives. Microbiology 146:1775–1795

Sharma SB, Davies KG (1996) Characterisation of *Pasteuria* isolated from *Heterodera cajani* using morphology, pathology and serology of endospores. Syst Appl Microbiol 19:106–112

Steichen CT, Chen P, Kearney JF et al (2003) Identification of an immunodominant protein and other proteins of the *Bacillus anthracis* exosporium. J Bacteriol 185:1903–1910

Stirling GR (1985) Host-specificity in *Pasteuria penetrans* within the genus *Meloidogyne*. Nematologica 31:203–209

Stirling GR (1991) Biological control of plant-parasitic nematodes: progress problems and prospects. CAB International, Wallingford

Stirling GR, Wachtel MF (1980) Mass production of *Bacillus penetrans* for the biological control of root-knot nematodes. Nematologica 26:308–312

Strous GJ, Dekker J (1992) Mucin-type glycoproteins. Crit Rev Biochem Mol Biol 27:57–92

Sylvestre P, Couture-Tosi E, Mock M (2002) A collagen-like surface glycoprotein is a structural component of the *Bacillus anthracis* exosporium. Mol Microbiol 45:169–178

Sylvestre P, Couture-Tosi E, Mock M (2003) Polymorphism in the collagen-like region of the *Bacillus anthracis* BclA protein leads to variation in length in the exosporium filament length. J Bacteriol 185:5155–5163

Sylvestre P, Couture-Tosi E, Mock M (2005) Contribution of ExsFA and ExsFB proteins to the localisation of BclA on the spoe surface and to the stability of the *Bacillus anthracis* exosporium. J Bacteriol 187:5122–5128

Takeuchi K, Taguchi F, Inagaki Y et al (2003) Flagellin glycosylation island in *Pseudomonas syringae* pv. Glycinea and its role in host specificity. J Bacteriol 185:6658–6665

Takumi K, Kinouchi T, Kawata T (1979) Isolation and partial characterization of exosporium from spores of a highly sporogenic mutant of *Clostridium botulinum* type A. Microbiol Immunol 28:443–454

The *C. elegans* Sequencing Consortium (1998) Science 282:2012–2018

Todd SJ, Moir AJ, Johnson MJ (2003) Genes of *Bacillus cereus* and *Bacillus anthracis* encoding proteins of the exosporium. J Bacteriol 185:3373–3378

Trudgill DL, Blok VC (2001) Apomictic, polyphagous root-knot nematodes: exceptionally successful and damaging biotrophic root pathogens. Annu Rev Phytopathol 39:53–77

Van der Beek JG, Poleij LM, Zulstra C et al (1998) Variation in virulence within *Meloidogyne chitwoodi, M. fallax*, and *M. hapla* on *Solanum* spp. Phytopathology 88:658–665

Van Valen L (1973) A new evolutionary law. Evol Theory 1:1–30

Wehrli E, Scherrer P, Kubler O (1980) The crystalline layer in spores of *Bacillus cereus* and *Bacillus thuringiensis* studied by freeze-etching and high resolution electron microscopy. Eur J Cell Biol 20:283–289

Welch M, Cummings MP, Hillis DM (2004) Divergent gene copies in the asexual class Bdelloidea (Rotifera) separated before the bdelloid radiation or within bdelloid families. Proc Natl Acad Sci USA 101:1622–1625

Williams AB, Stirling GR, Hayward AC (1989) Properties and attempted culture of *Pasteuria penetrans*, a bacterial parasite of root-knot nematode (*Meloidogyne javanica*). J Appl Microbiol 67:145–156

Williamson VM (1999) Plant nematode resistance genes. Curr Opin Plant Biol 2:327–331

Wilson AC, Soyer M, Hoch JA et al (2008) The bicarbonate transporter is essential for *Bacillus anthracis* lethality. PLoS Pathog 4(11):e1000210

Wishart J, Blok VC, Phillips MS et al (2004) *Pasteuria penetrans* and *P. nischizawae* attachment to *Meloidogyne chitwoodi, M. fallax* and *M.hapla*. Nematology 6:507–510

Zapf J, Hoch JA, Whitely JM (1996) A phosphotransferase activity of the *Bacillus subtilis* sporulation protein Spo0F that employs phosphoramidate substrates. Biochemistry 35:2926–2933

Zarkower D (2006) Somatic sex determination. In *Wormbook*, ed. The *C. elegans* research community, doi/10.1895/wormbook1.84.1. http://www.wormbook.org

Zawadzki JL, Presidente PJA, Meeusen EN et al (2006) RNAi in *Haemonchus contortus*: a potential method for target validation. Trends Parasitol 22:495–499

Chapter 5
Plant Nematode Surfaces

Rosane H.C. Curtis, John T. Jones, Keith G. Davies, Edna Sharon, and Yitzhak Spiegel

Abstract The outer surface of nematodes act as an external skeleton and is covered by a tough, but flexible, multi-layered, extracellular cuticle which protects them from the external environment, maintains body shape and is involved in locomotion and defence against their host or microorganism attack. This chapter highlights the role of the nematode surface cuticle, during the various life-stages, with their environment, including their host and other microorganism. A comprehensive appraisal is presented of the complex interactions between nematodes and microbial antagonists, as the surface cuticle is believed to be involved in the host-recognition events determining the specificity of such interactions.

5.1 Introduction

The nematode cuticle is an extracellular coating that is secreted by the hypodermis and has a variety of important roles in nematode biology. The cuticle maintains the body shape, provides a strong layer against which muscles can act during locomotion and protects the nematode from the external environment. The cuticle is overlaid with a fuzzy coating material – the surface coat (SC). Substances on the surface

R.H.C. Curtis (✉) and K.G. Davies
Department of Plant Pathology and Microbiology, Rothamsted Research, Harpenden, Hertfordshire AL5 2JQ, UK
e-mail: rosane.curtis@bbsrc.ac.uk; keith.davies@rothamsted.ac.uk

J.T. Jones
SCRI, Invergowrie, Dundee, DD2 5DA, Scotland, UK
e-mail: John.Jones@scri.ac.uk

E. Sharon* and Y. Spiegel
Division of Nematology, Institute of Plant Protection, ARO, The Volcani Center, Bet Dagan 50250, Israel
(*Part of the Ph.D. thesis supervised by Spiegel & Chet)
e-mail: ednash2@gmail.com; spiegely@volcani.agri.gov.il

K. Davies and Y. Spiegel (eds.), *Biological Control of Plant-Parasitic Nematodes: Building Coherence between Microbial Ecology and Molecular Mechanisms*, Progress in Biological Control 11, DOI 10.1007/978-1-4020-9648-8_5,

of the cuticle are in direct contact with the outside world and therefore, also have important roles in terms of the interactions of nematodes during the various life-stages with their environment, including their host and other microorganism. Nematodes eggs generally are the most resistant nematode life-stage and many have a remarkable capacity for survival in stressful environments; therefore, the eggshell is a most important barrier at this stage.

5.2 Eggshell

Most nematode eggs are morphologically very similar and are similar in size (average of 53–133 μm in length and 17–79 μm in width) irrespective of the size of the adult. They are ellipsoidal in shape with a transparent shell (Bird and Bird 1991). The eggshell in nematodes is formed after fertilization and usually contains four layers. The outermost layer (uterine) consists of material that is secreted by the uterine epithelial cells and can be absent in some nematodes. In plant-parasitic nematodes the composition of the uterine layer resembles the gelatinous matrix (gm) secreted from various organs (Mackintosh 1960; Maggenti 1962; Bird and Rogers 1965). The next layer, vitelline, originates from the vitelline membrane (oolemma) which is formed after fertilization of the oocyte. In many eggs this retains a unit membrane-like structure and forms the outer layer of the eggshell. Carbohydrate residues have been detected on the surfaces of the eggs (Rao et al. 1988; Taylor et al. 1986; Spiegel and McClure 1991).

The underlying chitinous layer is made-up of chitin microfibrils embedded in a protein coat (Wharton 1983). Chitin is formed from the polymerization of N-acetyl glucosamine which is itself synthesized from glycogen (Preston and Jenkins 1985). The chitinous layer is often the thickest layer and provides structural rigidity to the eggshell. Protein is frequently present in this layer; it has been estimated that the eggshell of *Globodera rostochiensis* contains 59% protein and 9% chitin while that of *Meloidogyne javanica* contains 50% protein and 30% chitin (Clarke et al. 1967; Bird and McClure 1976). Perry and Trett (1986) suggest the sub-division of the chitinous layers of cyst nematodes into distinctive outer and inner components due to differences in their chemical composition on the basis of ultrastructural studies that showed that the external outer layer is thin and amorphous followed by a predominantly tetra-pentalaminate inner layer. Chitin is synthesised by chitin synthase and the gene encoding this enzyme has been shown to be expressed in the egg-producing adult stages and fertilized eggs of various nematodes including *Caenorhabditis elegans*, *M. artiellia*, *Ascaris suum*, *Brugia malayi* and *Dirofilaria immitis* (Veronico et al. 2001; Harris and Fuhrman 2002; Dubinsky et al. 1986a, b). These observations are consistent with a role for this enzyme in producing chitin for the eggshell. This has been confirmed with RNAi-mediated ablation of the *C. elegans* chitin synthase gene function which resulted in sterile hermaphodites that lay defective eggs (Hanazawa et al. 2001). RNAi of a chitin synthase gene expressed in the eggs of root-knot nematode *M. artiella* was achieved by soaking intact eggs within their gelatinous matrix in a solution containing dsRNA and led to a reduction in stainable chitin in eggshells and a delay in hatching of juveniles from treated eggs (Fanelli et al. 2005). So far, the eggshell is the only structure in nematodes in which the presence of chitin has been conclusively demonstrated (Bird and Bird 1991). A chitin synthase gene has been shown to be expressed in the cells that form the pharynx of *C. elegans* at a time that precedes a moult. It has also been suggested that this gene might be involved in the synthesis of the feeding apparatus which is replaced during each moult (Veronico et al. 2001). A secreted chitinase has been identified in the perivitelline fluid surrounding the infective larva of *A. suum* prior to hatching indicating that this enzyme might be responsible for the digestion of the eggshell during hatching of this nematode (Geng et al. 2002). The potential of using bacterial and fungal chitinases and chitin synthase inhibitors to control root-knot nematodes have been demonstrated (Spiegel and Chet 1985; Jung et al. 2002; van Nguyen et al. 2007).

The most internal layer is the lipid layer which is responsible for the extreme impermeability of the nematode eggshell. It is formed in the middle region of the uterus in *M. javanica* (Bird and McClure 1976) where proline-containing proteins are incorporated into both lipid and chitinous layers and the synthesis of the eggshell is then completed (Bird and Bird 1991). Nematode eggs are permeable to chemicals prior to the formation of the lipid layer and when this layer is broken down before hatching. Egg permeability changes are central to the hatching of cyst nematodes (Perry and Clarke 1981; Bird and McClure 1976) and work by Twomey et al. (2000) corroborated these studies and showed that the biological nematicide DiTera (fermentation product of killed *Myrothecium verrucaria*) induced a significant inhibition of hatch of *G. rostochiensis* by possibly preventing eggshell permeability change by competitively blocking the Ca^{2+} binding sites on the eggshell. DiTera showed a lack of inhibition of hatch on *M. incognita* and therefore do not affect second-stage juveniles (J2) directly during the hatch process or inhibit the action of enzymes. It has been suggested that the breakdown of the lipid layer is enzymatic and secretions have been shown to emanate from various structures of J2s of *M. incognita* and *M. javanica*, including the amphids, secretory-excretory pore and from around the mouth while still inside the egg. An increase in size of the nucleolus of the dorsal pharyngeal glands of *G. rostochiensis* J2 inside the eggs that were stimulated with root diffusate was demonstrated indicating that the oesophageal glands are activated and leucine amonipeptidase activity was identified in the supernatant of eggs of *Heterodera glycines* (Atkinson et al. 1987; Premachandran et al. 1988; Bird and Bird 1991). Perry and Trett (1986) observed that the eggs within *H. glycines* cysts internally contaminated with fungus, had no inner lipid layer and it was suggested that fungal lipases might have contributed to disrupt both inner and outer lipid layers.

As both the lipid layer and the juvenile epicuticle are derived directly from the secondary vitelline membrane of the embryo, both probably share the same protein moieties. Cross-reactivity of polyclonal and monoclonal antibodies (Fig. 5.1) produced against *M. incognita* J2 with eggshells support this suggestion (R. Curtis unpubl; Sharon et al. 2002, 2009a).

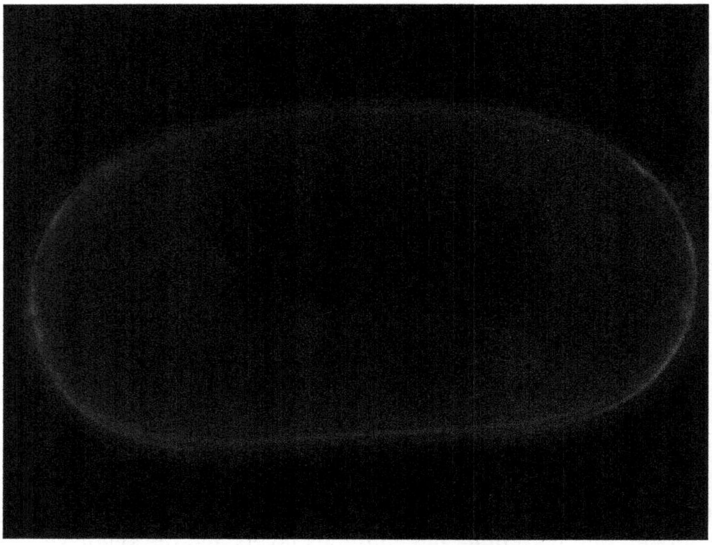

Fig. 5.1 Monoclonal antibody raised to second-stage juveniles of *Meloidogyne incognita* shows reactivity with the egg-shell

Cysteine proteinases are involved in a variety of biological functions and have been implicated in tissue remodeling in free-living and animal-parasitic nematodes. Cathepsin L has been shown to be present in the eggshell surrounding the embryos of *C. elegans*, *Onchocerca volvulus* and *B. pahangi* and is possibly involved in eggshell remodeling by processing of nutrients responsible for the synthesis and/or degradation of the eggshell in these nematodes (Hashmi et al. 2002; Guiliano et al. 2004). Transmission electron microscopy showed that the eggshells and cuticles layers of *C. elegans* and of the parasitic stages of *Haemonchus contortus* have an ABC transporter, P-glycoprotein. This might function as a membrane efflux 'pump' and may play a major role in the transport of antihelmintic drugs in parasitic nematodes of ruminants (Riou et al. 2005).

5.3 Cuticle Structure and Function

A detailed review describing the structure of the cuticle and contrasting the cuticular features of a wide range of nematode species has been published previously (Lee 2002) and only the salient points are summarized here. The cuticle of most nematodes can be subdivided into three main zones covered with an epicuticle. In addition, the cuticle is overlaid with a surface coat composed of mucins and other proteins (Fig. 5.2). The innermost cuticle zone, the basal zone, often has a striated appearance when sections through the cuticle are viewed under an electron microscope and is thought to be composed largely of collagens (see below), arranged in oriented layers of fibrils. The median zone varies in appearance

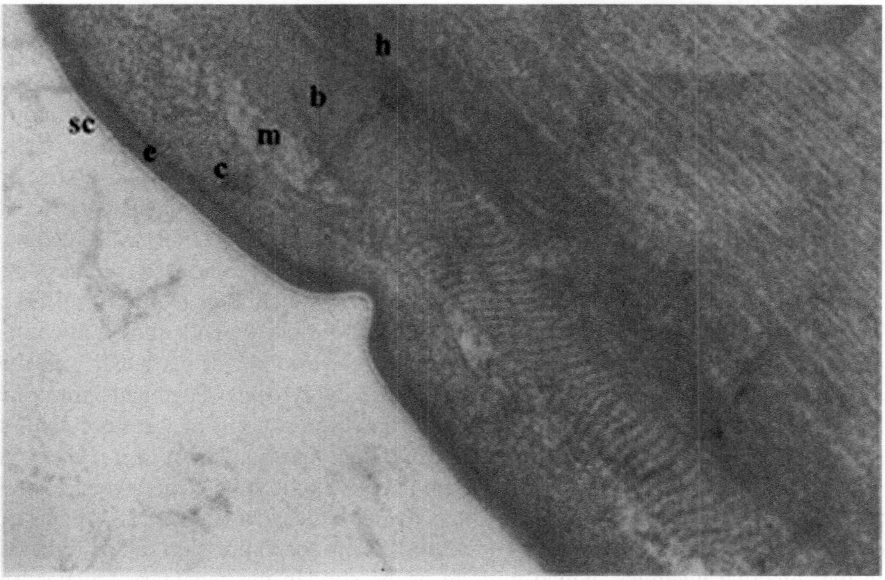

Fig. 5.2 Transmission electron micrograph of a longitudal section of *Meloidogyne javanica* second-stage juvenile (X 40,000), showing surface coat (*sc*), epicuticle (*e*), cortical zone (*c*), median zone (*m*), basal zone (*b*) and hypodermis (*h*)

in electron micrographs but often appears less electron-dense than the basal and cortical zones. It may appear to contain vacuoles and struts and has often been described as fluid-filled or gel-like. The appearance of this layer sometimes varies between nematode life-stages. For example, the median zone of unhatched second-stage juveniles (J2s) of *G. rostochiensis* appears to contain an electron dense material which is lost on hatching (Jones et al. 1993). It has been suggested that this zone is composed of a combination of collagens and soluble proteins, possibly destined for the nematode surface (Blaxter and Robertson 1998). The cortical zone varies enormously in thickness and may have regions that have different electron densities when viewed under an electron microscope. The cortical zone is composed of collagens and extremely insoluble cuticulins. The cuticle of many nematodes is annulated and may also carry a wide range of projections including hooks, bristles and papillae. Many of these projections are associated with mechanoreceptive neurons (Jones 2002). Lateral projections (alae) may also be present that extend along the length of the nematode body.

The best characterised, and most abundant, of the cuticle proteins are the collagens. Collagens from a wide range of animal species have a conserved and characteristic triple helical tertiary structure. This region is conserved to such an extent that the first nematode collagen gene was identified using a probe derived from a vertebrate sequence (Kramer et al. 1982). The triple helical region of the protein is formed from large Gly-X-Y repeat regions in which every third amino acid is a glycine residue and where Y is frequently proline, which forces the helical turn.

Collagen genes are present in extremely large gene families in a wide range of nematode species; over 170 genes are present in *C. elegans* (Page and Johnstone 2007) and 122 are predicted from the genome of *M. incognita* (Abad et al. 2008). This large suite of collagen proteins allows variation in the collagens present in different life-stages and thus variation in the structural and physical properties of the cuticle of different life-stages (Koltai et al. 1997; Liu et al. 2001). A number of conserved cysteine residues are present in the non Gly-X-Y, N- and C-terminal regions and within short stretches of amino acids that interrupt the Gly-X-Y regions. The number and spacing of these residues, along with other conserved residues, has been used to subdivide the *C. elegans* collagens into subfamilies (Johnstone 1994). The other major protein component of the cuticle is formed by cuticulins. Cuticulins contain cysteine rich regions and are extensively cross linked, particularly in the outer cortical layers and in the dauer cuticle, through tyrosine (Page and Johnstone 2007). It has been suggested on the basis of expression patterns of various cuticulin genes in *C. elegans* that cuticulins are also important in formation of lateral alae and other cuticular annulations and ridges (Sapio et al. 2005). Studies on the expression patterns during the development of *M. artiellia* indicate that there is a burst of expression of the cut-1 gene during moulting. Then, the expression rate is reduced in the infective juveniles, which migrate in the soil. In the sedentary females, in contrast, no expression is detected, while in the males which move freely through the soil, the gene is expressed and the transcript fully processed. These data strongly suggest that the gene is developmentally regulated. It is proposed that the production of cuticlin plays an important role in determining the mechanical properties of the cuticle (de Giorgi et al. 1997).

The cuticle itself is synthesized during a series of moults in the hypodermis, a syncytial cell layer immediately below the cuticle. The moulting process can be subdivided into several distinct phases and has been analysed in detail in *C. elegans*. Once the new cuticle has been resynthesised the nematode shows a decrease in activity and feeding (lethargus) followed by separation of the old cuticle from the new structure (apolysis). The nematode then moves rapidly in order to loosen the old cuticle which is shed during ecdysis. The new cuticle is synthesised at the end of each juvenile stage and the timing of the expression patterns of collagen genes during this process has been examined in detail (Johnstone and Barry 1996). Cuticle collagen genes are expressed in distinct waves and in an order that is repeated at each moult, with some genes always expressed early in the moulting cycle and others at a later stage. It is possible that these waves of gene expression exist to ensure that the collagen components required for each cuticle zone or substructure are expressed simultaneously, with different zones synthesised at different times in the moulting cycle.

5.4 The Surface Coat and Excreted/Secreted Antigens

The surface coat (SC) of nematodes contains various proteins, carbohydrates and lipids, as individual components or as glycoproteins/mucins, glycolipids, or lipoproteins. Studies have used antisera, lectins, biotin and neoglycoproteins to characterize and localize surface components and to learn about its nature. Binding of lectins indicated the presence of specific carbohydrates on the surface and amphids of different life-stages of plant-parasitic nematodes. Carbohydrate-recognition domains (CRDs) were demonstrated on the surface of plant-parasitic nematodes, suggesting the presence of lectins on the surface of the nematodes (Spiegel and McClure 1995; Sharon and Spiegel 1996). Carbohydrates or CRDs on nematodes surface and/or secretions might be involved in nematode-plant and nematode-microorganisms interactions (Spiegel and McClure 1995; Koltai et al. 2002). Surface coat proteins and glycoproteins from pre-parasitic J2s and adult females of *Meloidogyne* were labeled, extracted and partially characterized (Robinson et al. 1989; Lin and McClure 1996; Spiegel et al. 1997). Specific antibodies were raised against surface antigens and excretory-secretory (E-S) products of plant-parasitic nematodes and have been used to localize and characterize antigens on SC regions on the nematodes (Fig. 5.3), on amphids and secretory-excretory products. Antibodies served as a tool for studying interactions with plant hosts (*e.g.* Curtis 1996; Gravato-Nobre and Evans 1998; Lopez de Mendoza et al. 1999) and microorganisms (*e.g.* Spiegel et al. 1996; Davies and Danks 1992; Sharon et al. 2009a).

One of the most interesting features of the nematode SC is its labile and dynamic nature and there is a continuous turn-over of surface-associated antigens that involves shedding and replacing of the antigens (Blaxter and Robertson 1998). Studies have shown that surface coat of plant-parasitic nematodes is shed (Fig. 5.4) both in vitro (Bird et al. 1988; Lin and McClure 1996; Spiegel et al. 1997; Robertson et al. 2000) and in the host (Curtis 1996; Gravato-Nobre et al. 1999;

Fig. 5.3 Scaning electron micrograph of immunogold-labeled surface of *Meloidogyne javanica* second-stage juvenile; labeling was visualized by silver enhancement reaction, bar = 1 µm

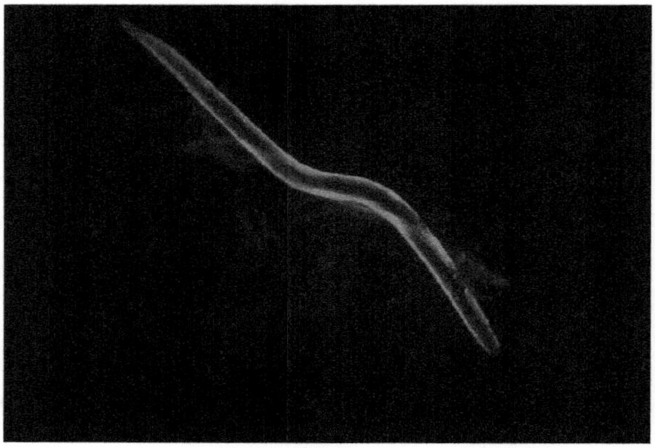

Fig. 5.4 Antiserum (anti-rGR-TpX) binds to the surface of *Globodera rostochiensis* and to material shed in great quantities from the parasite surface

Sharon et al. 2002). Unlike other cuticle proteins surface coat molecules are readily secreted/released into the environment and this was demonstrated for the pre-parasitic juveniles of *Meloidogyne* spp., as the surface coat proteins of *M. incognita* were released when the J2 were incubated in water (Lin and McClure 1996). Also, when J2 of *M. javanica* were treated with detergents there was a reduction in the binding of red blood cells to the nematode surface but the binding properties were completely renewed after 24 h at 25°C, but not at 4°C, indicating that the sloughing-off and replacement of the nematode's surface coat is an active event. This phenomenon was visualized also with detergent-extracted SC proteins, using gel-electrophoresis (Spiegel et al. 1997).

Interestingly, binding of antibodies to *M. javanica* and *G. pallida* J2s affected their normal movement pattern on agar plates, regardless the binding pattern (i.e. SC regions, head, amphids, E-S products); this was reversed within several hours, probably due to renewal of the SC. However, continuous binding of antibodies did not enable the recovery and nematodes stopped moving after 2–3 days, subsequently inhibiting plants inoculation by the nematodes (Fioretti et al. 2002; Sharon et al. 2002). Some antibodies caused even more drastic lethal effects after less than one day of contact with antibodies (Sharon et al. 2009a).

There is evidence to suggest that some components of the surface coat are synthesised in the hypodermis (*e.g.* Jones et al. 2004). Further evidence to support this idea comes from the fact that antibodies reactive with the SC of *M. incognita* and *M. javanica* also show reactivity with the hypodermis (Sharon et al. 2002). The origin of surface-associated antigens on nematodes may differ for various antigens and is still not clear in most cases (Spiegel and McClure 1995; Blaxter and Robertson 1998). These non-structural proteins can originate from gland cells such as excretory cells, pharyngeal glands, amphids and phasmids as well as from the hypodermis (Blaxter and Robertson 1998; Page and Johnstone 2007).

In *Meloidogyne* spp., as well as in other species like *Rotylenchulus*, *Tylenchulus* and *Heterodera*, eggs are enveloped with a gelatinous matrix (gm) that contains glycoproteins (Sharon and Spiegel 1993; Agudelo et al. 2004). The eggs and the emerging J2s are exposed to these components and some of it probably attach to the surface of the J2s and affect their interactions with the environment. Labelling of *M. javanica* egg mass-originated eggs and J2s with the monoclonal antibody (MAb), MISC, presented different patterns and was more intense than on gm-free ones (from hypochlorite-treated eggs); labellings were inhibited by fucose (Sharon et al. 2009a). This MAb had also labelled *M. incognita* J2s SC, the gm and the rectal glands, where the gm originates, which suggests that there are mutual epitopes in the gm and SC (Hu et al. 2000). Actually, when gm-originated J2s are used, some of the surface components can be of gm origin. The gm plays a key role in attachment and parasitism of microorganisms, such as *Trichoderma* (see Sect. 5.5.3 and Chap. 8) on *Meloidogyne* J2s and eggs. The role of gm-originated components on the surface of nematodes and their fate during SC turn-over should still be further investigated.

5.4.1 Role of the Surface Coat in Host-Nematode Interactions

Nematodes can rapidly change their surface composition in response to environmental signals, which may enable animal-parasitic nematodes to escape host immune responses and free-living nematodes to escape pathogenic infections (Grenache et al. 1996; Olsen et al. 2007; Proudfoot et al. 1993). A growing body of evidence indicates that some molecules present at the nematode surface of parasitic nematodes serve as an active defense against host responses and are therefore important for nematode survival (Blaxter et al. 1992; Jones et al. 2004; Olsen et al. 2007).

5.4.2 Changes in the Surface Coat in Response to Host Derived Signal

The idea that nematodes switch surface composition in response to environmental signals has been based on rapid changes in surface lipophilicity (Modha et al. 1995; Proudfoot et al. 1993) or surface antigenicity (Philipp and Rumjaneck 1984; Politz and Philipp 1992) that occur during parasitic nematode infections. Therefore, the surface composition can change within a single stage of the life-cycle, during the entry of parasitic nematodes into a new host or host tissue; these surface changes are different from the moulting process as they occur more rapidly (Modha et al. 1995; Proudfoot et al. 1993). Some surface proteic epitopes present in the preparasitic *Meloidogyne* J2s were shown to be abundantly shed during root invasion (De Mendoza et al. 2002; Sharon et al. 2002; Curtis 2007b). *In vitro*, plant signals present in root exudates trigger a rapid modification of the surface cuticle of *M. incognita* and *G. rostochiensis* (de Mendoza et al. 2000; Akhkha et al. 2002). Increase in the surface lipophilicity was also induced by phytohormones, in particular indole-acetic acid (IAA) and kinetin in *M. incognita* but not *G. rostochiensis* (Akhkha et al. 2002, 2004). It has been suggested that the ability of *M. incognita* to respond to a general plant compound as opposed to a specific root diffusate is related to the broad host-range of this species. The increase in the lipophilicty of the SC of *M. incognita* J2, induced by plant signals might allow this nematode to adapt to and survive plant defence processes. By contrast, more specific host cues from root exudates of Solanaceous plants increase the lipophilicity of the surface cuticle of infective J2 of *Globodera* species (Akhkha et al. 2002; Curtis 2007a).

The composition of the nematode surface is also important for the survival of free-living nematodes (see chapter on *C. elegans* for more details) as specific surface-altered mutants of *C. elegans* are resistant to pathogen infections (Mendoza de Gives et al. 1999; Gravato-Nobre et al. 2005; Hodgkin et al. 2000; Ewbank 2002). *C. elegans* responds to environmental conditions by modifying its surface, a process similar to surface switching in parasitic nematodes and these environmental signals are detected by the nematode's chemosensory organs (Grenache et al. 1996; Olsen et al. 2007). These studies suggest that surface switching in plant-parasites might also rely on chemosensation and it can be speculated that free-living and parasitic nematodes use their sensilla to detect environmental signals that lead to changes in the surface composition. Olsen et al. (2007) hypothesized that, like dauer formation, surface antigen switching in *C. elegans* is guided by chemical signals sensed by the amphid neurons. This behavioral adaptation may protect the nematodes from biological attack.

5.4.3 Protection of the Plant Nematode from Host Defence Responses

Endoparasitic nematodes spend a significant portion of their life-cycles within their hosts and are therefore exposed to host defence responses. All endoparasites will aim to minimise the effects of defence responses. In addition, many plant-parasitic nematodes, including cyst and root-knot nematodes, are biotrophic and it is particularly important for these nematodes to mask their presence from their host, or suppress any defence response that is mounted, as detection of the pathogen will lead to destruction of the feeding site and death of the pathogen. The nematode surface is in intimate contact with the host. Materials present on the surface of the nematode are therefore targets for detection by the plant and substances can be secreted to the surface that mask the presence of the nematode from the plant, suppress host defences or modulate the effects of any defence response that is mounted.

The ability of nematodes to continuously shed and renew the SC may help the nematode avoid recognition in the host-plant. It has been suggested that shedding of SC components may cause a defence response to be mounted against a region

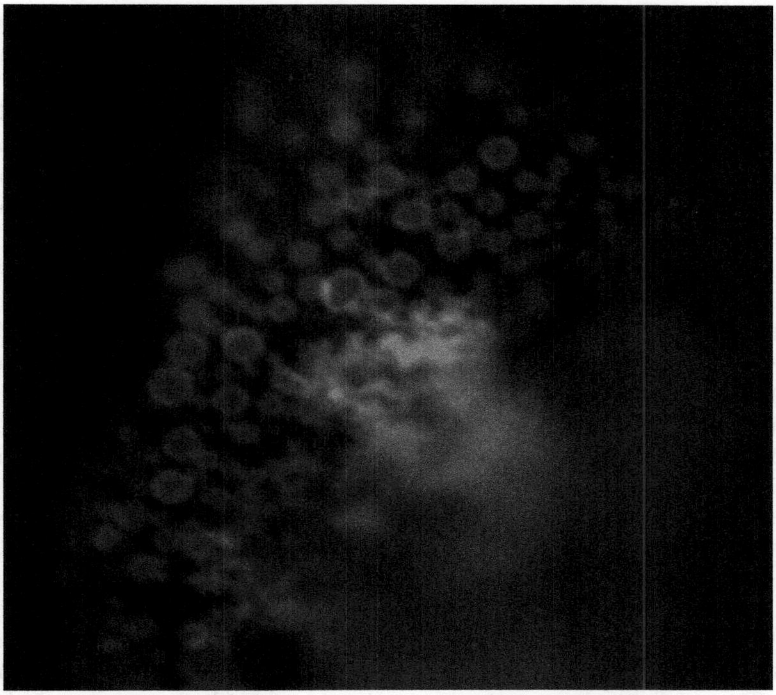

Fig. 5.5 Monoclonal antibody (IACR-CCNj.2a.15) shows reactivity with globules of different sizes exuded from the cuticle of adult female of *Meloidogyne incognita*

that the nematode has vacated, protecting the parasite from early host defences. However, this process of exudation continues once the nematode is established at the feeding site (*e.g.* Jones et al. 1993; de Mendoza et al. 2002; Curtis 2007b) as fibrillar exudates and secretion vesicles exuded as globules of different sizes (Fig. 5.5) have been observed on the cuticle surface of feeding nematodes, suggesting that other processes are also used by the nematode to mask its presence from the host. An abundant cuticular secretion was shown to envelop the adult females and the giant cells and it might have a role in protecting the nematodes from harmful root compounds (de Mendoza et al. 2002; Curtis 2007b).

Antibodies raised against nematode surface components cross react with host tissues; therefore it has been suggested that the SC may mimic host tissues in order to prevent a host reaction (Bird and Wilson 1994; Curtis 1996). Similarly, it has also been suggested that the presence of carbohydrate residues in glycoproteins of plant-parasitic nematodes SC, and therefore the binding of lectins to the nematode surface, is masking other components which may be recognised by the host. Numerous studies have shown that a range of lectins can bind to the surface of many plant-parasitic nematodes (reviewed by Lee 2002) but this idea has yet to be tested *in vitro*. The nematode surface may also play a role in suppressing or modulating the effects of host defences and striking parallels have been found between the proteins used by plant- and animal-parasites to suppress defence signalling. For plant-parasites these processes have been most thoroughly investigated in cyst nematodes. A range of enzymes that may neutralise active oxygen species produced by the host have been described on the nematode surface. A peroxiredoxin (thioredoxin peroxidase) secreted in great quantities from the surface of the potato cyst nematode *G. rostochiensis* was found to catalyse the breakdown of hydrogen peroxide but, unlike most proteins of this type, did not metabolise larger lipid hydroperoxides (Robertson et al. 2000). It is known that hydrogen peroxide is produced as part of the host response to cyst nematode infection (Waetzig et al. 1999) and it is therefore possible that the nematode peroxiredoxin has become modified to allow it to target host-derived hydrogen peroxide efficiently (Robertson et al. 2000). Another peroxidase, glutathione peroxidase, was also found on the surface of this nematode and this protein was shown to metabolise a wider range of hydroperoxides, including larger hydroperoxides (Jones et al. 2004). Like animal-parasites, plant-parasitic nematodes therefore have a range of peroxidases within their surface secretions that can metabolise a wide range of active oxygen species produced as part of the host defence response.

In contrast to root-knot nematodes which migrate between root cells, minimising disruption to host tissues, cyst nematodes migrate destructively through root cells to their chosen feeding site. This process and the damage caused to plant tissues are likely to provoke defence pathways that provide protection against herbivores. These defence pathways lead to production of antifeedants and are triggered by the plant hormone jasmonic acid (Kunkel and Brooks 2002). Jasmonic acid is produced via the octadecanoid pathway and peroxidation of linoleic and linolenic acid by lipoxygenase is an early step in this process. A surface localised retinol and fatty acid binding protein (GpFAR1) has been described from *G. pallida* which binds both linoleic and linolenic acids and it has been shown that FAR1 inhibits lipoxygenase mediated peroxidation of these fatty acids, presumably due to sequestration of the ligands (Prior et al. 2001). Cyst nematodes, therefore, contain proteins at their surface that can inhibit jasmonate signalling pathways and which can also metabolise active oxygen moieties produced as part of any defence response that is mounted. The genome sequencing projects for *M. incognita* and *M. hapla* (Abad et al. 2008; Opperman et al. 2008) show that similar proteins are also present in root-knot nematodes but functional studies examining localisation of these proteins and biochemical activities remain to be carried out.

5.4.4 Cross Reactivity of Surface Coats of Animal- and Plant-Parasitic Nematodes

Monoclonal antibodies (MAb) produced to excretory-secretory (E-S) products of plant-parasitic nematodes were shown to cross-react with E–S products and the SC of the animal-parasites *Trichinella spiralis* and *Haemonchus contortus* (de Mendoza et al. 1999). Glycosylated peptides have been reported to be present in abundance on the SC and in E–S products of several parasitic nematodes (Robertson et al. 1989; Schallig et al. 1994, 1995) and, in fact, most of the MAbs tested recognized carbohydrate epitopes. However, 2 out of 7 MAbs recognized proteic epitopes present in the SC and oral exudate of *M. incognita*, *T. spiralis* and *H. contortus*. One of these cross-reactive antigens was detected in the exudate present during ecdysis of *H. contortus* (de Mendoza et al. 1999). Whether this antigen plays any role in mediating ecdysis remains to be determined.

A cross reactive cuticular proteic epitope was identified in *T. spiralis* and *M. incognita* and this antigen might play a role in the interaction of *M. incognita* with its host-plant and in the interaction of *T. spirallis* inside the nurse cell as they are secreted *in planta* and *in vivo* (Lopez-Arellano and Curtis 2002). *In planta* this proteic antigen was immunolocalized surrounding the cuticle of the adult females of *M. incognita* and in the plasma membrane of root cells of *Arabidopsis thaliana*, close to the feeding cell formed during infection with *M. incognita*. *In vivo* this antigen was localised on the nematode surface and as secreted droplets close to the collagen capsule surrounding the nematode nurse cell (Curtis 1996; De Mendoza et al. 2002). These common antigens might represent immunodominant epitopes which are secreted inside the hosts and may perform related functions in these parasitic nematodes.

5.4.5 The Role of the Surface Coat in Immune Evasion by Animal-Parasitic Nematodes

Animal-parasitic nematodes have evolved a mutiplicity of evasive strategies to survive in immunologically competent host. The parasite's ability to exist for long periods of time in their host, has been attributed to a rapid turnover of their cuticle surface, shedding of surface antigens and membrane rigidity, which are likely to render the parasite less susceptible to immune attack (Simpson et al. 1984; Kusel and Gordon 1989). The mechanisms underlying surface antigen switching mechanisms are presently unknown but nematodes can alter their SC protein compositions at the moults between developmental stages or in response to host/environmental changes. As a rapid change in the surface lipophilicity of various animal-parasitic nematodes occurs during their transition from pre- to post-parasitic forms, these surface alterations may enable parasitic nematodes to evade host immune defenses during the course of infection (Jungery et al. 1983). Intracellular signalling and second messenger pathways involving cyclic nucleotides, calcium and intracellular alkalinisation participate in bringing about these surface changes (Modha et al. 1995, 1997).

Disguise of the parasite cuticle surface with the acquisition of host derived antigens (Smithers and Doenhoff 1982) and the action of parasite surface proteases that can cleave the Fc region of Immunoglobulins (Auriault et al. 1981) are some other mechanisms that may also help parasites to evade the host immune response by inhibiting important cellular functions.

Evasion of host immunity by *Toxocara canis* infective larvae is mediated by the nematode SC, as this nematode is able to shed the entire SC in response to binding antibodies or eosinophils, thus permitting parasites to physically escape immune attack (Maizels and Loukas 2001a). The major constituent of the SC of this nematode is the O-linked TES-120 (*Toxocara* excretory/secretory) glycoproteins series, which has a typical mucin domain and may explain a generally non-adhesive property of this parasite. Membrane associated mucins are closely concerned with the adhesion status of cells through electrostatic charge and due to steric effects of long chains protruding from the surface. It has been shown that the inhibition of T-cell adhesion can interfere with the ability of eosinophils to bind to the surface cuticle and kill schistosome parasites in *in vitro* tests (Hayes et al. 1990). TES-120 is secreted in internal excretory glands and ducted to the surface via the oesophagus and excretory pore and it is also released from *Toxocara* surface. The overexpression of some membrane-associated mucins suggests a possible model for the role of SC in immune evasion by parasitic nematodes, through changing the nematode surface cuticle adherence to defence cells and/or by releasing soluble mucins that might interact with host cells and blocks defence responses (Gems and Maizels 1996). *Toxocara canis* also secretes large quantities of a C-type lectin thought to compete with host innate immune system receptors (Loukas and Maizels 2000).

Another important group of surface proteins which may act to promote immune evasion in *B. malayi* are the anti-oxidant products glutathione peroxidase and superoxide dismutase. *Bm*-GPX-1 is the major 29 kDa surface glycoprotein of adult *Brugia* (Cookson et al. 1992; Maizels et al. 1989), which is believed to act as a lipid hydroperoxidase, protecting parasite membranes from peroxidation caused by free-oxygen radicals (Tang et al. 1996). A minor surface-associated protein of similar molecular weight is a superoxide dismutase, allowing the parasite to detoxify superoxide radicals (Tang et al. 1994). Many other surface-associated molecules may contribute to immune escape in a less obvious manner. For example, the polyprotein antigen (variously named gp15/400 or *Bm*-NPA-1) has a very high affinity for fatty acids (Kennedy et al. 1995; Smith et al. 1998) which could sequester substrate required for host leukotriene synthesis. Non-protein filarial products are also likely to play a significant role: a novel lipid found in the cuticle of *B. malayi* acts as a sink absorbing oxidative attack, perhaps protecting essential membrane lipids and proteins from degradation (Smith et al. 1998). A further component prominently expressed by *B. malayi* is phosphorylcholine (PC). Not only has this been suggested as an immunosuppressive moiety in lymphatic filariasis (Lal et al. 1990), but it has been possible to demonstrate direct down-regulation of both B (Deehan et al. 1998) and T cell (Harnett et al. 1998) function by a PC-bearing protein secreted from the filarial parasite *Acanthocheilonema viteae*.

A proteinase inhibitors member of the cystatin (cysteine protease inhibitor) family located on the surface of both L3 and adult *B. malayi*, and secreted by these parasites *in vitro* blocks conventional cysteine proteases but also the aspariginyl endopeptidase involved in the Class II antigen processing pathway in human B cells (Maizels et al. 2001b)

5.5 Interactions of Nematode Surfaces with Microorganisms

Complex interactions are formed between microorganisms, nematodes, plants and the environment. Some of the pant-parasitic nematodes microbial antagonists are also root colonizers and this may affect their activity against the nematodes (Kerry 2000; Bordallo et al. 2002; Sharon et al. 2007). Microorganisms have a wide range of suppressive activities on different nematode species. In direct interactions, nematode surface is believed to be important in recognition events and determining the specificity of interactions or the defence mechanisms involving microbial antagonists (Spiegel and McClure 1995; Kerry and Hominick 2001; Morton et al. 2004). Interactions between nematodes and microorganisms (fungi and bacteria) have been described in several reviews (Kerry and Hominick 2001; Bird 2004; Chen and Dickson 2004a; Chen and Dickson 2004b; Morton et al. 2004; Davies 2005; Tian et al. 2007; Lopez-Llorca et al. 2008). We will refer to some examples of such interactions, regarding attachment and penetration aspects.

5.5.1 *Interactions with Fungi*

Fungal antagonists of nematodes can be grouped into predacious fungi (nematode - trapping fungi), endoparasites of vermiform nematodes, parasites of sedentary females and eggs, and fungi that produce antibiotic substances. Nevertheless, some fungi can belong to more than one category. Attachment of fungi to nematodes is either specifically to head and tail regions, or all over the body, or very sparse. Zoospores usually attach near natural body openings (Chen and Dickson 2004a).

5.5.2 *Predacious Fungi*

Nematophagous fungi can capture, kill and consume their prey, and have evolved special devices for capturing vermiform nematodes: adhesive hyphae, branches, nets or knobs, non-constricting or constricting rings, and stephanocyts. Adhesive hyphae or branches are usually produced by lower fungi and Deuteromycetes such as *Arthrobotrys* and *Dactylaria* (Chen and Dickson 2004a), (see also Chap. 6).

In nematode-trapping fungi, parasitism begins with the induced formation of traps or other parasitism structures. Despite the variation in trap morphology, majorities of nematode-trapping fungi are closely related and the infection mechanism appears to be rather similar. Following traps development, the infection process proceeds through a sequence of events: attachment of the trap cells to nematode surface, penetration of the cuticle, digestion and assimilation of the nutrients from the killed nematode (Fekete et al. 2008). Free-living nematodes such as *Panagrellus redivivus* and *C. elegans* serve as model host nematodes in research of these fungi. Involvement of a Gal-NAc-specific lectin of *A. oligospora* (Nordbring-Hertz and Mattiasson 1979) in nematode recognition has been suggested. Subsequently, a carbohydrate-binding protein from the capture organs of the fungi, not present on hyphae, was isolated and partially characterized (Borrebaeck et al. 1984). These early infection events lead to signaling cascades necessary for penetration and colonization of the nematode prey (Tunlid et al. 1992). After contact, an extracellular material, or adhesive, is formed which adhere the fungus to the nematode surface. The adhesive layer has a fibrillar structure containing residues of neutral sugars, uronic acid and proteins (Tunlid et al. 1991). The adhesive on the traps of *A. oligospora* changes from an amorphous to a fibrillar appearance after contact with a nematode (Jansson and Nordbring-Hertz 1988).

Initial contact with the host cuticle is probably followed by a cascade of interactions with specific receptors, reorganization of surface polymers to strengthen the adhesions, changes in morphology, and the secretion of specific enzymes. Trapped nematodes become immobilized after adhesion, when the fungus starts to penetrate the nematode cuticle (Dijksterhuis et al. 1994). A narrow penetration tube develops from the trap cells and the nematode cuticle is breached by a combination of physical force and enzymatic degradation (Kerry and Hominick 2001). The nematode-trapping fungus, *A. oligospora*, produces a subtilisin-like serine protease (designated PII) that immobilized *P. redivivus* and hydrolyzed proteins of purified cuticle (Tunlid et al. 1994; Åhman et al. 2002). Once inside the nematode, the penetration tube swells to form an infection bulb from which hyphae are growing inside the infected nematodes. At this stage the internal tissues of the nematode are rapidly degraded (Dijksterhuis et al. 1994). An extracellular 35 kDa alkaline serine protease (Ds1) was purified and characterized from the nematode-trapping fungus *Dactylella shizishanna*. The purified protease could degrade purified cuticle of *P. redivivus* and a broad range of protein substrates. It showed a high homology with Aozl and PII, two serine proteases purified from *A.oligospora* (Wang et al. 2006).

Monacrosporium haptotylum traps nematodes using a spherical structure called knob, which develops on the apex of a hyphal branch. The transcriptional response in the parasitic fungus *M. haptotylum* and its nematode host *C.elegans* were analyzed during infection using cDNA microarrays (Fekete et al. 2008). Among the infection-induced *C. elegans* genes were those encoding antimicrobial peptides, protease inhibitors and lectins. *C. elegans* mount protective responses against bacterial and fungal pathogens by activating several intracellular signalling pathways that lead to the production of compounds that limit the infection or destroy invading microorganisms (Gravato-Nobre and Hodgkin 2005). C-type lectin domains (CTLD) might act as pathogen recognition molecules, or may mask the virulence factors of the pathogen. CTLD-containing genes were induced by bacterial infections of *C. elegans* (Mallo et al. 2002; O'Rourke et al. 2006), but were down-regulated during infection with *M. haptotylum* (Fekete et al. 2008).

5.5.3 Endoparasites of Vermiform Nematodes, Sedentary Females and Eggs

The fungal species, most commonly isolated from sedentary stages, include *Acremonium*, *Fusarium*, *Gliocladium*, *Nematophthora*, *Paecilomyces*, *Penicillium*, *Phoma* and *Pochonia* (*Verticillium*) (Chen and Dickson 2004a). The infection of root-knot and cyst nematode eggs by parasitic fungi involves formation of appressoria when the hyphae encounter the eggshell. This depends on recognition of host surface, where the hydrophobicity of the surface is an important factor (Morton et al. 2004).

Pochonia chlamydosporia (*Verticillium chlamydosporium*), which colonizes cysts or egg-masses, can also attack J2s. The infection of nematode eggs by *Pochonia rubescens* and *P. chlamydosporia* starts with contact of the hyphae and formation of an appressorium. Extracellular material, or an adhesive, formed on the appressorium, could be labeled with the lectin Concanavalin A (Con A), indicating the presence of glucose/mannose residues (Lopez-Llorca et al. 2002). The fungus penetrates the nematode eggshell mechanically and by enzymatic hydrolysis. *P. rubescens* protease P32 was immunolocalized on appressoria that were infecting eggs of the beet cyst nematode *H. schachtii* (Lopez-Llorca and Robertson 1992).

Conidia of the endoparasitic fungus, *Drechmeria coniospora*, adhere to the chemosensory organs natural body openings of the nematodes (Jansson and Nordbring-Hertz 1983). Conidial adhesion to *P. redivivus* was suggested to involve a sialic acid-like carbohydrate since treatment of nematodes with the lectin Limulin, and treatment of the spores with sialic acid, decreased adhesion (Jansson and Nordbring-Hertz 1984). Adhesion of conidia to *C. elegans* was reduced after treatment of the nematodes with Pronase E. The process was reversible within 2 h, indicating that the proteinaceous material emanating from the sensory structures was rapidly replaced (Jansson 1994). Conidia of *D. coniospora* adhere to the chemosensory organs of *Meloidogyne* spp., but do not penetrate and infect the nematodes. Nevertheless, the fungus reduced root galling in tomato (Jansson et al. 1985), probably due to behavior interference. A chymotrypsin-like, was partially characterized from *D. coniospora* conidia (Jansson and Friman 1999).

The fungus *Clonostachys rosea* (syn. *Gliocladium roseum*) is a widely distributed facultative saprophyte in the soil. Observations reveal that the pathogenesis on *P. redivivus* started from the adherence of conidia to nematode cuticle for germination, followed by the penetration of germ tubes into the nematode body and subsequent death and degradation of the nematodes. The conidia excrete a mucous liquid that can stick to the cuticle of nematodes. Microscope observations suggested that the glutinous substance could prevent the nematodes escape from the adhesive areas formed by the secretions of the conidia (Zhang et al. 2008). Two extracellular serine proteases (Lmz1 and PrC) were isolated from *C. rosea* and identified as important factors in fungal pathogenicity. Chemicals isolated from the fungus showed strong nematicidal activities against several nematodes: *C. elegans*, *P. redivivus*, and *Bursaphelenchus xylophilus* (Zhang et al. 2008).

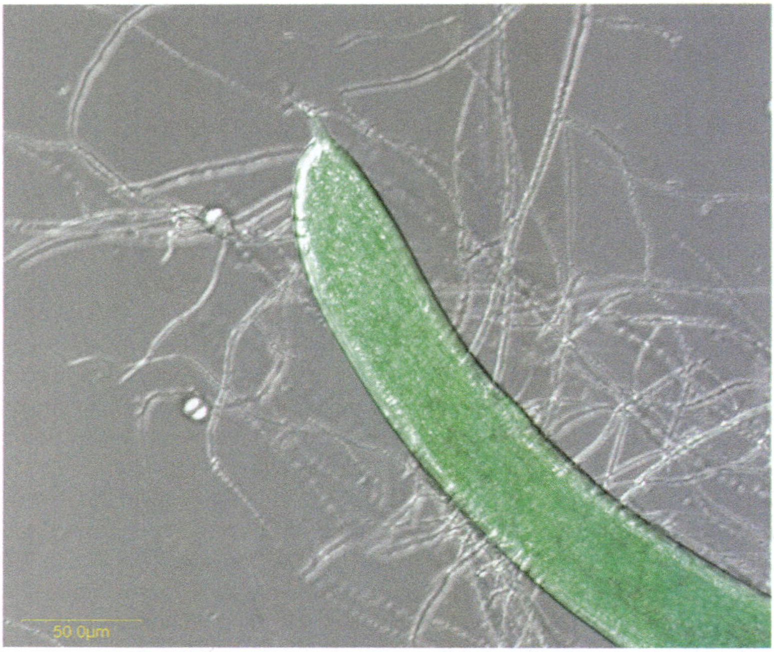

Fig. 5.6 Parasitism of *Trichoderma asperellum*-203 (constitutively expressing green fluorescent protein (GFP) construct) on *Xipinema index*. The image shows the posterior part of a female, colonized by the fungus, 4 days after inoculation with conidia, bar = 50 μm

Species of *Trichoderma* (see Chap. 8) are also facultative saprophytic fungi in soil and posses parasitism abilities against plant-pathogenic fungi and plant-parasitic nematodes, such as *Meloidogyne* spp. (Sharon et al. 2007, 2009b) and *Xipinema* spp. (Fig. 5.6) (Spiegel, Sharon, Chet unpubl.). Mechanisms involved in the attachment and parasitism processes on *M. javanica* were investigated, mainly with *T. asperellum*-203 and *T. atroviride*. It was found that the gelatinous matrix (gm) enables fungal attachment and enhances parasitic abilities of most isolates, which could also utilize it as a nutrient source. Fungal conidia can attach to nematode egg masses and to eggs and J2s that had contact with the gm, whereas gm-free J2s and eggs are almost unattached by fungal conidia; those were penetrated only by few fungal hyphae and colonized. Observations showed typical fungal parasitic behavior, including tight attachment of spores and hyphae to the J2s, coiling of hypae around J2s and appressoria-like structures formation upon egg penetration. Conidia were agglutinated by a gm suspension (enhanced in presence of Ca^{2+}) and their germination was improved. A model for fungal conidia attachment to nematodes suggests that carbohydrate-lectin-like interactions might be involved in this process (Sharon et al. 2007, 2009a).

Fungi of *Catenaria* spp. are known as obligate parasites of vermiform nematodes; however, *C. anguillulae* is a facultative parasite of females and eggs of root-knot nematodes (Wyss et al. 1990). Uniflagellate zoospores are attracted to natural openings of the nematode and show "amoeboid movement" upon contact with the nematode cuticle, before encystment. During encystment a cell wall is formed, covered by an adhesive, and the flagellum is withdrawn. The encysted zoospore forms an infection peg which penetrates the nematode cuticle and the nematode content is digested.

5.5.4 Fungi That Produce Antibiotic Substances and Toxins

Toxic effects of fungal culture filtrates on vermiform nematodes and eggs have been reported for several fungi, such as species of *Paecilomyces*, *Verticillium*, *Fusarium*, *Aspergillus*, *Trichoderma*, *Myrothecium* and *Penicillium* (Chen and Dickson 2004a; Morton et al. 2004). Toxins are important for parasitic microorganisms because they facilitate infection by debilitating the host. The combination of lytic enzymes and nematicidal compounds can improve the efficacy of the biocontrol agent by increasing the permeability of the nematode surfaces and eggshells. In biocontrol, our interest is to avoid damage to non-target organisms, such as beneficial nematodes. For example, nematicidal activity of *T. atroviride* culture filtrates was restricted to several plant parasitic species, and did not harm non-target and beneficial nematode species (Spiegel, Sharon, Chet unpubl.). The wide differences among the structures and compositions of the various nematode surfaces might determine the different permeability of the nematodes.

5.5.5 Interactions with Bacteria

Due to the chemical nature of the nematode's cuticle, few bacteria and fungi could utilize and degrade the extracellular exoskeleton. The physical body design of plant-parasitic nematodes prevents bacteria from entering through body openings; therefore, to become an endoparasite of nematodes, the bacterium must overcome the nematode surface barriers (Chen and Dickson 2004b). Most nematophagous bacteria, except for obligate parasites, are saprophytes that can also penetrate the nematodes and use them as a nutrition resource. Members of the genera *Pasteuria*, *Pseudomonas* and *Bacillus* have shown great potential for biological control of nematodes (Tian et al. 2007).

5.5.6 Obligate Bacterial Parasites

Species of *Pasteuria* are bacterial obligate hyperparasites of nematodes. The attachment process of endospores and its host specificity have been extensively studied mainly with *P. penetrans* on root-knot nematodes (recently reviewed by Bird 2004; Davies 2005, 2009; Tian et al. 2007). The involvement of surface coat components in the attachment has been demonstrated and antibodies raised against *Meloidogyne* surfaces, could inhibit spore attachment (Davies and Danks 1992; Spiegel et al. 1996). Monoclonal antibodies, raised against *H. cajani*, could also reduce spore attachment, whereas one antibody increased the attachment (Sharma and Davies 1997). There is high heterogeneity in bacterial surfaces, both within and among different *P. penetrans* populations (Davies and Redden 1997; Davies et al. 2000). The initial binding of endospores to their nematode hosts has been studied (Stirling et al. 1986; Persidis et al. 1991; Davies and Danks 1993; Spiegel et al. 1996). The results suggested a model in which a carbohydrate ligand on the surface of the endospore binds to a lectin-like receptor on the cuticle of the nematode host; however, binding of lectins to nematode surface also reduced the spore attachment (Spiegel et al. 1996). The fibres on the *Pasteuria* endospore are thought to be responsible for the adhesion to the host cuticle through a *Velcro*-like mechanism (Davies 2009). These fibres were shown to be glycoproteins, containing a high level of N-acetyglucosamine (Persidis et al. 1991). Collagen on the nematode's cuticle has been suggested to be involved in recognition process, because attachment was reduced by trypsin and endoglycosidase F, and because gelatin (denatured collagen) could inhibit spore attachment (Persidis et al. 1991; Mohan et al. 2001). However, treatments of J2s with collagenases did not inhibit endospore attachment (Davies and Danks 1993). Davies and Opperman (2006) identified several collagen-like proteins in *P. penetrans* that contain G-x-y repeats. Pretreatment of endospores with collagenase or the collagen-binding domain of fibronectin inhibited endospore binding to nematode cuticle (Mohan et al. 2001), suggesting that collagen-like proteins are also present on the *P. penetrans* exosporium surface and are involved in attachment.

5.5.7 Opportunistic Bacterial Parasites

Various isolates of *Brevibacillus laterosporus* have been reported as parasites of nematodes. The plant-parasites *Heterodera glycines*, *Trichostrongylus colubriformis* and *Bursaphelenchus xylophilus*, and the saprophytic nematode *P. redivius* could be killed by the bacterium (Tian et al. 2007). After attachment to the cuticle, the bacterium can multiply and form a single clone in the epidermis of the nematode cuticle. A circular hole can appear following the continuous degradation of the cuticle and tissue, until bacteria enter the nematode's body and digest it all for nutrition (Huang et al. 2005). The degradation of all the nematode cuticle components around the holes suggests the involvement of hydrolytic enzymes; the major pathogenic activity could be attributed to an extracellular alkaline serine protease, designated BLG4 (Huang et al. 2005; Tian et al. 2006).

A neutral protease (npr) (designated Bae16) of 40 kDa, toxic to nematodes, was purified to homogeneity from the strain *Bacillus nematocida*. The activity was tested against *P. redivivus* and *B. xylophilus*. This purified protease could destroy the nematode cuticle and its hydrolytic substrates included gelatin and collagen. The gene encoding Bae16 was cloned and expressed in *Escherichia coli*, confirming its nematicidal activity (Niu et al. 2006).

5.5.8 Other Bacterial Interactions

Bacterial antagonists that have been studied, such as *Telluria chitinolytica* (*Pseudomonas chitinolytica*) (Spiegel et al. 1991; Bowman et al. 1993) were able to attach to *M. javanica* juveniles, especially to head and tail regions, but no direct parasitism or degradation was observed. The bacteria and its culture filtrates could immobilize the nematodes and produce nematicidal compounds and hydrolytic enzymes such as chitinases, proteases and collagenases, which might contribute to the biocontrol process. These bacteria can colonize root surface and this might affect the nematode-plant interactions.

Bacillus cereus had also presented biocontrol activity against the root-knot nematode (Oka et al. 1993). An enzyme with collagenolytic/proteolytic activities that could degrade *M. javanica* cuticular collagens was purified from this bacterium (Sela et al. 1998).

Nematodes can serve as a vector of microorganisms that attach to their surface and enter the plants. Nematode species of *Anguina* are parasites of cereal grasses. They are vectors of *Rathaybacter* bacteria (formerly referred as *Clavibacter* and *Corynebacterium*). These bacteria produce toxins that can harm the animals feeding on the cereals. A positive correlation was found between a bacteriophage presence in *R. toxicus* and the toxin production (reviewed by Bird 2004). McClure and Spiegel (1991) showed that the bacteria are attached to the surface coat on the nematode's cuticle; however, the mechanism of the attachment remains unknown. Normal adhesion leads to fusion of the bacterium capsular material with the surface coat of the nematode. Slight swelling of nematode's epicuticle and cuticular damage, with loss of nematode's mobility may occur (Bird 2004).

Anguina funesta is also a vector of the fungus *Dilophosphora alopecuri*, a pathogen of cereals and grasses (reviewed by Bird 2004). The fungus has an inhibitory effect on nematode development in the plant. It shows also antagonism to *R. toxicus* and was suggested as a potential biocontrol agent. During adhesion, the fungal conidium secrets adhesive material that covers the surface of the nematode and fills the surface annulations, but no morphological changes were observed in the cuticle.

5.5.9 Role of Lytic Enzymes in Nematode Penetration

Enzymes degrade the host's barriers to infection and, therefore, play an important role in infection of nematodes by fungi and bacteria. The structure of nematode's cuticle and eggshells indicates that proteases and chitinases are necessary for their degradation. Such enzymes that have been identified and are associated with anti-nematode activity were described (Morton et al. 2004; Casas-Flores and Herrera-Estrella 2007; Lopez-Llorca et al. 2008; Tian et al. 2007; Gortari and Hours 2008). A cooperative effect of proteolytic and chtinolytic enzymes in eggshells degradation was demonstrated (Tikhonov et al. 2002). In nematophagous fungi, Pr1-like alkaline serine proteases act on the collagen-like proteins of the nematode cuticle and the protein-containing elements of the eggshell; such enzymes have been identified in several fungi. The major proteases identified from nematophagous fungi belong to the proteinase K family of subtilases (from Peptidase S8 Subtilase family) (Morton et al. 2004). This family includes also the proteinase Prb1, produced by *T. atroviride* and is also active against nematodes (Sharon et al. 2009b). Bacterial serine protease genes from nematophagous bacteria (*Brevibacillus* and *Bacillus* strains) have shown high sequence identity, indicating that these genes are highly conserved in these bacteria (Tian et al. 2007). Collagenases might also be important in cuticle degradation, since the majority of the proteins in the nematode cuticle are collagens. Collagenase was identified in *Arthrobotrys* spp (Tosi et al. 2002). Galper et al. (1991) evaluated a collagenolytic fungus, *Cunninghamella elegans*, for biocontrol activity against plant-parasitic nematodes. Lipases have been implicated in the infection of nematode eggs: *Heterodera schachtii* eggs, infected by fungi, appeared to have their inner lipid layers degraded (Perry and Trett 1986).

5.6 Concluding Remarks and Future Directions

Plant nematodes must survive the hostile environment that runs from the bulk soil through to the rhizosphere which contains numerous microorganisms and then onto the host plant. Plant nematode surfaces are targets for passive and active environmental attack by the plant immune system, the attachment of bacterial spores and fungal traps. The nematode egg surface is a target for fungal attack and the understanding of this fungus-nematode interaction might lead to the development of biological control methods for soil nematodes. The cuticle and its surface coat have a central importance not only in the biosynthesis and maintenance, but also play a part in defence against their host innate immunity and pathogens. The cuticle, and in particular its surface, can be looked upon as a major part of the nematodes immune system as can be seen from the specifity observed between *Pasteuria penetrans* and root-knot nematodes. It is likely that this variation, as observed by endospore attachment to the nematode cuticle may also play an important role in nematode-plant interactions.

Nematodes respond to environmental conditions by modifying their surfaces, a process formally similar to surface antigen switching in animal-parasitic nematodes. Now that the full genome of various plant-parasitic nematodes and their host plants are available it should be possible to perform detailed study of the genes involved in these interactions and the molecular nature of the surface modification response. It is now possible to study the tritrophic interactions between both the plant and nematode, and the nematode and the microbe, so including all sides of the pathosystem (see Chap. 12). It is also important to identify the environmental factors in the soil, rhizosphere and plants inducing changes in the nematode surface and behaviour.

Another important aspect of research in plant nematology is the identification of nematode effector proteins present in the surface cuticle and also the plant compounds being suppressed by the nematode. There is a great interest in the development of high throughput method for identifying putative nematode effector proteins, which are essential parasitism genes.

This knowledge should lead to a better understanding of interactions between plants and would undoutdly contribute to the development of novel environmentally friendly methods to control plant nematodes which could be incorporated into an integrated pest management programme. Plant nematode surfaces are easily accessible to control measures and constitute an important target for any anti-nematode gene therapy.

References

Abad P, Gouzy J, Aury JM et al (2008) Genome sequence of the metazoan plant-parasitic nematode *Meloidogyne incognita*. Nat Biotechnol 26:909–915

Agudelo P, Robbins RT, JMcD S et al (2004) Glycoproteins in the gelatinous matrix of *Rotylenchulus reniformis*. Nematropica 34:229–234

Åhman J, Johansson T, Olsson M et al (2002) Improving the pathogenicity of a nematode-trapping fungus by genetic engineering of a subtilisin with nematotoxic activity. Appl Environ Microbiol 68:3408–3415

Akhkha A, Kusel J, Kennedy M et al (2002) Effects of phytohormones on the surfaces of plant-parasitic nematodes. Parasitology 125:165–175

Akhkha A, Curtis R, Kennedy M et al (2004) The potential signalling pathways which regulate surface changes induced by phytohormones in the potato cyst nematode (*Globodera rostochiensis*). Parasitology 128:533–539

Atkinson HJ, Taylor JD, Fowler M (1987) Changes in the 2nd stage juveniles of *Globodera rostochiensis* prior to hatching in response to potato root diffusate. Ann Appl Biol 110:105–114

Auriault C, Ouaissi MA, Torpier G et al (1981) Proteolytic cleavage of IgG bound to the Fc receptor of *Schistosoma mansoni* schistosomula. Parasite Immunol 3:33–44

Bais HP, Weir TL, Perry LG et al (2006) The role of root exudates in rhizosphere interactions with plants and other organisms. Annu Rev Plant Biol 57:233–266

Bird AF (2004) Surface adhesion to nematodes and its consequences. In: Chen ZX, Chen SY, Dickson DW (eds) Nematology: advances and perspectives, vol I, Nematode morphology, physiology and ecology. CABI Publishing, Cambridge

Bird AF, Bird J (1991) The structure of nematodes, 2nd edn. Academic, San Diego

Bird AF, McClure MA (1976) Tylenchid (Nematoda) egg-shell, structure, composition and permeability. Parasitology 72:19–27

Bird AF, Rogers GE (1965) Ultrastructure of the cuticle and its formation in *Meloidogyne javanica*. Nematologica 11:224–230

Bird DM, Wilson MA (1994) Plant molecular and cellular responses to nematode infection. In: Lamberti F (ed) Advances in molecular plant nematology. Plenum Press, New York

Bird AF, Bonig I, Bacic A (1988) A role for the excretory-secretory in senernentean nematodes. J Nematol 20:493–496

Blaxter ML, Robertson WM (1998) The cuticle. In: Perry RN, Wright DJ (eds) The physiology and biochemistry of free-living and plant-parasitic nematodes. CABI Publishing, Wallingford

Blaxter ML, Page AP, Rudin W et al (1992) Nematode surface coats, actively evading immunity. Parasitol Today 8:243–247

Bordallo JJ, Lopez-Llorca LV, Jansson H-B et al (2002) Colonization of plant roots by egg-parasitic and nematode-trapping fungi. New Phytol 154:491–499

Borrebaeck CA, Mattlasson B, Nordbring-Hertz B (1984) Isolation and partial characterization of a carbohydrate-binding protein from a nematode-trapping fungus. J Bacteriol 159: 53–56

Bowman JP, Sly LI, Hayward AC et al (1993) *Telluria mixta* (*Pseudomonas mixta* Bowman, Sly, and Hayward 1988) gen. nov., comb. nov., and *Telluria chitinolytica* sp. nov., soil-dwelling organisms which actively degrade polysaccharides. Int J Syst Bacteriol 43:120–124

Casas-Flores S, Herrera-Estrella A (2007) Antagonism of plant parasitic nematodes by fungi. In: Kubicek CP, Druzhinina IS (eds) The mycota IV: environmental and microbial relationships, 2nd edn. Springer, Berlin

Chen S, Dickson DW (2004) Biological control of nematodes by fungal antagonists. In: Chen ZX, Chen SY, Dickson DW (eds) Nematology: advances and perspectives, vol II, Nematode management and utilization. CABI Publishing, Cambridge

Clarke AJ, Cox PM, Shepherd AM (1967) Chemical composition of egg shells of potato cyst-nematode *Heterodera rostochiensis* woll. Biochem J 104:1056–1060

Cookson E, Blaxter ML, Selkirk ME (1992) Identification of the major soluble cuticular glycoprotein of lymphatic filarial nematode parasites (gp29) as a secretory homolog of glutathione peroxidase. Proc Natl Acad Sci USA 89:5837–5841

Couillault C, Ewbank JJ (2002) Diverse bacteria are pathogens of *Caenorhabditis elegans*. Infect Immun 70:4705–4707

Curtis RHC (1996) Identification and *in vitro* and *in vivo* characterisation of secreted proteins produced by plant-parasitic nematodes. Parasitology 113:589–597

Curtis RHC (2007a) Do phytohormones influence nematode invasion and feeding site establishment? Nematology 9:155–160

Curtis RHC (2007b) Plant-parasitic nematodes and the host-parasite interactions. Brief Funct Genomics Proteomics 6:50–58

Davies KG (2005) Interactions between nematodes and microorganisms: bridging ecological and molecular approaches. Adv Appl Microbiol 57:53–78

Davies KG (2009) Understanding the interaction between an obligate hyperparasitic bacterium, *Pasteuria penetrans* and its obligate plant-parasitic nematode host, *Meloidogyne* spp. Adv Parasitol 68:211–245

Davies KG, Danks C (1992) Interspecific differences in the nematode surface coat between *Meloidogyne incognita* and *M. arenaria* related to the adhesion of the bacterium *Pasteuria penetrans*. Parasitology 105:475–480

Davies KG, Danks C (1993) Carbohydrate/protein interactions between the cuticle of infective juveniles of *Meloidogyne incognita* and spores of the obligate hyperparasite *Pasteuria penetrans*. Nematologica 39:53–64

Davies KG, Opperman CH (2006) A potential role for collagen in the attachment of *Pasteuria penetrans* to nematode cuticle. In: Raaijmakers JM, Sikora RA (eds) Multitrophic interactions in the soil. IOBC/WPRS Bull 29:11–16

Davies KG, Redden M (1997) Diversity and partial characterization of putative virulence determinants in *Pasteuria penetrans*, the hyperparasitic bacterium of root-knot nematodes (*Meloidogyne* spp.). J Appl Microbiol 83:227–235

Davies KG, Spiegel Y (2011) Biological control of plant parasitic nematodes: towards understanding field variation through molecular mechanisms. In: Jones J, Gheysen G, Fenoll C (eds) Exploiting genomics to understand plant-nematode interactions. Springer (in press)

Davies KG, Fargette M, Balla G et al (2000) Cuticle heterogeneity as exhibited by *Pasteuria* spore attachment is not linked to the phylogeny of parthenogenetic root-knot nematode (*Meloidogyne* spp.). Parasitology 122:111–120

De Giorgi C, De Luca F, Di Vito D et al (1997) Modulation of expression at the level of splicing of *cut-1* RNA in the infective second stage juvenile of the plant parasitic nematode *Meloidogyne artiella*. Mol Gen Genet 253:589–598

De Mendoza MEL, Curtis RHC, Gowen S (1999) Identification and characterization of excreted–secreted products and surface coat antigens of animal and plant-parasitic nematodes. Parasitology 118:397–405

De Mendoza MEL, Modha J, Roberts MC et al (2000) Changes in the lipophilicity of the surfaces of *Meloidogyne incognita* and *Haemonchus contortus* during exposure to host signals. Parasitology 120:203–209

De Mendoza MEL, Abrantes IMO, Rowe J et al (2002) Immunolocalisation *in planta* of secretions from parasitic stages of *Meloidogyne incognita* and *M. hispanica*. Int J Nematol 12:149–154

Deehan MR, Frame MJ, Parkhouse RME et al (1998) A phosphorylcholine-containing filarial nematode-secreted product disrupts B lymphocyte activation by targeting key proliferative signaling pathways. J Immunol 160:2692–2699

Dijksterhuis J, Veenhuis M, Harder W et al (1994) Nematophagous fungi: Physiological aspects and structure-function relationships. In: Rose AH, Tempest DW (eds) Advances in microbial physiology, vol 36. Academic, New York

Dubinsky P, Rybos M, Turcekova L (1986a) Properties and localization of chitin synthase in *Ascaris suum* eggs. Parasitology 92:219–225

Dubinsky P, Rybos M, Turcekova L et al (1986b) Chitin synthesis in zygotes of *Ascaris suum*. J Helminthol 60:187–192

Ewbank JJ (2002) Tackling both sides of the host-pathogen equation with *Caenorhabditis elegans*. Microb Infect 4:247–256

Fanelli E, Di Vito M, Jones JT et al (2005) Analysis of chitin synthase function in a plant parasitic nematode, *Meloidogyne artiellia*, using RNAi. Gene 349:87–95

Fekete C, Tholander M, Rajashekar B et al (2008) Paralysis of nematodes: shifts in the transcriptome of the nematode-trapping fungus *Monacrosporium haptotylum* during infection of *Caenorhabditis elegans*. Environ Microbiol 10:364–375

Fioretti L, Porter A, Haydock PJ et al (2002) Monoclonal antibodies reactive with secreted-excreted products from the amphids and the cuticle surface of *Globodera pallida* affect nematode movement and delay invasion of potato roots. Int J Parasitol 32:1709–1718

Galper S, Cohn E, Spiegel Y et al (1991) A collagenolytic fungus, *Cunninghamella elegans*, for biological control of plant-parasitic nematodes. J Nematol 23:269–274

Gems D, Maizels RM (1996) An abundantly expressed mucin-like protein from *Toxocara canis* infective larvae: The precursor of the larval surface coat glycoproteins. Proc Natl Acad Sci USA 93:1665–1670

Geng JM, Plenefisch J, Komuniecki PR et al (2002) Secretion of a novel developmentally regulated chitinase (family 19 glycosyl hydrolase) into the perivitelline fluid of the parasitic nematode, *Ascaris suum*. Mol Biochem Parasitol 124:11–21

Gortari MC, Hours RA (2008) Fungal chitinases and their biological role in the antagonism onto nematode eggs. Annu Rev Mycol Prog 7:221–238

Gravato-Nobre MJ, Evans K (1998) Plant and nematode surfaces: Their structure and importance in host-parasite interactions. Nematologica 44:103–124

Gravato-Nobre MJ, Hodgkin J (2005) *Caenorhabditis elegans* as a model for innate immunity to pathogens. Cell Microbiol 7:741–751

Gravato-Nobre MJ, McClure MA, Dolan L et al (1999) *Meloidogyne incognita* surface antigen epitopes in infected *Arabidopsis* roots. J Nematol 31:212–223

Gravato-Nobre MJ, Nicholas HR, Nijland R et al (2005) Multiple genes affect sensitivity of *Caenorhabditis elegans* to the bacterial pathogen *Microbacterium nematophilum*. Genetics 171:1033–1045

Grenache DG, Caldicott I, Albert PS et al (1996) Environmental induction and genetic control of surface antigen switching in the nematode *Caenorhabditis elegans*. Proc Natl Acad Sci USA 93:12388–12393

Guiliano DB, Hong XQ, McKerrow JH et al (2004) A gene family of cathepsin L-like proteases of filarial nematodes are associated with larval molting and cuticle and eggshell remodeling. Mol Biochem Parasitol 136:227–242

Hanazawa M, Mochii M, Ueno N et al (2001) Use of cDNA subtraction and RNA interference screens in combination reveals genes required for germ-line development in *Caenorhabditis elegans*. Proc Natl Acad Sci USA 98:8686–8691

Harnett MM, Deehan MR, Williams DM et al (1998) Induction of signalling energy via the T-cell receptor in cultured Jurkat T cells by pre-exposure to a filarial nematode secreted product. Parasite Immunol 20:551–563

Harris MT, Fuhrman JA (2002) Structure and expression of chitin synthase in the parasitic nematode *Dirofilaria immitis*. Mol Biochem Parasitol 122:231–234

Hashmi S, Britton C, Liu J et al (2002) Cathepsin L is essential for embryogenesis and development of *Caenorhabditis elegans*. J Biol Chem 277:3477–3486

Hayes DF, Silberstein DS, Rodrique SW et al (1990) DF3 antigen, a human epithelial-cell mucin, inhibits adhesion of eosinophils to antibody-coated targets. J Immunol 145:962–970

Hill DE, Fetterer RH, Urban JF (1991) *Ascaris suum*, stage-specific differences in lectin binding to the larval cuticle. Exp Parasitol 73:376–383

Hodgkin J, Kuwabara PE, Corneliussen B (2000) A novel bacterial pathogen, *Microbacterium nematophilum*, induces morphological change in the nematode *C. elegans*. Curr Biol 10:1615–1618

Hu GG, McClure MA, Schmitt ME (2000) Origin of a *Meloidogyne incognita* surface coat antigen. J Nematol 32:174–182

Huang XW, Tian BY, Niu QH et al (2005) An extracellular protease from *Brevibacillus laterosporus* G4 without parasporal crystal can serve as a pathogenic factor in infection of nematodes. Res Microbiol 156:719–727

Jansson H-B (1994) Adhesion of conidia of *Drechmeria coniospora* to *Caenorhabditis elegans* wild type and mutants. J Nematol 26:430–435

Jansson H-B, Friman E (1999) Infection-related surface proteins on conidia of the nematophagous fungus *Drechmeria coniospora*. Mycol Res 103:249–256

Jansson H-B, Nordbring-Hertz B (1983) The endoparasitic fungus *Meria coniospora* infects nematodes specifically at the chemosensory organs. J Gen Microbiol 129:1121–1126

Jansson H-B, Nordbring-Hertz B (1984) Involvement of sialic acid in nematode chemotaxis and infection by an endoparasitic nematophagous fungus. J Gen Microbiol 130:39–43

Jansson H-B, Nordbring-Hertz B (1988) Infection events in the fungus-nematode system. In: Poinar GO, Jansson H-B (eds) Diseases of nematodes, vol II. CRC Press, Boca Raton

Jansson H-B, Jeyaprakash A, Zuckerman BM (1985) Control of root-knot nematodes on tomato by the endoparasitic fungus *Meria coniospora*. J Nematol 17:327–330

Johnstone IL (1994) The cuticle of the nematode *Caenorhabditis elegans*, a complex collagen structure. Bioessays 16:171–178

Johnstone IL, Barry JD (1996) Temporal reiteration of a precise gene expression pattern during nematode development. EMBO J 15:3633–3639

Jones JT (2002) Nematode sense organs. In: Lee DL (ed) The biology of nematodes. Taylor & Francis, London

Jones JT, Perry RN, Johnston MRL (1993) Changes in the ultrastructure of the cuticle of the potato cyst-nematode, *Globodera rostochiensis*, during development and infection. Fundam Appl Nematol 16:433–445

Jones JT, Reavy B, Smant G et al (2004) Glutathione peroxidases of the potato cyst nematode *Globodera rostochiensis*. Gene 324:47–54

Jung WJ, Jung SJ, An KN et al (2002) Effect of chitinase-producing *Paenibacillus illinoisensis* KJA-424 on egg hatching of root-knot nematode (*Meloidogyne incognita*). J Microbiol Biotechnol 12:865–871

Jungery M, Clark NWT, Parkhouse RM (1983) A major change in surface-antigens during the maturation of newborn larvae of *Trichinella spiralis*. Mol Biochem Parasitol 7:101–109

Kennedy MW, Allen JE, Wright AS et al (1995) The gp15/400 polyprotein antigen of *Brugia malayi* binds fatty-acids and retinoids. Mol Biochem Parasitol 71:41–50

Kerry BR (2000) Rhizosphere interactions and exploitation of microbial agents for the biological control of plant-parasitic nematodes. Annu Rev Phytopathol 38:423–441

Kerry BR, Hominick WM (2001) Biological control. In: Lee DL (ed) Biology of nematodes. Taylor & Francis, London

Koltai H, Chejanovski N, Raccha B et al (1997) The first isolated collagen gene of the root-knot nematode *Meloidogyne javanica* is developmentally regulated. Gene 196:191–199

Koltai H, Sharon E, Spiegel Y (2002) Root-nematode interactions: recognition and pathogenicity. In: Waisel Y, Eshel A, Kafkafi U (eds) Plant roots: the hidden half, 3rd edn. Marcel Dekker, New York

Kramer JM, Cox GN, Hirsh D (1982) Comparisons of the complete sequences of 2 collagen genes from *Caenorhabditis elegans*. Cell 30:599–606

Kunkel BN, Brooks DM (2002) Cross talk between signaling pathways in pathogen defense. Curr Opin Plant Biol 5:325–331

Kusel JR, Gordon JF (1989) Biophysical studies of the *Schistosome* surface and their relevance to its properties under immune and drug attack. Parasite Immunol 11:431–451

Labrousse A, Chauvet S, Couillault C et al (2000) *Caenorhabditis elegans* is a model host for *Salmonella typhimurium*. Curr Biol 10:1543–1545

Lal RB, Kumaraswami V, Steel C et al (1990) Phosphocholine-containing antigens of *Brugia malayi* nonspecifically suppress lymphocyte function. Am J Trop Med Hyg 42:56–64

Lee DL (2002) Cuticle, moulting and exsheathment. In: Lee DL (ed) The biology of nematodes. Taylor & Francis, London

Lin HJ, McClure MA (1996) Surface coat of *Meloidogyne incognita*. J Nematol 28:216–224

Liu J, Koltai H, Chejanovski N et al (2001) Isolation of a novel collagen gene in *Meloidogyne javanica* and studies on its expression pattern. J Parasitol 87:801–807

Lopez-Arellano ME, Curtis RHC (2002) Immunolocalization of *Trichinella spiralis* L1 surface and excreted/secreted antigens *in situ*. Int J Nematol 12:55–58

Lopez-Llorca LV, Robertson WM (1992) Immunocytochemical localization of a 32-kDa protease from the nematophagous fungus *Verticillium suchlasporium* in infected nematode eggs. Exp Mycol 16:261–267

Lopez-Llorca LV, Olivares-Bernabeu C, Salinas J et al (2002) Prepenetration events in fungal parasitism of nematode eggs. Mycol Res 106:499–506

Lopez-Llorca LV, Macia-Vicente JG, Jansson H-B (2008) Mode of action and interactions of nematophagous fungi. In: Ciancio A, Mukerji KG (eds) Integrated pest and disease management and biocontrol of vegetable and grain crops nematodes. Springer Publishers, Dordrecht

Loukas A, Maizels RM (2000) Helminth C-type lectins and host-parasite interactions. Parasitol Today 16:333–339

Mackintosh GM (1960) The morphology of the *Brassica* root eelworm *Heterodera cruciferae* Franklin, 1945. Nematologica 5:158–165

Maggenti AR (1962) Production of gelatinous matrix and its taxonomic significance in *Tylenchulus* (Nematoda, Tylenchulinae). Proc Helminth Soc Washington 29:139–144

Maizels RM, Loukas A (2001) The surface and secreted antigens of *Toxocara canis*: genes, protein structure and function. In: Kennedy MW, Harnett W (eds) Parasitic nematodes – molecular biology, biochemistry and immunology. CABI Publishing, Wallingford

Maizels RM, Gregory WF, Kwanlim GE et al (1989) Filarial surface-antigens, the major 29 Kilodalton glycoprotein and a novel 17–200 Kilodalton complex from adult *Brugia malayi* parasites. Mol Biochem Parasitol 32:213–227

Maizels RM, Gomez-Escobar N, Gregory WF et al (2001) Immune evasion genes from filarial nematodes. Int J Parasitol 31:889–898

Mallo GV, Kurz CL, Couillault C et al (2002) Inducible antibacterial defense system in *C. elegans*. Curr Biol 12:1209–1214

McClure MA, Spiegel Y (1991) Role of the nematode surface coat in the adhesion of *Clavibacter sp.* to *Anguina funesta* and *Anguina tritici*. Parasitology 103:421–427

Mendoza de Gives P, Davies KG, Clark SJ, Behnke JM (1999) Predatory behavior of trapping fungi against *srf* mutants of *Caenorhabditis elegans* and different plant- and animal-parasitic nematodes. Parasitology 119:95–104

Modha J, Kusel JR, Kennedy MW (1995) A role for 2nd messengers in the control of activation-associated modification of the surface of *Trichinella spiralis* infective larvae. Mol Biochem Parasitol 72:141–148

Modha J, Roberts MC, Kennedy MW et al (1997) Induction of surface fluidity in *Trichinella spiralis* larvae during penetration of the host intestine: simulation by cyclic AMP *in vitro*. Parasitology 114:71–77

Mohan S, Fould S, Davies KG (2001) The interaction between the gelatin-binding domain of fibronectin and the attachment of *Pasteuria penetrans* endospores to nematode cuticle. Parasitology 123:271–276

Morton CO, Hirsch PR, Kerry B (2004) Infection of plant-parasitic nematodes by nematophagous fungi – a review of application of molecular biology to understand infection processes and to improve biological control. Nematology 6:161–170

Neuhaus B, Bresciani J, Peters W (1997) Ultrastructure of the pharyngeal cuticle and lectin labelling with wheat germ agglutinin-gold conjugate indicating chitin in the pharyngeal cuticle of *Oesophagostomum dentatum* (Strongylida, Nematoda). Acta Zool 78:205–213

Niu Q, Huang X, Zhang L et al (2006) A neutral protease from *Bacillus nematocida*, another potential virulence factor in the infection against nematodes. Arch Microbiol 185:439–448

Nordbring-Hertz B, Mattiasson B (1979) Action of a nematode-trapping fungus shows lectinmediated host-microorganism interaction. Nature 281:477–479

O'Rourke D, Baban D, Demidova M et al (2006) Genomic clusters, putative pathogen recognition molecules, and antimicrobial genes are induced by infection of *C. elegans* with *M. nematophilum*. Genome Res 16:1005–1016

Oka Y, Chet I, Spiegel Y (1993) Control of the root-knot nematode *Meloidogyne javanica* by *Bacillus cereus*. Biocontrol Sci Technol 3:115–126

Olsen DP, Phu D, Libby LJM et al (2007) Chemosensory control of surface antigen switching in the nematode *Caenorhabditis elegans*. Genes Brain Behav 6:240–252

Opperman CH, Bird DM, Williamson VM et al (2008) Sequence and genetic map of *Meloidogyne hapla*: A compact nematode genome for plant parasitism. Proc Natl Acad Sci USA 105:14802–14807

Page AP, Johnstone IL (2007) The cuticle. In: Wormbook. The *C. elegans* research community, doi/10.1895/wormbook.1.7.1, http://www.wormbook.org

Perry RN, Clarke AJ (1981) Hatching mechanisms of nematodes. Parasitology 83:435–449

Perry RN, Trett MW (1986) Ultrastructure of the eggshell of *Heterodera schachtii* and *Heterodera glycines* (Nematode, Tylenchida). Rev Nematol 9:399–403

Persidis A, Lay JG, Manousis T et al (1991) In: Wormbook. Characterization of potential adhesions of the bacterium *Pasteuria penetrans*, and of putative receptors on the cuticle of *Meloidogyne incognita*, a nematode host. J Cell Sci 100:613–622

Philipp M, Rumjaneck FD (1984) Antigenic and dynamic properties of helminth surface structures. Mol Biochem Parasitol 10:245–268

Politz SM, Philipp M (1992) *Caenorhabditis elegans* as a model for parasitic nematodes: a focus on the cuticle. Parasitol Today 8:6–12

Premachandran D, Vonmende N, Hussey RS et al (1988) A method for staining nematode secretions and structures. J Nematol 20:70–78

Preston CM, Jenkins T (1985) *Trichuris muris* – structure and formation of the egg polar plugs. Z Parasitenkunde – Parasitol Res 71:373–381

Prior A, Jones JT, Blok VC et al (2001) A surface-associated retinol- and fatty acid-binding protein (Gp-FAR-1) from the potato cyst nematode *Globodera pallida*: lipid binding activities, structural analysis and expression pattern. Biochem J 356:387–394

Proudfoot L, Kusel JR, Smith HV et al (1993) Rapid changes in the surfce of parasitic nematodes during transition from pre-parasitic to post-parasitic forms. Parasitology 107:107–117

Pujol N, Link EM, Liu LX et al (2001) Reverse genetic analysis of components of the Toll signaling pathway in *Caenorhabditis elegans*. Curr Biol 11:809–821

Rao UR, Chandrashekar R, Subrahmanyam D (1988) Developmental-changes in the surface carbohydrates of filariae. Indian J Med Res 87:9–14

Riou M, Koch C, Delaleu B et al (2005) Immunolocalisation of an ABC transporter, P-glycoprotein, in the eggshells and cuticles of free-living and parasitic stages of *Haemonchus contortus*. Parasitol Res 96:142–148

Robertson WM, Spiegel Y, Jansson H-B et al (1989) Surface carbohydrates of plant-parasitic nematodes. Nematologica 35:180–186

Robertson L, Robertson WM, Sobczak M et al (2000) Cloning, expression and functional characterisation of a peroxiredoxin from the potato cyst nematode *Globodera rostochiensis*. Mol Biochem Parasitol 111:41–49

Robinson MP, Delgado J, Parkhouse RMW (1989) Characterisation of stage-specific cuticular proteins of *Meloidogyne incognita* by radio-iodination. Physiol Mol Plant Pathol 35:135–140

Sapio MR, Hilliard MA, Cermola M et al (2005) The Zona Pellucida domain containing proteins, CUT-1, CUT-3 and CUT-5, play essential roles in the development of the larval alae in *Caenorhabditis elegans*. Dev Biol 282:231–245

Schallig H, Vanleeuwen MAW, Hendrikx WML (1994) Immune-responses of texel sheep to excretory/secretory products of adult *Haemonchus contortus*. Parasitology 108:351–357

Schallig H, Vanleeuwen MAW, Hendrikx WML (1995) Isotype-specific serum antibody-responses of sheep to *Haemonchus contortus* antigens. Vet Parasitol 56:149–162

Sela S, Schickler H, Chet I et al (1998) Purification and characterization of a B*acillus cereus* collagenolytic/proteolytic enzyme and its effect on *Meloidogyne javanica* cuticular proteins. Eur J Plant Pathol 104:59–67

Sharma SB, Davies KG (1997) Modulation of spore adhesion of the hyperparasitic bacterium *Pasteuria penetrans* to nematode cuticle. Lett Appl Microbiol 25:426–430

Sharon E, Spiegel Y (1993) Glycoprotein characterization of the gelatinous matrix in the root-knot nematode *Meloidogyne javanica*. J Nematol 25:585–589

Sharon E, Spiegel Y (1996) Gold-conjugated reagents for the labelling of carbohydrate-recognition domains and glycoconjugates on nematodes surfaces. J Nematol 28:124–127

Sharon E, Spiegel Y, Salomon R et al (2002) Characterization of *Meloidogyne javanica* surface coat with antibodies and their effect on nematode behaviour. Parasitology 125:177–185

Sharon E, Chet I, Viterbo A et al (2007) Parasitism of *Trichoderma* on *Meloidogyne javanica* and role of the gelatinous matrix. Eur J Plant Pathol 118:247–258

Sharon E, Chet I, Spiegel Y (2009a) Improved attachment and parasitism of *Trichoderma* on *Meloidogyne javanica in vitro*. Eur J Plant Pathol 123:291–299

Sharon E, Chet I, Bar-Eyal M et al (2009b) Biocontrol of root-knot nematodes by *Trichoderma* – modes of action. In: Steinberg C, Edel-Hermann V, Friberg H et al (eds) Multitrophic interactions in soil. IOBC/WPRS Bull 42:159–163

Simpson AJG, Payares G, Walker T et al (1984) The modulation of expression of polypeptide surface-antigens on developing schistosomula of *Schistosoma mansoni*. J Immunol 133:2725–2730

Smith VP, Selkirk ME, Gounaris K (1998) *Brugia malayi*: Resistance of cuticular lipids to oxidant-induced damage and detection of alpha-tocopherol in the neutral lipid fraction. Exp Parasitol 88:103–110

Smithers SR, Doenhoff MJ (1982) Schistosomiasis. In: Cohen S, Warren KS (eds) Immunology of parasitic infections. Blackwell Scientific Publications, Oxford

Spiegel Y, Chet I (1985) Chitin synthetase inhibitors and their potential to control the root-knot nematode, *Meloidogyne javanica*. Nematologica 31:480–482

Spiegel Y, McClure MA (1991) Stage-specific differences in lectin binding to the surface of *Anguina tritici* and *Meloidogyne incognita*. J Nematol 23:259–263

Spiegel Y, McClure MA (1995) The surface coat of plant-parasitic nematodes: chemical composition, origin and biological role: a review. J Nematol 27:127–134

Spiegel Y, Cohn E, Galper S et al (1991) Evaluation of a newly isolated bacterium, *Pseudomonas chitinolytica* sp. nov. for controlling the root-knot nematode *Meloidogyne javanica*. Biocontrol Sci Technol 1:115–125

Spiegel Y, Mor M, Sharon E (1996) Attachment of *Pasteuria penetrans* endospores to the surface of *Meloidogyne javanica* second-stage juveniles. J Nematol 28:328–334

Spiegel Y, Kahane I, Cohen L et al (1997) *Meloidogyne javanica* surface proteins: characterization and lability. Parasitology 115:513–519

Stirling GR, Bird AF, Cakurs AB (1986) Attachment of *Pasteuria panetrans* spores to the cuticle of root-knot nematodes. Rev Nematol 9:251–260

Tang L, Ou X, Henkleduhrsen K et al (1994) Extracellular and cytoplasmic superoxide dismutases from *Brugia* lymphatic filarial nematode parasites. Infect Immun 62:961–967

Tang L, Smith VP, Gounaris K et al (1996) *Brugia pahangi*: the cuticular glutathione peroxidase (gp29) protects heterologous membranes from lipid peroxidation. Exp Parasitol 82:329–332

Taylor DW, Goddard JM, McMahon JE (1986) Surface components of *Onchocerca volvulus*. Mol Biochem Parasitol 18:283–300

Tefft PM, Bone LW (1985) Plant-induced hatching of eggs of the soybean cyst nematode *Heterodera glycines*. J Nematol 17:275–279

Tian B, Li N, Lian L et al (2006) Cloning, expression and deletion of the cuticle-degrading protease BLG4 from nematophagous bacterium *Brevibacillus laterosporus* G4. Arch Microbiol 186:297–305

Tian B, Yang J, Zhang K-Q (2007) Bacteria used in the biological control of plant-parasitic nematodes: populations, mechanisms of action, and future prospects. FEMS Microbiol Ecol 61:197–213

Tikhonov VE, Lopez-Llorca LV, Salinas J et al (2002) Purification and characterization of chitinases from the nematophagous fungi *Verticillium chlamydosporium* and *V. suchlasporium*. Fungal Genet Biol 35:67–78

Tosi S, Annovazzi L, Tosi I et al (2002) Collagenase production in an Antarctic strain of *Arthrobotrys tortor jarowaja*. Mycopathologia 153:157–162

Tunlid A, Johansson T, Nordbring-Hertz B (1991) Surface polymers of the nematode-trapping fungus *Arthrobotrys oligospora*. J Gen Microbiol 137:1231–1240

Tunlid A, Jansson H-B, Nordbring-Hertz B (1992) Fungal attachment to nematodes. Mycol Res 96:401–412

Tunlid A, Rosen S, Ek B et al (1994) Purification and characterization of extracellular serine protease from the nematode-trapping fungus *Arthrobotrys oligospora*. Microbiology 140:1687–1695

Twomey U, Warrior P, Kerry BR et al (2000) Effects of the biological nematicide, DiTera, on hatching of *Globodera rostochiensis* and *G. pallida*. Nematology 2:355–362

Van Nguyen N, Kim Oh KT, Jung W et al (2007) The role of chitinase from *Lecanicillium antillanum* B-3 in parasitism to root-knot nematode *Meloidogyne incognita* eggs. Biocontrol Sci Technol 17:1047–1058

Veronico P, Gray LJ, Jones JT et al (2001) Nematode chitin synthases: gene structure, expression and function in *Caenorhabditis elegans* and the plant parasitic nematode *Meloidogyne artiellia*. Mol Genet Genomics 266:28–34

Waetzig GH, Sobczak M, Grundler FMW (1999) Localization of hydrogen peroxide during the defence response of *Arabidopsis thaliana* against the plant-parasitic nematode *Heterodera glycines*. Nematology 1:681–686

Wang RB, Yang JK, Lin C et al (2006) Purification and characterization of an extracellular serine protease from the nematode-trapping fungus *Dactylella shizishanna*. Lett Appl Microbiol 42:589–594

Wharton DA (1983) The production and functional-morphology of helminth egg-shells. Parasitology 86:85–97

Wyss U, Voss B, Jansson H-B (1990) *In vitro* observations on the infection of *Meloidogyne incognita* eggs by the zoosporic fungus *Catenaria anguilulae* Sorokin. Fundam Appl Nematol 15:133–139

Zhang L, Yang J, Niu Q et al (2008) Investigation on the infection mechanism of the fungus Clonostachys rosea against nematodes using the green fluorescent protein. Appl Microbiol Biotechnol 78:983–990

Chapter 6
Molecular Mechanisms of the Interaction Between Nematode-Trapping Fungi and Nematodes: Lessons From Genomics

Anders Tunlid and Dag Ahrén

Abstract Soil contains a diverse range of fungi that are parasites on nematodes. These fungi include the nematode-trapping fungi that are dependent on specific hyphal structures on or in which nematodes can be trapped mechanically or by adhesion. The interests of studying these fungi come from their potential use as biological control agents against plant- and animal-parasitic nematodes. Studies on the molecular mechanisms of the interaction between nematode-trapping fungi and nematodes were initiated already in the 1950s. Recently, the infection process, including the differentiation of trap cells and the penetration and digestion of nematodes, has been examined using tools of genomics and functional genomics. The results from these studies are reviewed. We discuss how genomic approaches can provide insights into the infection process, the environmental factors that influence survival and activity in soils, and the mechanisms that could account for the variation in parasitic activity within and between species of nematode-trapping fungi.

6.1 Introduction

Fungi which prey or parasitize on nematodes were first reported more than a century ago. The first attempt to use these fungi as biological control agents against plant-parasitic nematodes were in the late 1930s (Linford 1937). Although, the nematophagous fungi attracted the attention from numerous mycologists during the coming decades (Dudddington 1955; Drechsler 1941), the interest of using them for biological control was rather low until the 1980s. Research in this area was then reinforced as a response to the public demands for environmentally friendly ways to control plant-parasitic nematodes.

A. Tunlid (✉) and D. Ahrén
Department of Microbial Ecology, Ecology Building, SE 224 62 Lund, Sweden
e-mail: anders.tunlid@mbioekol.lu.se; dag.ahren@biol.lu.se

K. Davies and Y. Spiegel (eds.), *Biological Control of Plant-Parasitic Nematodes: Building Coherence between Microbial Ecology and Molecular Mechanisms*, Progress in Biological Control 11, DOI 10.1007/978-1-4020-9648-8_6,

There are more than 200 species of nematophagous fungi described. Based on the infection mechanism, they are commonly subdivided into three main groups: the nematode-trapping fungi that captures free-living nematodes using specialized morphological structures (i.e. traps), the endoparasitic fungi that infects nematodes using adhesive spores, and the egg- and cyst-parasitic fungi that infect these stages with their hyphal tips (Barron 1977). The nematophagous fungi that have received most attention for biological control of plant-parasitic nematodes include various species of the nematode-trapping fungi *Arthrobotrys* spp. (Stirling and Smith 1998), and the egg-parasitic fungi *Pochonia chlamydosporium* (Kerry 2001) and *Paecilomyces lilacinus* (Gaspard et al. 1990). In addition, nematode-trapping fungi, in particular strains of *Duddingtonia flagrans*, have successfully been used to control animal-parasitic nematodes (Larsen 2000).

Despite the fact that significant levels of control have been obtained in many experiments, biological control based on nematophogaous fungi have not yet become a reliable and economically viable method for nematode control in agriculture. There are many reasons for the limited success (Jansson and Lopez-Llorca 2004; Kerry 2000). Among them are the problem of producing commercially accepted formulations for field applications and the low ability of added fungi to become established and active in the complex soil environment. Furthermore, strains of nematophagous fungi can differ significantly in their capacity for biological control and survival in different habitats. Thus, a strain that can cause significant control in one trial may fail in another environment.

It has been proposed that a better knowledge on the infection process of nematophagous fungi could lead to the development of more efficient and reliable methods for biological control of parasitic nematodes (Tunlid and Ahrén 2001; Morton et al. 2004; Davies 2005). Through application of biochemistry and molecular biology, the molecular mechanisms of the interaction between nematophagous fungi and their hosts could be elucidated in detail. Based on such information, novel molecular markers that reflect the growth and parasitic activity of nematophagous fungi could be developed. With such markers in hand, the biotic and abiotic factors regulating the population levels and parasitic activity of nematophagous fungi in soils could be identified. A better understanding of the infection process can also be used to develop new screening procedures of nematophagous fungi with high and consistent biocontrol activity.

Studies on the molecular background to the interactions between nematophagous fungi and nematodes were initiated already in the 1950s. The pioneer investigations were done on nematode-trapping fungi, and this is still the best examined group when it comes to mechanistic studies at cellular and molecular levels. Recently, the infection process of nematode-trapping fungi has been examined using tools of functional genomics. In this paper, the result of these studies will be reviewed and their implications for understanding the ecology of nematode-trapping fungi and improving the biological control activity will be discussed. First, some general aspects of the biology of nematode-trapping fungi and earlier biochemical and molecular studies (done during the "pre-genomic era") will be presented.

6.2 Biology of Nematode-Trapping Fungi

Nematode-trapping fungi are found in all regions of the world, from the tropics to Antarctica. They are commonly found in soils and decaying leaf litter, decaying wood, dung, compost and mosses. When grown in soils, nematode-trapping fungi can grow as saprophytes as vegetative hyphae (mycelium). Traps are initiated either spontaneously or in response to signals from the environment, including peptides and other compounds secreted by the host nematode (Dijksterhuis et al. 1994). There is a large variation in the morphology of trapping structures, even between closely related species (Barron 1977) (Fig. 6.1). In some species, the trap consists of an erect branch that is covered with an adhesive material. In other species, such as the well-studied *Arthrobotrys oligospora*, the trap is a complex three-dimensional net. A third type of trap is the adhesive knob. The knob is a morphologically distinct cell, often produced on the apex of a slender hyphal stalk. A layer of adhesive polymers covers the knob, which is not present on the support stalks. Finally, there are some species of nematode-trapping fungi that capture nematodes in mechanical traps called "constricting rings". When a nematode enters this type of trap, the three ring cells are triggered to swell rapidly and close around the nematode (Barron 1977).

Despite the large morphological variation in trapping structures, phylogenies inferred from molecular data have shown that a majority of nematode-trapping fungi belong to a monophyletic group placed in the family of *Orbiliales*, Ascomycota. These studies have shown that the trapping mechanisms have evolved along two major lineages, one leading to the constricting rings, and the other into adhesive traps. Among species with adhesive traps, those with adhesive networks separated early from the species with adhesive knobs and branches (Liou and Tzean 1997; Ahrén et al. 1998; Hagedorn and Scholler 1999; Li et al. 2005; Yang et al. 2007b) (Fig. 6.1).

The phylogenetic relationships of the various types of nematode-trapping fungi are supported by ultra structural studies. For example, one feature, which is common to the traps of all adhesive species, is the presence of numerous cytosolic organelles, the so-called "dense bodies". Although the function of these organelles is not yet clear, the fact that they exhibit catalase and D-amino acid oxidase activity indicates that the dense bodies are peroxisomal in nature. Furthermore, adhesive trap cells have extensive layers of extracellular polymers, which are thought to be important for attachment of the traps to the surface of the nematode. The trap cells of the inflating constricting rings have a unique, highly ordered structure that are not observed in cells of adhesive traps (Dijksterhuis et al. 1994).

Detailed microscopic studies of net-forming species, in particular *A. oligospora*, by Birgit Nordbring-Hertz, Martin Veenhuis and colleagues, have revealed that the infection of nematodes occurs by a sequence of events (Dijksterhuis et al. 1994). Following a physical contact between the trap cells and the nematode, the nematodes become attached to the trap surface. In the electron microscope, the trap cells are surrounded by a layer of extracellular fibrils. After contact, these fibrils become directed perpendicularly to the nematode surface (i.e. cuticle). Subsequently, the fungus pierces the cuticle by forming a penetration tube. This step probably

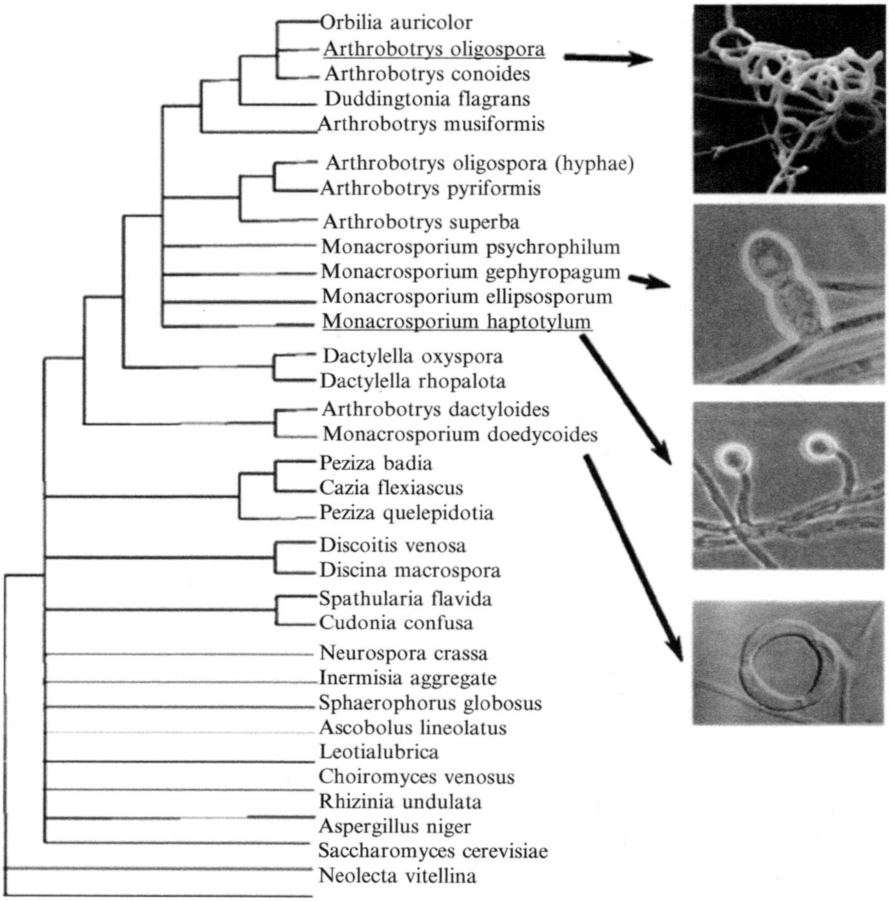

Fig. 6.1 A cladistic tree showing the relationship between various nematode-trapping fungi and other ascomycetes based on 18SrDNA sequences. Only branches with bootstrap support values above 50 are shown. Note that the nematode-trapping fungi form a monophyletic clade among an unresolved cluster of apothecial ascomycetes. The phylogenetic pattern within the clade of nematode-trapping fungi is concordant with the morphology of the traps (The tree is redrawn from Ahrén et al. (1998). The pictures are reproduced from Norbring-Hertz et al. (1995), courtesy of Birgit Nordbring-Hertz and IWF, Göttingen)

involves both the activity of hydrolytic enzymes solubilizing components of the cuticle and the activity of a mechanical pressure. Concomitant with penetration, the nematodes become paralyzed (immobilized). Once inside the nematode, the penetration tube swells to form an infection bulb from which trophic hyphae develops that colonizes the dead nematode. The infection bulb can be considered as an intermediate morphological structure, between the highly differentiated trap cells and the trophic hyphae which develops from the bulb. Upon maturation of the bulb the dense bodies are degraded. At the same time, the endoplasmic reticulum proliferates extensively. When the nematode cavity becomes invaded by the trophic hyphae, the internal tissues of the nematode are rapidly degraded. Some of the nematode content is converted to lipid droplets, which can be metabolized to support growth of new vegetative mycelium that develops outside the nematode. The infection process is usually completed in 48–60 h (Dijksterhuis et al. 1994).

6.3 Biochemical and Molecular Studies

6.3.1 Peptides Inducing Trap Formation

Many nematode-trapping fungi only form trapping structures after induction by external stimuli like living nematodes, while others produce traps spontaneously. Already, in the late 1950s, David Pramer and co-workers showed that the external signals included substances released from living nematodes. A compound named "nemin" that stimulated formation of traps (nets) by *Arthrobotrys conoides* was extracted from culture filtrates of nematodes (Pramer and Stoll 1959). Subsequent experiments showed that nemin is of peptide nature and that it is a mixture of different compounds (Pramer and Kuyama 1963). Further insights into the structure of nemin were obtained by Nordbring-Hertz and colleagues. They isolated and characterized a number of peptides from enzyme-hydrolyzed casein which significantly stimulated trap formation in *A. oligospora*. The active peptides had a high proportion of non-polar and aromatic amino acid residues and such peptides were also detected in exudates from the nematode *Panagrellus redivivus* (Nordbring-Hertz 1973; Nordbring-Hertz and Brinck 1974; Nordbring-Hertz 1977). A prerequisite for the morphogenic activity was that the fungus was grown on low-nutrient mineral salts medium which suggests that the trap formation in *A. oligospora* is favoured during nutrient limiting conditions (Nordbring-Hertz 1973).

The fact that trap formation could be stimulated by the presence of specific peptides have been used for developing a method for growing trap-containing mycelium of *A. oligospora* in liquid cultures (Friman et al. 1985). The procedure has been used in numerous studies for obtaining enough biomass of *A. oligospora* for the purification and characterization of proteins that are produced during the infection of nematodes. Among these proteins are lectins and extracellular serine proteases.

6.3.2 Lectins – Recognition and Storage of Nutrients

Lectins are carbohydrate-binding proteins that function as recognition molecules in cell-molecule and cell-cell interactions in a variety of organisms (Sharon and Lis 2004). The involvement of lectins in the interaction between nematode-trapping fungus and nematodes was proposed by Nordbring-Hertz and Mattiasson (1979). Based on sugar-inhibition experiments, it was suggested that the infection process in *A. oligospora* is initiated by a binding between an N-acetylgalactosamine (GalNac)-specific lectin present on the traps and a carbohydrate ligand found on the nematode surface (Nordbring-Hertz and Mattiasson 1979). However, the specificity for GalNac was not complete, which was also shown when red blood cells (RBC) were used as model prey. RBC type A which are characterized by a terminal GalNac, tended to adhere more easily than did Types B and O. Subsequently, a GalNac-binding protein was isolated from *A. oligospora* by affinity chromatography (Borrebaeck et al. 1984; Premachandran and Pramer 1984). Pre-treatment of nematodes with the lectin purified by Premachandran and Pramer (1984) reduced the entrapment, which further supported the action of protein-carbohydrate recognition event. Evidences for lectin-mediated interactions were also obtained in several other species of nematode-trapping fungi. The binding specificities of these lectins appear to be different (Nordbring-Hertz and Chet 1988).

To gain a more detailed understanding on the role of lectins in the capture of nematodes, larger amount of a lectin named AOL was purified and characterized from *A. oligospora* (Rosén et al. 1992). AOL is an abundant, saline-soluble, dimeric protein with a molecular mass of 36 kDa. AOL is most probably identical to the previously identified GalNac-binding protein of *A. oligospora* (Rosén et al. 1992). The gene encoding AOL was cloned, and the deduced primary sequence showed a high sequence similarity to a lectin (designated ABL) from the basidiomycete *Agaricus bisporus* but not to any other fungal, plant or animal lectins (Rosén et al. 1996b). Furthermore, AOL and ABL have similar binding specificities. Both proteins are multispecific lectins binding to N- and O-linked oligosaccharide chains of glycoproteins (including Galβ1-3GalNAcα-, the Thomsen-Friedenreich or T antigen), sulfated glycoconjugates and some phospholipids (Rosén et al. 1996a).

The observations that AOL can be abundant in both trap-containing and vegetative mycelium and that AOL has similar structure and binding specificity to a lectin (ABL) from a non-parasitic fungus, suggest that these lectins can have functions not only related to parasitic growth. Based on a study on the expression and localization of AOL during a number of different growth conditions, it was demonstrated that AOL can function as a storage protein (Rosén et al. 1997). During saprophytic growth in liquid cultures, the levels of AOL were regulated depending on the C/N ratio of the medium. In media with low C/N ratios, AOL comprised a large fraction (5–20%) of the soluble proteins present in the mycelium. Under conditions of nitrogen starvation, AOL was preferentially degraded to a higher degree compared with other proteins. During infection of nematodes, AOL is rapidly synthesized in *A. oligospora*, once nematodes have been penetrated and digestion started. Large amounts of AOL accumulate in the trophic hyphae growing inside the nematode. Later, the lectin is transported to other parts of the mycelium, where it can degrade and supports the growth of the fungus (Rosén et al. 1997).

The fact that AOL binds to sugar structures that are common in glycoproteins of animals, but not of fungi, suggests that the lectin can have a role in the interaction with nematodes as first proposed by Nordbring-Hertz and Mattiasson (1979). This hypothesis was examined by constructing AOL deletion mutants using a transformation system for *A. oligospora* (Tunlid et al. 1999; Balogh et al. 2003).

Notably, there was no significant difference between the deletion mutants and the wild-type strain in the ability to infect nematodes. Furthermore, the deletion mutants did not express any other lectins with hemagluttinating activity. Thus, saline soluble lectins like AOL do not appear to have a major role in the infection of nematodes. However, it cannot be ruled out that *A. oligospora* produces GalNac binding proteins that cannot be extracted with the saline buffers used for isolating AOL.

AOL and ABL are members of a growing family of T antigen specific lectins that are unique to fungi (Carrizo et al. 2005; Leonidas et al. 2007; Birck et al. 2004; Iijima et al. 2002). Recent studies on some of these lectins imply that AOL can bind to endogenous receptor molecules, and have other functions apart from being a storage protein. ABL has anti proliferative effects on human cancer cells, and in such cells, ABL selectively block the nuclear localization sequence (NLS) – dependent nuclear protein import system (Yu et al. 2002). Notably, AOL is located in the cytoplasm as well as in the nucleus (Rosén et al. 1997). Another function has been suggested for the lectin of the soil borne pathogen *Scelrotium rolfsii*. This lectin is developmentally regulated and expressed at high levels in sclerotia. At maturation, the lectin associates with cell wall-associated endogenous receptors. Preliminary characterization indicates that these receptors are glycosphingolipids, and it was proposed that the lectin-receptor complex function as signalling molecule in the germination of sclerotia (Swamy et al. 2004).

6.3.3 Proteases – Penetration and Digestion

The nematode cuticle consists mainly of proteins including collagens (Cox et al. 1981), and it can be assumed that nematode-trapping fungi penetrate the host cuticle with the aid of extracellular proteases. The first report on the production of proteases by nematode-trapping fungi was by Schenck et al. (1980). They partially purified a protease from a culture filtrate of *Arthrobotrys amerospora* that could hydrolyze components in collagen prepared from different organisms like chicken, fish and earthworms. Studies on the structure and function of extracellular proteases from nematode-trapping fungi continued during the 1990s. Bioassay experiments showed that the paralysis of captured nematodes by *A. oligospora* was severely affected by treating trap-bearing mycelium with inhibitors against serine proteases. A limited effect was observed using inhibitors of metalloproteases, but none of cysteine or aspartic proteases (Tunlid and Jansson 1991). The serine protease inhibitors did not affect the adhesion of the nematodes to the traps. Microscopic studies have indicated that nematodes captured by *A. oligospora* become paralysed (immobilized) at the time when the fungus starts to penetrate the nematode cuticle (Dijksterhuis et al. 1994). Accordingly, the added serine protease inhibitors most probably affect the stage of penetration.

This hypothesis was further examined by isolating and characterizing extracellular proteases from *A. oligospora*. When grown in liquid culture allowing the formation of nematode traps, *A. oligospora* produces at least two different serine proteases (designated PI and PII). In bioassays, PII, but not PI, immobilized free-living nematodes which suggested a role for this protease in infection. PII was purified from the culture filtrate. The enzyme belongs to the subtilisin family of serine proteases, it has a molecular weight of ca. 35 kDa, and can digest a number of different protein substrates including proteins present in the cuticle of the nematode *Panagrellus redivivus* (Tunlid et al. 1994). The *PII* gene displays significant sequence similarity to that of other fungal subtilisins (Åhman et al. 1996). Furthermore, during saprophytic growth, the expression of *PII* is induced by various soluble and insoluble protein sources. The expression is repressed in the presence of more easily metabolized forms of nitrogen like ammonia, nitrate and amino acids. Thus, the expression of PII is controlled by carbon and nitrogen metabolite repression, in a similar way as identified for subtilisins isolated from other ascomycetes (Åhman et al. 1996).

During the infection of nematodes, the *PII* gene is expressed by *A. oligospora* during the early phase of colonization, but high levels are also expressed at later stages when the nematode is digested by the trophic hyphae (Åhman et al. 2002). The importance of the activity of PII for the pathogenicity of *A. oligospora* was further studied by constructing various *PII* mutants (Åhman et al. 2002). *PII* deletion mutants were not significantly affected in pathogenesis. This could be explained by the presence of a significant residual proteolytic activity (including that of PI) in the deletion mutants. However, the role of PII in virulence was demonstrated by analyzing mutants containing additional copies of the PII gene. Multi-copy mutants had an increased speed of capture and killing nematodes compared to the wild type. The increased virulence of these mutants could be accounted for by at least two different mechanisms: First, the multi-copy mutants developed a higher number of traps than the wild type. The stimulation of trap formation can be explained by a rapid digestion of the infected nematodes by over-expressed PII. Second, the multi-copy mutants have enhanced nematotoxic activity. Such an activity was verified by demonstrating that heterologous-produced PII (in *Aspergillus niger*) immobilized free-living nematodes in bioassays (Åhman et al. 2002).

The factor(s) that contributes to the comparable rapid killing of nematodes captured by nematophagous fungi has intrigued scientists for a long time. Early studies indicated that toxic fungal metabolites might be involved in the immobilization of nematodes by *A. oligospora* (Olthof and Estey 1963). Based on experiments demonstrating that only extracts from nematodes infected by the fungus, but not extracts from the fungus or nematodes alone, were toxic to nematodes, it was proposed that *A. oligospora* secretes a nematotoxic substance which paralyzes nematodes following the capture and penetration of them (Olthof and Estey 1963). This paper stimulated a number of studies trying to identify toxic metabolites from nematode-trapping fungi. The only such compound yet identified are fatty acids including linoleic acid (Stadler and Sterner 1993). The studies of PII show that not only metabolites but also a protease of nematode-trapping fungi can have nematotoxic activity. In agreement with the observations of the killing of nematodes by Olthof and Estey (1963), the nematotoxic activity of PII is produced by the fungus after the nematodes are caught by the adhesive trapping nets. Experiments with the recombinant PII showed that its nematotoxic activity is significantly higher than that of other commercially available serine proteases, which suggest that this enzyme has some specific properties that contribute to its toxicity (Åhman et al. 2002).

During the last years, a number of extracellular subtilisins have been isolated and characterized from nematode-trapping fungi (Fig. 6.2). The biochemical properties

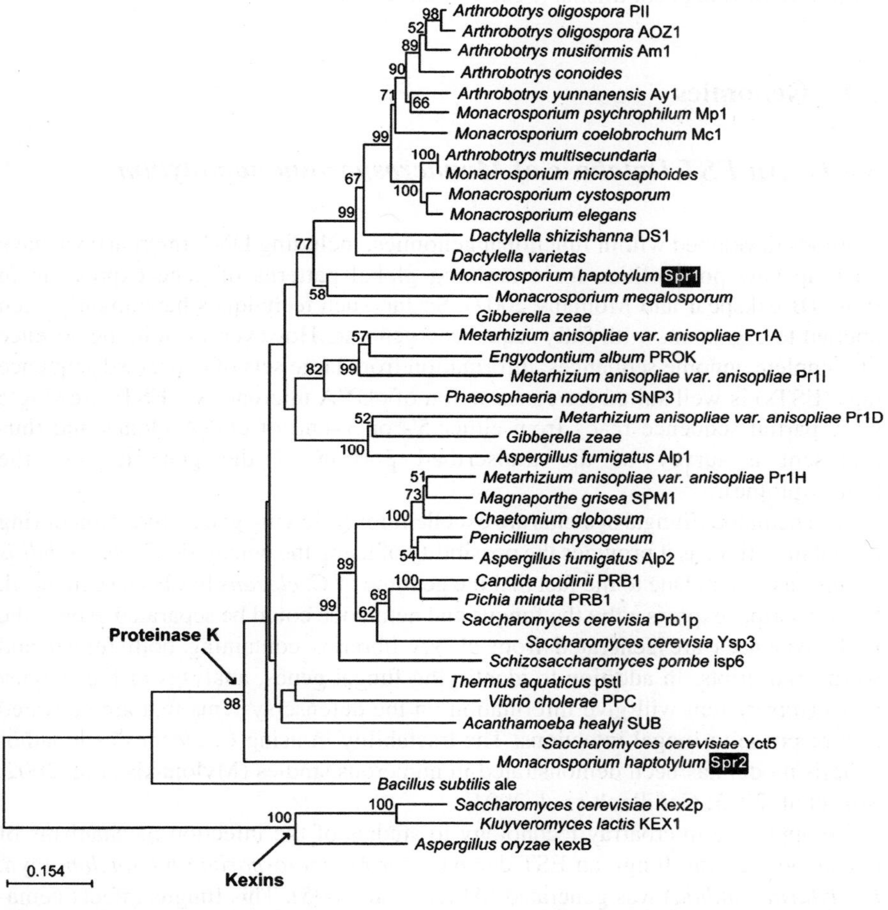

Fig. 6.2 Phylogenetic relationships of subtilisins identified in nematode-trapping fungi. The analysis was done by aligning the conserved peptidase_S8 domains (PF00082) using the neighbor-joining method. The values at the nodes indicate the bootstrap support in percent for 1,000 replicates (>50 are shown). The kexins group was used as outgroup to root the tree. Accession numbers for the above proteins from top to bottom are as follows: CAA63841.1, AAM93666.1, ABL74282.1, AAX54903.1, ABL74283.1, ABL74284.1, ABL74285.1, ABK54363.1, AAW21809.2, AAX54901.1, AAX54902.1, ABL74286.1, ABF72192.1, EF681769 (Spr1), BAD44716.1, XP_389558.1, CAC95049.1, P06873, CAB64346.1, AAP30889.1, CAB89873.1, XP_380982.1, XP_751651.1, CAD13274.1, BAB63284.1, XP_001222475.1, AAG44693.2, XP_753718.1, BAC75710.1, BAE91901.1, NP_010854.1, NP_014645.1, NP_593815.1, P08594, EAX56701.1, AAF32368.1, P25381, EF681770 (Spr2), AAA87324.1, NP_014161.1, XP_453942.1, BAC66791.1. Spr1 and Spr2 are two subtilisins identified in *M. haptotylum* (The figure is reprinted from Fekete et al. (2008) with permission)

and the primary structure of some of them have recently been reviewed (Yang et al. 2007a). In short, these subtilisins have similar molecular weights; they can degrade a broad range of protein substrates, and display a high sequence identity. A phylogenetic analysis of the subtilisins of nematode-trapping fungi shows that all of them, apart from a subtilisin recently isolated from *Monacrosporium haptotylum* (spr2, see below), cluster into a well-resolved clade among the proteinase K-like subfamily of fungal subtilisins (Fekete et al. 2008).

6.4 Genomics

6.4.1 An EST Database of Monacrosporium haptotylum

Methods developed within functional genomics, including DNA microarrays, have open-up new possibilities for examining global patterns of gene expression in fungi (Breakspear and Momany 2007). So far, such techniques have mainly been applied to organisms with fully sequenced genome. However, even in the absence of complete genome sequences, information from large sets of expressed sequence tags (ESTs) is well suited for construction of cDNA microarrays. ESTs are single pass, partial sequence reads from either 5′- or 3′-end of cDNA clones and thus represent a survey of the transcribed portion of the genome (i.e. the transcriptome).

The nematode/fungal systems are excellent for following gene expression during fungal infection, as it provides the possibility of using the nematode *Caenorhabditis elegans* as a host. Due to the fact that the genome of *C. elegans* has been sequenced, the transcripts expressed by the fungus and nematode could be separated even if the EST sequences are generated from cDNA libraries containing both fungal and worm transcripts. In addition to identifying fungal genes, analyses of the fungus/ *C. elegans* system will give information on the defense systems that are activated as a response to fungal infections. The tractability in using *C. elegans* as a pathogenesis model has been demonstrated in numerous studies (Mylonakis et al. 2002; Sifri et al. 2005; Couillault et al. 2004).

To apply the microarray technology to studies of the infection mechanisms of nematode-trapping fungi, an EST database for *Monacrosporium haptotylum* (syn. *Dactylaria candida*) was generated (Ahrén et al. 2005). This fungus infects nematodes using an adhesive knob (c.f. Fig. 6.1). The advantage of using *M. haptotylum* is that during growth in liquid cultures with heavy aeration, the connections between the traps (knobs) and mycelium can be broken easily and the knobs can be separated from the mycelium by filtration (Friman 1993). The isolated knobs retain their function as infection structures, i.e. they can "capture" and infect nematodes. We constructed four directional cDNA libraries from mycelium, knobs, and knobs infecting *C. elegans* for 4 and 24 h, respectively. In total, 8,463 ESTs were sequenced from the four cDNA libraries. The sequences were assembled into 3,121 contigs that putatively represent unique genes/transcripts. Between 5% and 37% of the assembled sequences displayed a high degree of similarity to sequences in the GenBank nr protein database. Based on the information obtained through these searches, ESTs were assigned into various functional categories. A large fraction (38–60%) of the assembled sequences was orphans that showed no homology to protein sequences in the GenBank database. A cDNA microarray was constructed by amplifying and spotting 3,518 EST clones including 2,822 of fungal, 540 of *C. elegans* and 156 of unknown origin (Ahrén et al. 2005).

6.4.2 *Gene Expression in Trap Cells Versus Vegetative Mycelium*

In the first cDNA array study, gene expression in trap cells were compared with that in vegetative mycelium (Ahrén et al. 2005). Despite the fact that the mycelium and knobs were grown at identical conditions, there were substantial differences in the patterns of genes expressed in the two cell types. In total, 23.3% (657 of 2,822) of the EST probes (putative genes) were differentially expressed in knobs versus mycelium. The proposed functions of the genes that were significantly up or down regulated in the knobs differed. Genes with putative roles in "transcription", "cellular transport and transport mechanisms", "cellular communication or signal transduction", "cell rescue, defence, cell death and ageing" and "cell growth, cell division and DNA synthesis" were expressed at lower levels in the knobs than in the mycelium (Fig. 6.3).

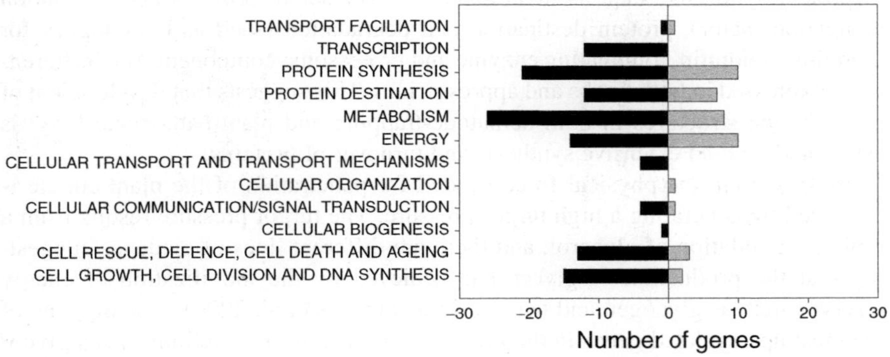

Fig. 6.3 The number of *down* (*left*) and *up* (*right*) regulated genes in knobs versus mycelium classified into various functional categories. Data are derived from cDNA microarray analysis using a significance level of $P < 0.05$. Annotation of putative functional roles is based on sequence similarity to information in the GenBank nr protein database. Of a total of 398 genes being regulated, 123 genes were annotated into the displayed functional categories, whereas the remaining (not shown) were either annotated as unclassified proteins (179 genes) or orphans with no sequence homology (96 genes). Among a total number of 259 up regulated genes, 43 were assigned functional annotation, whereas 119 clones were assigned as unclassified and 97 genes as orphans (The figure is reprinted from Ahrén et al. (2005) with permission)

The knob of *M. haptotylum* is a spherical cell that develops at the tips of an apically growing hyphal branch. The change in morphology represents a shift in the polarity of the cells. Notably, several of the genes being differentially expressed in knobs displayed sequence similarities to genes known to be involved in regulating morphogenesis and cell polarity in fungi. Among them were the actin-binding proteins profilin and cofilin, and several small GTPases including homologs of *rho*1, *rac*1 and *ras*1(Ahrén et al. 2005). Several of these genes were also differentially expressed during the infection of nematodes which suggests that they are important for regulating the morphogenetic changes occurring during penetration and digestion of the nematode (Fekete et al. 2008).

There are several similarities in the structure and function of knobs of nematode-trapping fungi and that of appressoria formed by plant-pathogenic fungi. Like a knob, an appressorium is a specialized infection structure, which develops as a spherical cell at the tip of a hypha (germ tube). Both structures contain an adhesive layer on the outside, which binds to the surface of the host. Furthermore, both appressoria and knobs form a hypha that penetrates the host using a combination of physical force and extracellular enzymatic activities (Tucker and Talbot 2001). Comparison of data from the transcriptional profiling of knobs in *M. haptotylum* and those of appressoria in *Magnaporthe grisea* and *Blumeria graminis* shows that there are many similarities in the patterns of gene regulation (Takano et al. 2004; Thomas et al. 2002; Rauyaree et al. 2004). For example, genes involved in stress and defence responses are one of the largest classes of genes that are differentially expressed during appressoria formation. Several such genes, including cyclophilins, peptidyl-prolyl cis-trans isomerases, metallothionein and thioredoxins, were differentially expressed in the knobs of *M. haptotylum*. A number of genes involved in protein synthesis (such as homologues for ribosomal proteins and translation elongation factor), protein destination and degradation (such as homologues for ubiquitin, ubiquitin-conjugating enzyme and proteasome components) are differentially expressed in both knobs and appressorium. This suggests that development of the infection structures in both nematode-trapping and plant-pathogenic fungi is associated with an extensive synthesis and turnover of proteins.

In *M. grisea*, the physical force needed for penetration of the plant cuticle is produced by generating a high turgor pressure. The turgor pressure results from a rapid accumulation of glycerol, and there are different lines of evidence suggesting that the production of glycerol is achieved by the mobilization of energy reserves such as glycogen and neutral lipids (Thines et al. 2000). Notably, one of the most up-regulated genes in the knobs, as compared to mycelium, was a glycogen phosphorylase (*gph*1) gene homologue. This enzyme catalyzes and regulates the degradation of glycogen to glucose-1-phosphate. This product is further metabolized in the glycolytic pathway and glycerol can be synthesized from several of the intermediates in this pathway. Notably, *gph*1 is among the most down-regulated genes following the infection of nematodes (Fekete et al. 2008). The pattern of regulation of the *gph*1 gene suggests that the penetration of nematodes by *M. haptotylum* could involve the action of a turgor pressure generated from the degradation of glycogen.

6.4.3 Shifts in the Fungal Transcriptome During Infection

In a more recent study, the changes in the transcriptome of *M. haptotylum* during the adhesion, penetration and digestion of the nematode *C. elegans* were examined. Worm genes that were activated in response to the fungal infection were also analyzed (Fekete et al. 2008). In the experiments, knobs were added to L1 larvae of *C. elegans*. The infection of nematodes was followed under a light microscope (Fig. 6.4a). The immobilization of nematodes was fairly synchronized: approximately 40% of the nematodes were paralyzed (immobilized) after 4 h, 70% after 16 h, and 80% after 24 h of infection (Fig. 6.4b). In comparison, approximately 90% of the nematodes were alive when incubated axenically for 24 h. Samples for cDNA microarray hybridizations were taken after 1, 4, 16 and 24 h of infection. The experimental design also included reference material of *M. haptotylum* and *C. elegans* grown axenically in parallel during the time course, with the purpose of detecting differentially regulation of genes not related to the infection process.

A significant fraction of the 2,684 *M. haptotylum* and 372 C. *elegans* gene reporters that were analyzed on the array was regulated during the infection. Hence, 1,562 of the fungal, and 299 of the worm reporters differed in the expression levels in at least one of the pair-wise comparisons made between the infected and non infected (reference) material. Analysis of the differentially expressed genes showed that there were dramatic changes in the transcriptome of *M. haptotylum* during the different stages of the infection. An initial transcriptional response was recorded after 1 h of infection when the traps adhered to the cuticle, but before the immobilization of the captured nematodes. The most highly up-regulated gene at this stage was a subtilisin that displayed a high sequence similarity to *PII* of *A. oligospora*. This gene (designated *spr*1) had a similar expression patterns as *PII*. The *spr*2 mRNA was up-regulated at 1 h, then down-regulated at 4 and 16 h, and then up-regulated at 24 h when the fungus was digesting the killed nematode. Up-regulated at 1 h was also a gene encoding another extracellular subtilisin (*spr*2) that had a similar expression profile as *spr*1. Although spr1 and spr2 had low sequence similarity (36%), a phylogenetic analysis showed that both of them belong to the proteinase K-like subfamily of subtilisins (Siezen and Leunissen 1997) (Fig. 6.3). Spr1 clustered into a clade containing subtilisins isolated from other species of nematode-trapping fungi including PII. Spr2 was distantly related to these proteases and other yet characterized fungal subtilisins (Fekete et al. 2008). Apart from *spr*1 and *spr*2, no other serine protease-encoding genes were identified among the ESTs of *M. haptotylum.* Thus, nematode-trapping fungi do not appear to have a large number of subtilisins similar to that observed in the entomopathogenic fungus *Metarhizium anisopliae*, which is among the best studied fungal pathogen of invertebrates (Bagga et al. 2004).

Several of the *M. haptotylum* genes that were regulated during the initial stage of the infection have homologs in phytopathogenic fungi which are known to be regulated during plant infection. Among them were two genes (designated *CFEM*1 and *CFEM*2) that encoded proteins with a CFEM domain. This is an eight cysteine-containing domain that is unique for fungal proteins (Kulkarni et al. 2003). CFEM1

a

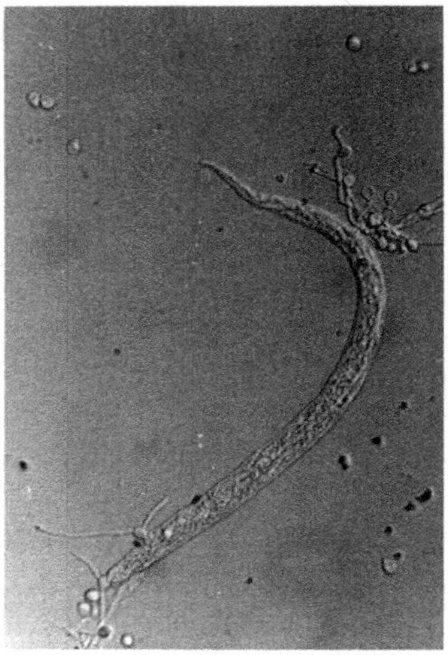

b

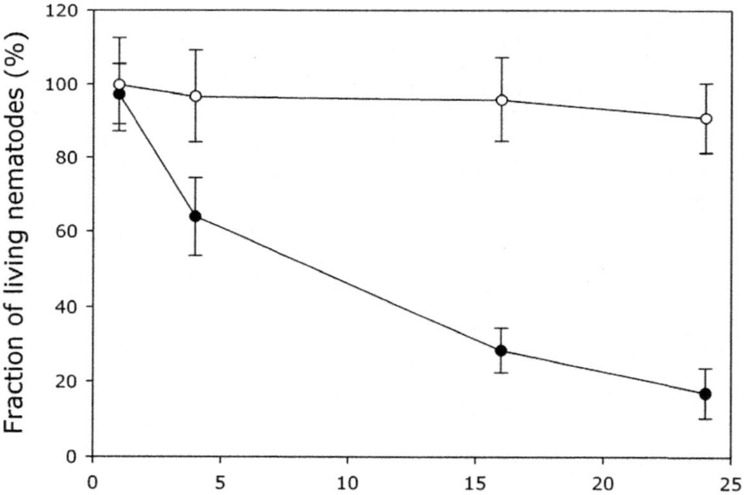

Fig. 6.4 Infection of the nematode *C. elegans* by the nematode-trapping fungus *M. haptotylum*. (**a**) Micrograph showing an infected nematode 24 h after adding the fungus. The nematode is infected by knobs adhered to the cuticle of the nematode. The fungus has penetrated the nematode and trophic hyphae are growing inside the killed nematode. Nutrients obtained from the partly degraded nematode support growth of external hyphae present outside the infected nematode. (**b**) The killing of *C. elegans* following the infection by *M. haptotylum*. Nematodes were considered as killed when they were not moving. Data (mean and SD) from three independent infection (●) and control (○) (axenically incubated nematodes) experiments are shown (The figure is reprinted from Fekete et al. (2008) with permission)

and CFEM2 displayed sequence similarities to several fungal proteins that have been proposed to have important roles during fungal pathogenesis. The *M. grisea* Pth1 protein is required for appressorium development (DeZwaan et al. 1999). ACI1 is an adenylate cyclase interacting protein which is a key regulator of appressoria formation (Choi and Dean 1997). The *Candida albicans* CSA1 protein and the *Coccidioides posadasii* proline-rich antigen are highly expressed in stages that are associated with infection (Peng et al. 1999).

By clustering the expression profiles, it was possible to identify a cohort of 372 genes that were transiently up-regulated at 4 h, concomitant with the onset of the paralysis of the captured nematodes. A large proportion (79%) of the genes in this cohort was orphans, i.e. they did not show any significant homology to gene and proteins in other organisms. Thus, the genes appear to be specific for *M. haptotylum*. To gain some more information on these infection-regulated orphans, the full-length cDNAs of 21 orphans were sequenced. These genes were of two different classes; those translating into presumably functional peptides and those with no protein coding potential. Transcription of non-coding RNAs has recently been identified in several eukaryotes including *Saccharomyces cerevisiae* and *Schizosaccharomyces pombe* (Miura et al. 2006; Wilhelm et al. 2008). The pool of non-coding RNAs includes transcripts derived from antisense strands, sense transcripts from internally transcribed open reading frames, and transcripts from intergenic regions. The biological roles of these non-coding RNAs are largely unknown, but the transcripts of many of them are regulated depending on the growth conditions.

Among the infection-regulated orphans were several transcripts that translated into short peptides (26–80aa). Recently, there has been a large interest in the role of small, secreted proteins for the virulence of plant pathogenic fungi (Rep 2005). Most of these proteins are highly species-specific, and the expressions of the corresponding genes are in many cases induced after infection. Although, the biochemical functions for most of these proteins are not known, a few of them have been shown to have a role in manipulating the host cell structure and function (Rep 2005; Kamoun 2006). Whether the early expressed orphans of *M. haptotylum* represent genes encoding short peptides with similar functions as proposed for the small secreted proteins in plant-pathogenic fungi deserves further studies.

6.4.4 Defence Response in C. elegans

In contrast to the patterns observed in *M. haptotylum*, no distinct shifts occurred in the global expression profile of the *C. elegans* genes when comparing the transcriptome at 1 and 4 h of the infection. Thus, a majority of the genes that were regulated at 1 h were also regulated at 4 h. In total, 177 of the *C. elegans* genes were regulated more than twofold at 1 and 4 h after infection.

C. elegans mount protective responses against bacterial and fungal pathogens by activating several intracellular signalling pathways. These pathways in turn mediate up-regulation of the expression of defensive gene products (effector molecules) which serve to limit the infection or destroy invading microorganisms (Gravato-Nobre and Hodgkin 2005). There are at least six different signal transduction cascades identified in *C. elegans* that may be activated in response to pathogen attack. One of them is the DAF2/DAF16 pathway (Gravato-Nobre and Hodgkin 2005). One of the targets of this pathway is members of the *dod* gene group. We identified *dod*-3 amongst the most up-regulated genes in the infected nematodes. Two other genes were identified (F31C3.6 and *sip*-1) that are known to be regulated by the transcription factor DAF16. A previous microarray study showed that a family of neuropeptide-like peptides was induced in *C. elegans* following the infection by the endoparasitic nematophagous fungus *Drechmeria coniospora* (Couillault et al. 2004). Indeed, such peptides including nlp-24, nlp-29 and nlp-34 are encoded by genes being highly up-regulated in *C. elegans* after infection by *M. haptotylum* (Fig. 6.4). Another peptide-encoding gene that was up-regulated in responses to both *D. coniospora* and *M. haptotylum* is the *cnc*-4 (CaeNaCin-4 (*Caenorhabditis* bacteriocin-4). Other strongly induced genes were two genes (*lec*8 and *lec*10) encoding galectins (galactose binding proteins). Members of this family are also regulated during bacterial infection of *C. elegans* (Mallo et al. 2002; Gravato-Nobre and Hodgkin 2005).

The *C. elegans* genome encodes 135 proteins containing the C-type lectin-like domain (CTLD) (Stein et al. 2003). C-type lectins have been proposed to have an important role in the defense response of *C. elegans* (O'Rourke et al. 2006). They might act as pathogen-recognition molecules, or may mask the virulence factors of the pathogen. Previous studies have shown that CTLD-containing genes are induced by the infection of bacteria (Mallo et al. 2002; O'Rourke et al. 2006). In contrast, all three genes encoding CTLD proteins identified in our study were all down-regulated during infection.

6.5 Linking Insights from Genomics to Organismal Biology and Ecological Research

6.5.1 Understanding the Infection Process

The exploitation of functional genomics and genomics in gaining insights into the infection biology of nematode-trapping fungi is in its infancy. Although, the cDNA microarray experiments of *M. haptotylum* have shown that several hundreds of genes are regulated during infection, there are major difficulties in interpreting this data due to the limited molecular background information of nematode-trapping fungi. A common approach to predict gene function in such organisms is to search for homologs in other organisms in which large numbers of genes have been functionally characterized. The rationale is that the rate of evolution of many genes with respect to both sequence and function has been so slow that characterization in one organism can suffice for many. Using this approach, the microarray data indicate that genes with roles in signal transduction pathways, stress and defence responses, protein synthesis/turnover and metabolism are differentially expressed in *M. haptotylum* during the infection of nematodes. In addition, the analysis suggests that many of the molecular components of the infection mechanisms of nematode-trapping fungi are similar to those in other well-characterized plant- and animal-parasitic fungi.

However, a majority of the infection-regulated genes in *M. haptotylum* display a low sequence similarity to genes in other organisms and their function must be examined by experimental methods. It must also be remembered that expression profiling using DNA arrays assays functionality in an indirect way. Transcripts are only transmitters of the instructions for synthesizing proteins, while it is the proteins and metabolites that are the functional entities in the cell. Thus, the function of the infection-regulated genes in *M. haptotylum* should be analyzed using a combination of different approaches including analysis of mutants, examination of proteins and recombinant gene products, and cellular localization of proteins. Notably, transformation protocols are available for nematode-trapping fungi that can be used for the construction of knock-out and over-expressing mutants (Tunlid et al. 1999; Jin et al. 2005). During the last years, numerous novel methods have also been developed that make it possible to perform functional analyses on larger scales than previously anticipated, and some of them can readily be adopted for studies on the infection biology of nematode-trapping fungi. For example, the first proteomic analysis of a nematode-trapping fungus was recently published (Khan et al. 2008). The proteome of the mycelium of the knob-forming fungus *Monacrosporium lysipagum* was analyzed using 2D gel electrophoresis and mass spectrometry. Out of the 250 proteins analyzed by 2D gel electrophoresis and mass spectrometry, 51(20%) were identified by cross-species matches.

6.5.2 Interactions with Other Organisms

A majority of the molecular studies of the interactions between nematode-trapping fungi and nematodes have been done in axenic systems. Experiments in such systems might overlook important aspects on the biology of nematode-trapping fungi that are significant for their activity and survival in natural environments. A striking example is the observation that *A. oligospora*, when grown in natural substrates such as cow dung and rhizosphere soils, does not only trap nematodes using three-dimensional nets but also by so-called conidial traps (Dackman and Nordbring-Hertz 1992). These traps are formed directly upon germination of conidia without an intermediate hyphal phase. Conidial traps (CTs) have been observed in several species of nematode-trapping fungi (Barron 1977; Persmark and Nordbring-Hertz 1997). CTs do not appear to be developed in response to nematodes. They are produced when conidia are germinated in soil extracts. Experiments have shown that the induction is completely lost when microorganisms are removed from the extracts and that the formation can be increased by pre-incubating the soil or soil extracts before adding conidia to the extracts. These results suggest that a certain level of competition for nutrients by microorganisms is necessary for CT formation. The ability to form CTs may significantly increase the survival potential and parasitic activity of nematode-trapping fungi in natural soils where germination and growth of fungi are commonly suppressed by fungistasis (Dackman and Nordbring-Hertz 1992; Persmark and Nordbring-Hertz 1997).

Another interaction that might influence the survival of nematode-trapping fungi in soils is the parasitism of other fungi (mycoparasitism). Thus, several species of nematode-trapping fungi, including *A. oligospora*, may form hyphal coil around the hyphae of other fungi. At the site of coiling, the host cell wall is partly degraded and the cell content is taken up by the attacking fungus (Persson et al. 1985). Nematode-trapping fungi are also capable of morphogenetic responses towards plant roots, resulting in the formation of appressoria (Bordallo et al. 2002). It was demonstrated that *A. oligospora* can colonize barley roots by growing inter- and intracellularly, and forming appressoria during penetration of the plant cell walls.

The above studies clearly demonstrate that nematode-trapping fungi display a large variety of morphological adaptations in the response to other organisms that are most probably crucial for their survival in soils. Accordingly, studies of the interaction between nematode-trapping fungi, nematodes, plants and other microorganisms are needed in more complex laboratory settings and soil microcosms. There are several difficulties associated with the application of DNA microarrays to mRNA analysis in such samples like low levels of high-quality mRNA and cross-hybridization from transcripts of non-target organisms. However, recent studies have demonstrated that cDNA microarrays can be used for analyzing gene expression patterns of fungi growing and interacting with other organisms in soil microcosms (Wright et al. 2005). Hence, we foresee that the cDNA microarray technology will provide insights into common mechanisms as well as the unique adaptations that are expressed in the interaction between nematode-trapping fungi, nematodes, and other organisms in soils.

6.5.3 Evolutionary Genomics

As discussed above, the commonly used methods for analyzing microarray data favour the analysis on genes, proteins and pathways that are conserved among organisms. With such approaches, it might be difficult to get insights into the unique molecular mechanisms that could account for the parasitic activity of nematode-trapping fungi. To elucidate such mechanisms, an evolutionary approach might be rewarding. Molecular phylogenies strongly support that the nematode-trapping life-style evolved once in the *Orbiliales*. Subsequently, several distinct trapping types evolved (Fig. 6.1). On the genomic level there are basically three compatible mechanisms that could account for such evolutionary patterns. They are as follows: first, parasitism is associated with the presence of novel genes. Such genes could be acquired by gene duplication or horizontal gene transfer. Second, adaptations to the parasitic habit may result from the differences in the regulation of gene expression. Third, parasitism is associated with gene loss and deletions (Tunlid and Talbot 2002).

The findings of genes in the EST database of *M. haptotylum* that exhibit no significant similarity to protein sequences in other organisms suggest that nematode-trapping fungi possess a set of unique genes. Notably, the expression analysis showed that some of these orphans appeared to be distinctly regulated during the penetration and paralysis of nematodes. Further insights into the genomic mechanisms associated with the evolution of the nematode-trapping habit will come from the analysis of complete genome sequences. With the arrival of the high throughput sequencing technology such as 454 pyrosequencing and Solexa, we anticipate that genome sequences of nematode-trapping fungi will be available in the near future. Comparisons of these sequences with those of other fungi will provide an unparalleled opportunity to develop a deeper understanding of the processes by which nematode-trapping fungi infect nematodes.

6.5.4 Molecular Markers for Ecological Studies

DNA based methods have increasingly been used for elucidating the taxonomic identity of fungi in soils (Peay et al. 2008; Anderson and Cairney 2004). Typically, DNA is extracted from soils, and specific region of ribosomal DNA is amplified by PCR and analyzed by gel electrophoresis or sequencing. Recently, *Orbiliales*-specific PCR primers for the internal transcribed spacer (ITS) and 28S ribosomal DNA were designed to directly detect nematode-trapping fungi without culturing in soils (Smith and Jaffee 2008). The primers were used to selectively amplify, clone and sequence *Orbiliales* DNA extracted from soil, litter and wood. The analyses showed that these habitats contained a number of sequences in the *Orbiliales* clade that could not be cultured using the classical culture based methods. These data suggest that there is a hidden, not yet characterized diversity of *Orbiliales* in soils. Members of the *Orbiliales* are predators on nematodes but also on other microbes, mites and insects. Some species even appear to lack predatory adaptations (Barron 1977). Hence, it is difficult to predict if the ribosomal sequences of non-cultured *Orbiliales* represent nematode-trapping fungi or species with other growth habits.

The complications might be solved with the development of genetic markers that are indicative of the parasitic activity of nematode-trapping fungi. Candidates for functional markers of nematode-parasitism may be selected among those genes that are highly over-expressed during the infection of nematodes. The translation of genome-wide expression data into ecologically useful biomarkers requires however, a number of carefully designed validation experiments. First, the expression of the candidate genes needs to be examined under a number of different growth conditions to make sure that they are uniquely expressed during parasitic growth. In many nematode-trapping fungi including *A. oligospora*, the shift from saprophytic to parasitic stages is triggered not only by external signals from the nematodes, but also by other signals reflecting the nutrient level in the environment. Notably, detailed studies on the regulation of *AOL* and *PII* in *A. oligospora* have shown that these genes are regulated both during infection, but also during saprophytic growth depending on the nutrient composition of the environment (c.f. Sects. 6.3.2 and 6.3.3). It has been proposed that the nematode-trapping habit evolved among cellulolytic or lignolytic fungi as a response to nutrient deficiency in nitrogen-limiting habitats. In such environments (like soils) with a high carbon:nitrogen ratio, nematodes might serve as an import source of nitrogen (Barron 1992). Hence, it can be expected that many of the genes that are regulated during nematode infection are also regulated in response to nutrient-related signals. Second, experiments need to be carried out in several fungus-nematode systems to assure that the biomarkers could indicate parasitic activity in many, if not all nematode-trapping fungi of the *Orbiliales*. Microarray expression data from a diverse array of species and environmental conditions should be compared and mined by bioinformatic methods to identify candidates of functional gene markers.

Although not yet applied to nematode-trapping fungi, quantitative PCR (QPCR) has been used in numerous studies to measure the abundance and expression of functional gene markers of microorganisms within the environment (Smith and Osborn 2009). So far, QPCR analyses have mainly been done on functional genes that code for enzymes catalyzing various biogeochemical processes including the carbon, nitrogen, and sulphur cycles, and for genes encoding key reactions in biodegradation pathways. Several recent studies have investigated how the numbers and expression of such genes relates to the activity of the processes that they encode, and how these measurements are affected by changes in the environment and experimental manipulations (Nicolaisen et al. 2008; Blackwood et al. 2007; McKew et al. 2007). In the future, similar approaches may be used for assessing the abundance and activity of nematode-trapping fungi in soils, identifying the environmental factors that determine their distribution and activity, and how they are affected by various farming practices and agrochemical usages. Furthermore, QPCR methods have the potential to detect and monitor the fate of released specific biological control agents; to what extent they are spread from the point of inoculation, colonize the habits of target nematodes (e.g. the rhizosphere), and are resting or active in killing nematodes.

6.5.5 Assessing Variability Within and Between Species

Nematode-trapping fungi vary extensively in phenotypic traits like the morphology and induction mechanisms of traps, growth rate at various temperatures, and production of hydrolytic enzymes. Many of these traits can be expected to affect the survival and parasitic activity of nematode-trapping fungi in soils. Insights into the genetic background to the variation of such traits will be useful for researchers that screen for more potent biological control agents. When searching for such genes, a clear distinction must here be made between genes that contribute to the *manifestation* of a phenotypic trait and genes that contribute to the *variation* in the traits. Although, there are yet some knowledge on the former class of genes, knowledge on the second class of genes is lacking in nematode-trapping fungi. The DNA microarray technology has opened up new possibilities for comparing transcript abundance between closely related species/and or strains, and identifying genes that are associated with morphological and physiological divergence (Ferea et al. 1999; Le Quéré et al. 2004) In addition, DNA microarray-based comparative genome hybridization can be used to assess genomic rearrangements like amplifications or deletion at single gene resolution which might play an important role in generating variation in virulence of fungal pathogens (Hu et al. 2008).

Acknowledgments Work done in the laboratory of the authors has been supported by grants from the Swedish Research Council (VR) and the Swedish Research Council for Agricultural Sciences and Spatial Planning (FORMAS).

References

Åhman J, Ek B, Rask L et al (1996) Sequence analysis and regulation of a gene encoding a cuticle-degrading serine protease from the nematophagous fungus *Arthrobotrys oligospora*. Microbiology 142:1605–1616

Åhman J, Johansson T, Olsson M et al (2002) Improving the pathogenicity of a nematode-trapping fungus by genetic engineering of a subtilisin with nematotoxic activity. Appl Environ Microbiol 68:3408–3415

Ahrén D, Ursing BM, Tunlid A (1998) Phylogeny of nematode-trapping fungi based on 18 S rDNA sequences. FEMS Microbiol Lett 158:179–184

Ahrén D, Tholander M, Fekete C et al (2005) Comparison of gene expression in trap cells and vegetative hyphae of the nematophagous fungus *Monacrosporium haptotylum*. Microbiology 151:789–803

Anderson IC, Cairney JW (2004) Diversity and ecology of soil fungal communities: increased understanding through the application of molecular techniques. Environ Microbiol 6:769–779

Bagga S, Hu G, Screen SE et al (2004) Reconstructing the diversification of subtilisins in the pathogenic fungus *Metarhizium anisopliae*. Gene 324:159–169

Balogh J, Tunlid A, Rosen S (2003) Deletion of a lectin gene does not affect the phenotype of the nematode-trapping fungus *Arthrobotrys oligospora*. Fungal Genet Biol 39:128–135

Barron GL (1977) The nematode-destroying fungi. Lancester Press, Guelph

Barron GL (1992) Lignolytic and cellulolytic fungi as predators and parasites. In: Carroll GC, Wicklow DT (eds) The fungal community. Marcel Dekker, New York, pp 311–326

Birck C, Damian L, Marty-Detraves C et al (2004) A new lectin family with structure similarity to actinoporins revealed by the crystal structure of *Xerocomus chrysenteron* lectin XCL. J Mol Biol 344:1409–1420

Blackwood CB, Waldrop MP, Zak DR et al (2007) Molecular analysis of fungal communities and laccase genes in decomposing litter reveals differences among forest types but no impact of nitrogen deposition. Environ Microbiol 9:1306–1316

Bordallo JJ, Lopez-Llorca LV, Jansson H-B et al (2002) Colonization of plant roots by egg-parasitic and nematode-trapping fungi. New Phytol 154:491–499

Borrebaeck CA, Mattiasson B, Nordbring-Hertz B (1984) Isolation and partial characterization of a carbohydrate-binding protein from a nematode-trapping fungus. J Bacteriol 159:53–56

Breakspear A, Momany M (2007) The first fifty microarray studies in filamentous fungi. Microbiology 153:7–15

Carrizo ME, Capaldi S, Perduca M et al (2005) The antineoplastic lectin of the common edible mushroom (*Agaricus bisporus*) has two binding sites, each specific for a different configuration at a single epimeric hydroxyl. J Biol Chem 280:10614–10623

Choi W, Dean RA (1997) The adenylate cyclase gene MAC1 of *Magnaporthe grisea* controls appressorium formation and other aspects of growth and development. Plant Cell 9:1973–1983

Couillault C, Pujol N, Reboul J et al (2004) TLR-independent control of innate immunity in *Caenorhabditis elegans* by the TIR domain adaptor protein TIR-1, an ortholog of human SARM. Nat Immunol 5:488–494

Cox GN, Kusch M, Edgar RS (1981) Cuticle of *Caenorhabditis elegans*: its isolation and partial characterization. J Cell Biol 90:7–17

Dackman C, Nordbring-Hertz B (1992) Conidial traps: a new survival structure of the nematode-trapping fungus *Arthrobotrys oligospora*. Mycol Res 96:194–198

Davies KG (2005) Interactions between nematodes and microorganisms: bridging ecological and molecular approaches. Adv Appl Microbiol 57:53–78

DeZwaan TM, Carroll AM, Valent B et al (1999) *Magnaporthe grisea* pth11p is a novel plasma membrane protein that mediates appressorium differentiation in response to inductive substrate cues. Plant Cell 11:2013–2030

Dijksterhuis J, Veenhuis M, Harder W et al (1994) Nematophagous fungi: physiological aspects and structure-function relationships. Adv Microb Physiol 36:111–143

Drechsler C (1941) Predacious fungi. Biol Rev Camb Philos Soc 16:265–290

Dudddington CL (1955) Fungi that attack microscopic animals. Bot Rev 7:377–439

Fekete C, Tholander M, Rajashekar B et al (2008) Paralysis of nematodes: shifts in the transcriptome of the nematode-trapping fungus *Monacrosporium haptotylum* during infection of *Caenorhabditis elegans*. Environ Microbiol 10:364–375

Ferea TL, Botstein D, Brown PO et al (1999) Systematic changes in gene expression patterns following adaptive evolution in yeast. Proc Natl Acad Sci USA 96:9721–9726

Friman E (1993) Isolation of trap cells from the nematode-trapping fungus *Dactylaria candida*. Exp Mycol 17:368–370

Friman E, Olsson S, Nordbring-Hertz B (1985) Heavy-trap formation by *Arthrobotrys oligospora* in liquid culture. FEMS Microbiol Ecol 31:17–21

Gaspard JT, Jaffe BA, Ferris H (1990) *Meloidogyne incognita* survival in soil infested with *Paecilomyces lilacinus* and *Verticillium chlamydosporium*. J Nematol 22:176–181

Gravato-Nobre MJ, Hodgkin J (2005) *Caenorhabditis elegans* as a model for innate immunity to pathogens. Cell Microbiol 7:741–751

Hagedorn G, Scholler M (1999) A reevaluation of predatory orbiliaceous fungi. I: Phylogentic analysis using rDNA sequence data. Sydowia 51:27–48

Hu G, Liu I, Sham A et al (2008) Comparative hybridization reveals extensive genome variation in the AIDS-associated pathogen *Cryptococcus neoformans*. Genome Biol 9:R41

Iijima N, Yoshino H, Ten LC et al (2002) Two genes encoding fruit body lectins of *Pleurotus cornucopiae*: sequence similarity with the lectin of a nematode-trapping fungus. Biosci Biotechnol Biochem 66:2083–2089

Jansson H-B, Lopez-Llorca LV (2004) Control of nematodes by fungi. In: Arora DK (ed) Fungal biotechnology in agricultural, food, and environmental applications. Marcel Dekker, New York, pp 205–215

Jin X, Ming-He M, Xiao-Wei H et al (2005) Improvement on genetic transformation in the nematode-trapping fungus *Arthrobotrys oligospora* and its quantification on dung samples. Mycopathologia 159:533–538

Kamoun S (2006) A catalogue of the effector secretome of plant pathogenic oomycetes. Annu Rev Phytopathol 44:41–60

Kerry BR (2000) Rhizosphere interactions and the exploitation of microbial agents for the biological control of plant-parasitic nematodes. Annu Rev Phytopathol 38:423–441

Kerry BR (2001) Exploitation of the nematophagous fungus *Verticillium chlamydosporium* Goddar for the biological control of root-knot nematodes (*Meloidogyne* spp.). In: Butt TM, Jackson C, Magan M (eds) Fungi as biocontrol agents; progress, problems and potential. CABI Publishing, Wallingford, pp 155–168

Khan A, Williams KL, Soon J et al (2008) Proteomic analysis of the knob-producing nematode-trapping fungus *Monacrosporium lysipagum*. Mycol Res 112:1447–1452

Kulkarni RD, Kelkar HS, Dean RA (2003) An eight-cysteine-containing CFEM domain unique to a group of fungal membrane proteins. Trends Biochem Sci 28:118–121

Larsen M (2000) Prospects for controlling animal parasitic nematodes by predacious micro fungi. Parasitology 120(Suppl):S121–S131

Le Quéré A, Schützendübel A, Rajashekar B et al (2004) Divergence in gene expression related to variation in host specificity of an ectomycorrhizal fungus. Mol Ecol 13:3809–3819

Leonidas DD, Swamy BM, Hatzopoulos GN et al (2007) Structural basis for the carbohydrate recognition of the *Sclerotium rolfsii* lectin. J Mol Biol 368:1145–1161

Li Y, Hyde KD, Jeewon R et al (2005) Phylogenetics and evolution of nematode-trapping fungi (Orbiliales) estimated from nuclear and protein coding genes. Mycologia 97:1034–1046

Linford MB (1937) Stimulated activity of natural enemies of nematodes. Science 85:123–124

Liou GY, Tzean SS (1997) Phylogeny of the genus *Arthrobotrys* and allied nematode-trapping-fungi based on rDNA sequences. Mycologia 89:876–884

Mallo GV, Kurz CL, Couillault C et al (2002) Inducible antibacterial defense system in *C. elegans*. Curr Biol 12:1209–1214

McKew BA, Coulon F, Yakimov MM et al (2007) Efficacy of intervention strategies for bioremediation of crude oil in marine systems and effects on indigenous hydrocarbonoclastic bacteria. Environ Microbiol 9:1562–1571

Miura F, Kawaguchi N, Sese J et al (2006) A large-scale full-length cDNA analysis to explore the budding yeast transcriptome. Proc Natl Acad Sci USA 103:17846–17851

Morton CO, Hirsch PR, Kerry BR (2004) Infection of plant-parasitic nematodes by nematophagous fungi - a review of the application of molecular biology to understand infectionprocesses and to improve biological control. Nematology 6:161–170

Mylonakis E, Ausubel FM, Perfect JR et al (2002) Killing of Caenorhabditis elegans by Cryptococcus neoformans as a model of yeast pathogenesis. Proc Natl Acad Sci USA 99:15675–15680

Nicolaisen MH, Baelum J, Jacobsen CS et al (2008) Transcription dynamics of the functional tfdA gene during MCPA herbicide degradation by *Cupriavidus necator* AEO106 (pRO101) in agricultural soil. Environ Microbiol 10:571–579

Norbring-Hertz B, Jansson H-B, Persson Y et al (1995) Nematophagous fungi. Film C1851. Institut für den Wissenschafltichen Film, Göttingen

Nordbring-Hertz B (1973) Peptide induced morphogenesis in the nematode-trapping fungus *Arthrobotrys oligospora*. Physiol Plant 29:223–233

Nordbring-Hertz B (1977) Nematode-induced morphogenesis in the nematode-trapping fungus *Arthrobotrys oligospora*. Nematologica 23:443–451

Nordbring-Hertz B, Brinck B (1974) Quantitative characterization of some peptides inducing morphogenesis in the nematode-trapping fungus *Arthrobotrys oligospora*. Physiol Plant 31:59–63

Nordbring-Hertz B, Chet I (1988) Fungal lectins and agglutinins. In: Mirelman D (ed) Microbial lectins and agglutinins: properties and biological activity. Wiley, New York, pp 393–408

Nordbring-Hertz B, Mattiasson B (1979) Action of a nematode-trapping fungus shows lectin-mediated host-microorganism interaction. Nature 281:477–479

O'Rourke D, Baban D, Demidova M et al (2006) Genomic clusters, putative pathogen recognition molecules, and antimicrobial genes are induced by infection of *C. elegans* with *M. nematophilum*. Genome Res 16:1005–1016

Olthof THA, Estey RH (1963) A nemtatotoxin produced by the nematode-trapping fungus *Arthrobotrys oligospora*. Nature 197:514–525

Peay KG, Kennedy PG, Bruns T (2008) Fungal comminity ecology: a hybrid beast with a molecular master. BioScience 58:799–810

Peng T, Orsborn KI, Orbach MJ et al (1999) Proline-rich vaccine candidate antigen of *Coccidioides immitis:* conservation among isolates and differential expression with spherule maturation. J Infect Dis 179:518–521

Persmark L, Nordbring-Hertz B (1997) Conidial trap formation of nematode-trapping fungi in soil and soil extracts. FEMS Microbiol Ecol 22:313–323

Persson Y, Veenhuis M, Nordbring-Hertz B (1985) Morphogenesis and significance of hyphal coling by nemtaode-trapping fungi in mycoparasitic relationships. FEMS Microbiol Ecol 31:283–291

Pramer D, Kuyama S (1963) Symposium on biochemical bases of morphogenesis in fungi II: Nemin and the nematode-trapping fungi. Bacteriol Rev 27:282–292

Pramer D, Stoll NR (1959) Nemin: a morphogenetic substance causing trap formation by predaceous fungi. Science 129:966–967

Premachandran D, Pramer D (1984) Role of N-Acetylgalactosamine-specific protein in trapping of nematodes by *Arthrobotrys oligospora*. Appl Environ Microbiol 47:1358–1359

Rauyaree P, Choi W, Fang E et al (2004) Genes expressed during early stages of rice infection with the rice blast fungus *Magnaporthe grisea*. Mol Plant Pathol 2:347–354

Rep M (2005) Small proteins of plant-pathogenic fungi secreted during host colonization. FEMS Microbiol Lett 253:19–27

Rosén S, Ek B, Rask L et al (1992) Purification and characterization of a surface lectin from the nematode-trapping fungus *Arthrobotrys oligospora*. J Gen Microbiol 138:2663–2672

Rosén S, Bergstrom J, Karlsson KA et al (1996a) A multispecific saline-soluble lectin from the parasitic fungus *Arthrobotrys oligospora*. Similarities in the binding specificities compared with a lectin from the mushroom agaricus bisporus. Eur J Biochem 238:830–837

Rosén S, Kata M, Persson Y et al (1996b) Molecular characterization of a saline-soluble lectin from a parasitic fungus Extensive sequence similarities between fungal lectins. Eur J Biochem 238:822–829

Rosén S, Sjollema K, Veenhuis M et al (1997) A cytoplasmic lectin produced by the fungus *Arthrobotrys oligospora* functions as a storage protein during saprophytic and parasitic growth. Microbiology 143:2593–2604

Schenck S, Chase T, Rosenzweig WD et al (1980) Collagenase production by nematode-trapping fungi. Appl Environ Microbiol 40:567–570

Sharon N, Lis H (2004) History of lectins: from hemagglutinins to biological recognition molecules. Glycobiology 14:53R–62R

Siezen RJ, Leunissen JA (1997) Subtilases: the superfamily of subtilisin-like serine proteases. Protein Sci 6:501–523

Sifri CD, Begun J, Ausubel FM (2005) The worm has turned–microbial virulence modeled in *Caenorhabditis elegans*. Trends Microbiol 13:119–127

Smith ME, Jaffee BA (2008) PCR Primers with Enhanced Specificity for Nematode-Trapping Fungi (Orbiliales). Microb Ecol 58:117–128

Smith CJ, Osborn AM (2009) Advantages and limitations of quantitative PCR (Q-PCR)-based approaches in microbial ecology. FEMS Microbiol Ecol 67:6–20

Stadler M, Sterner O (1993) Linoleic acid - The nematicidal principle of several nematophagous fungi and its production in trap-forming submerged culture. Arch Microbiol 160:401–405

Stein LD, Bao Z, Blasiar D et al (2003) The genome sequence of *Caenorhabditis briggsae*: a platform for comparative genomics. PLoS Biol 1:E45

Stirling GR, Smith LJ (1998) Filed tests of formulated products containing either *Verticillium chlamydoposporium* or *Arthrobotrys dactyloides* for biological control of root-knot nematodes. Biol Control 11:231–239

Swamy BM, Bhat AG, Hegde GV et al (2004) Immunolocalization and functional role of *Sclerotium rolfsii* lectin in development of fungus by interaction with its endogenous receptor. Glycobiology 14:951–957

Takano Y, Choi W, Mitchell TK et al (2004) Large scale parallel analysis of gene expression during infection-related morphogenesis of *Magnaporthe grisea*. Mol Plant Pathol 4:337–346

Thines E, Weber RW, Talbot NJ (2000) MAP kinase and protein kinase A-dependent mobilization of triacylglycerol and glycogen during appressorium turgor generation by *Magnaporthe grisea*. Plant Cell 12:1703–1718

Thomas SW, Glaring MA, Rasmussen SW et al (2002) Transcript profiling in the barley mildew pathogen *Blumeria graminis* by serial analysis of gene expression (SAGE). Mol Plant Microbe Interact 15:847–856

Tucker SL, Talbot NJ (2001) Surface attachment and pre-penetration stage development by plant pathogenic fungi. Annu Rev Phytopathol 39:385–417

Tunlid A, Ahrén D (2001) Application of genomics to the improvement of nematode-trapping fungi. In: Vurro M, Gressel J, Butt T et al (eds) Enhancing biocontrol agents and handling risks. Ios Press, Amsterdam, pp 193–200

Tunlid A, Jansson S (1991) Proteases and their involvement in the infection and immobilization of nematodes by the nematophagous fungus *Arthrobotrys oligospora*. Appl Environ Microbiol 57:2868–2872

Tunlid A, Talbot NJ (2002) Genomics of parasitic and symbiotic fungi. Curr Opin Microbiol 5:513–519

Tunlid A, Rosen S, Ek B et al (1994) Purification and characterization of an extracellular serine protease from the nematode-trapping fungus *Arthrobotrys oligospora*. Microbiology 140:1687–1695

Tunlid A, Åhman J, Oliver RP (1999) Transformation of the nematode-trapping fungus *Arthrobotrys oligospora*. FEMS Microbiol Lett 173:111–116

Wilhelm BT, Marguerat S, Watt S et al (2008) Dynamic repertoire of a eukaryotic transcriptome surveyed at single-nucleotide resolution. Nature 453:1239–1243

Wright DP, Johansson T, Le Quéré A et al (2005) Spatial patterns of gene expression in the extramatrical mycelium and mycorrhizal root tips formed by the ectomycorrhizal fungus *Paxillus involutus* in association with birch (*Betula pendula* Roth.) seedlings in soil microcosms. New Phytol 167:579–596

Yang J, Tian B, Liang L et al (2007a) Extracellular enzymes and the pathogenesis of nematophagous fungi. Appl Microbiol Biotechnol 75:21–31

Yang Y, Yang E, An Z et al (2007b) Evolution of nematode-trapping cells of predatory fungi of the Orbiliaceae based on evidence from rRNA-encoding DNA and multiprotein sequences. Proc Natl Acad Sci USA 104:8379–8384

Yu LG, Andrews N, Weldon M et al (2002) An N-terminal truncated form of Orp150 is a cytoplasmic ligand for the anti-proliferative mushroom *Agaricus bisporus* lectin and is required for nuclear localization sequence-dependent nuclear protein import. J Biol Chem 277:24538–24545

Chapter 7
Ecology of *Pochonia chlamydosporia* in the Rhizosphere at the Population, Whole Organism and Molecular Scales

Brian R. Kerry and Penny R. Hirsch

Abstract *Pochonia chlamydosporia* is a facultative egg parasite of cyst and root-knot nematodes. It is a widely distributed parasite and it has been developed as a biological control agent. It appears to have only limited growth in soil and chlamydospores are an important survival stage of the fungus. The fungus can however proliferate in the rhizosphere where it presumably grows on plant root exudates but in the presence of plant-parasitic nematodes it can switch its trophic state and become a parasite of nematode eggs. The molecular mechanism by which it becomes a nematode parasite will be reviewed in the context of biodiversity and how to identify nematode pathogenic strains. The relationship between parasitism and fungal abundance in relationship to infection processes will also be discussed.

7.1 Introduction

The nematophagous fungus, *Pochonia chlamydosporia* (Goddard) Zare & Gams (syn *Verticillium chlamydosporium* Goddard) is a facultative parasite mainly of the eggs and females of cyst and root-knot nematodes. It has been developed as a biological control agent for use in Cuba (Hernández and Hidalgo Díaz 2008) and India (Rao et al. 1997) against these pests and is being evaluated elsewhere as a tool for nematode management. The fungus was first reported as a parasite of nematode eggs in the UK (Willcox and Tribe 1974) and its role as one of the causal agents of the populations decline of cereal cyst nematode, *Heterodera avenae*, was demonstrated soon afterwards (Kerry and Crump 1977; Kerry et al. 1982). Since then it has been reported from many countries in most continents as an important pathogen in the regulation of nematode populations including those of *H. avenae* in Europe

B.R. Kerry and P.R. Hirsch (✉)
Nematode Interactions Unit, Plant Pathology & Microbiology Department,
Rothamsted Research, Harpenden, Hertfordshire AL3 2JQ, UK
e-mail: penny.hirsch@bbsrc.ac.uk

K. Davies and Y. Spiegel (eds.), *Biological Control of Plant-Parasitic Nematodes: Building Coherence between Microbial Ecology and Molecular Mechanisms*,
Progress in Biological Control 11, DOI 10.1007/978-1-4020-9648-8_7,

(Kerry 1975) and in Australia (Stirling and Kerry 1983), of *H. schachtii* in Holland (Heijbroek 1983), and Iran (Ayatollahy et al. 2008), of *H. glycines* in the USA (Gintis et al. 1983), and China (Meyer et al. 2004), of *Meloidogyne* spp. in California (Loffredo et al. 2007), China (Sun et al. 2006), Cuba (Hidalgo-Díaz et al. 2000), Pakistan (Zaki and Maqbool 1993), and of *Nacobbus aberrans* in Mexico (Flores-Camacho et al. 2007). These nematode genera include the most important nematode pests affecting world crop production and are characterised by sedentary, saccate females, which produce several 100 eggs in gelatinous matrices; egg masses are colonised by *P. chlamydosporia* and the eggs are destroyed. In contrast to the sedentary plant-parasites, most nematodes remain active throughout their development and females lay their eggs singly: whether or not the fungus has a role in parasitizing these eggs would be technically difficult to determine and thus to date, remains unknown.

There appears to be only limited growth of *Pochonia chlamydosporia* in soil and the chlamydospores produced by the fungus are important for its survival. Fungal propagules applied to soil at planting time survive in sufficient numbers to infect significant numbers of nematode eggs produced 1–3 months later (Bourne and Kerry 1999). The fungus is more prolific in the rhizosphere where presumably it derives its nutrition from nutrients released by the roots. When contact is made with a nematode egg mass, the fungus switches from the saprophytic phase of its development to the parasitic phase and the colonisation of nematode eggs. The conditions required to induce the switch in the trophic phases of the fungus are poorly understood. However, investigation of factors affecting fungal gene expression, in particular the regulation of genes encoding enzymes involved in egg infection, may help to explain the switch (see below).

There is limited endophytic growth of the fungus within roots (Lopez-Llorca et al. 2002a; Maciá-Vicente et al. 2009) but the fungus is mostly confined to the rhizosphere (de Leij and Kerry 1991). A number of key factors affect the interactions between the plant, the nematode and the fungus that affect its activity as a biological control agent (Table 7.1). The observations that led to the identification of these factors were based on the use of dilution plating techniques onto a selective medium for the fungus (Kerry et al. 1993). The methods are useful to detect relative changes in abundance and the extent of colonisation of the soil and rhizosphere, but as colonies that develop on the plates may develop from a hyphal fragment, a conidium or a resting spore, little can be deduced about the physiological state of the fungus. Indeed, large increases in the number of colonies observed may be caused by the production of spores (Bourne et al. 1994; Mauchline et al. 2002). Similarly, use of DNA-based methods may enable quantification of the fungus in small soil samples but they do not separate viable from dead fungal propagules. Studying the ecology of *P. chlamydosporia* and other fungi in soil requires the use of a number of methods to measure changes in the abundance and structure of fungal populations and their activity against nematode hosts.

Changes in the abundance of chlamydospores of *P. chlamydosporia* in cereal cyst nematode suppressive soils, under different cropping regimes, have been followed (Kerry and Crump 1998) and the establishment of the fungus in rotations of vegetable

Table 7.1 Factors that affect the tri-trophic interaction between the plant and nematode and *Pochonia chlamydosporia* in the rhizosphere

Factor	Reference
Colonisation of the rhizosphere is essential for nematode control	Bourne et al. (1994)
Fungal isolates differ in their saprotrophic and parasitic abilities	Morton et al. (2003a, b), Mauchline et al. (2004), and Siddiqui et al. (2009)
Plant species differ in their ability to support the fungus in their rhizospheres	Bourne et al. (1994, 1996) and Atkins et al. (2003)
Fungal abundance may not be related to the extent of parasitism of nematode eggs	Bourne and Kerry (1999), Atkins et al. (2009), and Siddiqui et al. (2009)
Fungus more abundant on nematode infected than healthy roots	Bourne et al. (1996)
Fungal biotypes exist with nematode host preferences	Morton et al. (2003a), Mauchline et al. (2004), Manzanilla-López et al. (2009), and Siddiqui et al. (2009)
Presence of *P. chlamydosporia* in the rhizosphere does not reduce nematode invasion of roots or the abundance of non-target nematodes and microbes but may change the structure of microbial communities.	O'Flaherty et al. (2003) and Bailey et al. (2008)

crops monitored after application to a tropical soil (Atkins et al. 2003). However, little is known of the epidemiology of the fungus in soil. Despite the *in vitro* production of nematicidal products such as the antibiotic phomalactone (Khambay et al. 2000), the presence of the fungus in the rhizosphere did not affect the invasion of tomato roots by infective juveniles of *M. incognita* (Bailey et al. 2008) or have direct effects on the microbial and nematode communities in the rhizosphere (O'Flaherty et al. 2003; Tahseen et al. 2005). Nevertheless, the fungus can affect other soil microorganisms, as it is a parasite of oospores of the fungus-like oomycete *Phytopthera* (Sneh et al. 1977) and has been used as a control agent for plant diseases caused by *Phytopthera* (Sutherland and Papavizas 1991). Abiotic factors such as temperature (Dackman et al. 1989) and water stress (Esteves et al. 2009a) affect fungal growth *in vitro* and are likely to affect the growth and spread of the fungus in soil.

Pochonia chlamydosporia infects nematode eggs through the development of appressoria at the hyphal tip or laterally, which appear to attach tightly to the surface of eggshells, which are penetrated by an infection peg. A post-infection bulb leads to the development of a mycelium within the egg and the destruction of its contents (Segers et al. 1996). All eggs that are colonised by the fungus are destroyed within a few days of infection. The fungus produces a range of enzymes *in vitro* (Dackman et al. 1989; Dupont et al. 1999; Esteves et al. 2009b) but their role as host range and virulence determinants is poorly understood. A serine proteinase designated VCP1 is responsible for the degradation of the outer vitelline membrane of nematode eggs (Segers et al. 1996). Polymorphisms in this enzyme were correlated with host preference and those fungal biotypes with a preference for root-knot nematode eggs produced VCP1 with a specific difference in substrate affinity, arising from amino acid substitutions at active sites within the enzyme, compared to those biotypes that preferred cyst nematodes (Morton et al. 2003a).

Molecular studies of *P. chlamydosporia* have and will continue to be essential for further understanding of (a) the functional diversity of the fungus, (b) gene expression during the saprophytic and parasitic phases of the fungus, (c) the infection process, and (d) will be important in developing diagnostic methods for the detection of specific isolates and biotypes in soil. This chapter concentrates on the information gleaned from using a range of molecular methods to improve our understanding of the dynamics and genetic structure of *P. chlamydosporia* populations and of gene expression during the infection process.

7.2 Diversity of *P. Chlamydosporia*

Isolates of *P. chlamydosporia* differ greatly in their growth, development and virulence, as has been observed with most microbial natural enemies. Much variation within the fungus has been observed in different isolates from the same soil (Hidalgo-Díaz et al. 2000; Morton et al. 2003a, b; Flores-Camacho et al. 2008; Manzanilla-López et al. 2009) and may be correlated with changes in the host nematode in accordance with Red Queen dynamics but such interrelationships are unproven. However, differences between isolates from amphimictic cyst nematodes were generally greater than those between isolates from parthenogenetic root-knot nematodes (Fig. 7.1). The role of variation in the regulation of nematode populations is not understood and may await a detailed understanding of the infection process at the molecular scale. Biotypes of the fungus have also been recognised which exhibit different host preferences (Morton et al. 2003a, b; Mauchline et al. 2004; Siddiqui et al. 2009). A range of molecular diagnostic methods have been established to discriminate between isolates and enable quantification of populations of the fungus in the soil, rhizosphere and nematode egg masses (see "Molecular Diagnostics" section, below). The recent development of methods to visualise the fungus in the rhizosphere by transformation with a fluorescent marker gene offers exciting prospects (Maciá-Vicente et al. 2009).

There is now sufficient ribosomal gene sequence data to identify, and to design primers and probes for use in PCR assays, to differentiate between the different species and varieties of *Pochonia* (Zare and Gams 2004). The polymorphisms in the serine protease VCP1 correlating to host nematode preference demonstrate both functional and genetic diversity at the level of individual isolates but there is not yet corresponding sequence data for other genes that may be involved in nematode infection. However, different *P. chlamydosporia* isolates grown in liquid culture vary in the production of extracellular enzymes (chitinases, lipases and esterases) that are required for egg infection in addition to proteases (Esteves et al. 2009b). The differential responses to potential inducers and variations in enzyme activity indicate underlying genetic differences that may be revealed by further study.

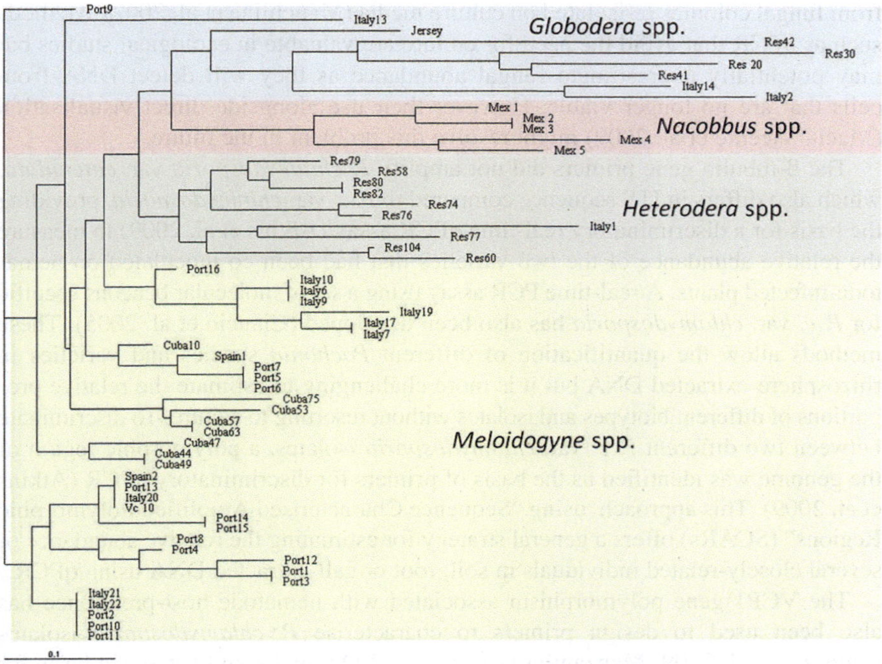

Fig. 7.1 Phylogram showing the genetic variation in isolates of *Pochonia chlamydosporia* from different geographical regions and different nematode hosts as determined by ERIC-PCR. The phylogram is a neighbour-joining tree generated using FreeTree (After Morton et al. 2003b)

7.3 Molecular Diagnostic Techniques

Different isolates of *P. chlamydosporia* that are impossible to differentiate by morphology can be discriminated using a variety of molecular methods based on PCR. Restriction digest fingerprints of the products from PCR amplification of the ribosomal intergenic spacers (IGS), and the patterns obtained using PCR with arbitrary ERIC and REP primers, could differentiate isolates although digests of the ribosomal internal transcribed spacers (ITS) did not (Arora et al. 1996). Subsequently, ERIC fingerprinting was used to infer a phylogenetic tree of isolates, which grouped according to both their geographical and the host nematode origin (Morton et al. 2003b). PCR fingerprinting has been used to identify fungal colonies re-isolated from soil and plants during mixed infection experiments (Mauchline et al. 2004; Flores-Camacho et al. 2008).

The β-tubulin gene of *P. chlamydosporia* var. *chlamydosporia* was found to have an intron not present in other fungi, which provided the basis for designing specific primers to positively identify fungal cultures (Hirsch et al. 2001). Subsequently, these were used for quantitative PCR (qPCR) estimates of fungal abundance in the rhizosphere, using DNA extracted from soil, roots and nematode galls rather than from fungal colonies re-isolated on culture media (Mauchline et al. 2002). Methods such as qPCR that avoid the need for culture are valuable in ecological studies but may potentially overestimate fungal abundance as they will detect DNA from cells that are no longer viable. However their use alongside direct visualisation (Maciá-Vicente et al. 2009) might resolve this problem in the future.

The β-tubulin gene primers did not amplify *P. chlamydosporia* var. *catenulata,* which also differs in ITS sequence compared to *P. c.* var. *chlamydosporia,* providing the basis for a discriminatory real-time qPCR assay (Atkins et al. 2009) to measure the relative abundance of the two varieties that had been co-inoculated on nematode-infected plants. A real-time PCR assay using a set of molecular beacons specific for *P. c.* var. *chlamydosporia* has also been developed (Ciancio et al. 2005). These methods allow the quantification of different *Pochonia* species and varieties in rhizosphere-extracted DNA but it is more challenging to estimate the relative proportions of different biotypes and isolates without resorting to culture. To discriminate between two different *P. c.* var. *chlamydosporia* isolates, a polymorphic region of the genome was identified as the basis of primers for discriminatory qPCR (Atkins et al. 2009). This approach, using "Sequence Characterised Amplified Polymorphic Regions" (SCARs) offers a general strategy for estimating the relative abundance of several closely-related individuals in soil, root or gall-extracted DNA using qPCR.

The VCP1 gene polymorphism associated with nematode host-preference has also been used to design primers to characterise *P. chlamydosporia* isolates (Siddiqui et al. 2009; Manzanilla-López et al. 2009) and could act as the basis for qPCR assays in the future.

7.4 Relationship Between Activity and Abundance

Biological control depends on the relative abundance of host and pathogen populations but the quantitative relationships between nematodes and their natural enemies are poorly understood. Although there are a range of methods to assess soil microbial populations, all have limitations and there is a need to use more than one approach to estimate active populations in soil. Standard approaches using dilution plating onto selective media have been used to estimate changes in the abundance of *P. chlamydosporia* but the data produced may be difficult to interpret. For example, the development of resting spores in sterile soil could not be detected using dilution plate techniques whereas estimates of the ATP content of the fungus were more sensitive (Kerry et al. 1993). It is therefore not possible to relate changes in abundance to changes in vegetative growth using dilution plates.

Density dependent interactions are common between hosts and their parasites and provide a feedback mechanism for regulating populations. In research on interactions between *H. schachtii* and the nematophagous fungus *Hirsutella rhossiliensis,* Jaffee et al. (1992) demonstrated density dependent parasitism, host threshold densities and low transmission rates. Hence, epidemics in *H. schachtii* populations were expected to develop slowly and natural control was unlikely to develop within a single growing season. Density dependent dynamics have been demonstrated for *P. chlamydosporia* and may indicate greater dependence on their nematode hosts than their status as facultative parasites suggests but the development of suppressive soils and epidemics within nematode populations are again slow to develop (Kerry and Crump 1998). However, unlike *H. rhossiliensis* that has limited growth outside its host, *P. chlamydosporia* is thought to have a significant saprotrophic phase in soil and can develop in the rhizosphere in the absence of nematode hosts providing more options for its manipulation as a biological control agent. This offers the possibilities of managing nematode infestations throughout a crop rotation rather than only during the growth of the susceptible crop. Indeed, the strategy for root-knot nematode management with *P. chlamydosporia* relies on the use of poor hosts for *Meloidogyne* spp., which support extensive growth of the fungus on their roots, enabling fungal abundance to be maintained whilst nematode multiplication is limited (Atkins et al. 2003). Hence, nematode management is optimised through a combination of appropriate crop rotation and biological control.

Isolates of the fungus differ in their ability to grow in soil and the rhizosphere (Bourne et al. 1994, 1996) and tend to be more abundant in organic than in mineral soils but abundance does not necessarily relate to the parasitic activity of the fungus and the number of nematode eggs destroyed (Atkins et al. 2009; Siddiqui et al. 2009). Such a lack of relationship presents a problem for the exploitation of *P. chlamydosporia* as a biological control agent. Successful establishment of the fungus in soil relies on the provision of an energy source to help overcome competition from the resident soil microbial community. The fungus has been successfully established in soil from inocula of chlamydospores, which contain sufficient internal resources for fungal growth, or from conidia and hyphae formulated with an external nutrient medium (Kerry et al. 1993). Hence, chlamydospores that may be applied at rates of 5,000 spores g^{-1} soil in aqueous suspension have been the preferred inoculum to test the efficacy of the fungus as a biological control agent (Bourne and Kerry 1999; Kerry and Hidalgo-Díaz 2004). However, these spores are slow to be produced *in vitro* and as yields of only 10^7 g^{-1} medium mean are obtained after 3 weeks culture, large amounts (30 $g\,m^{-2}$) are needed to treat soil. Applications of organic soil amendments as additional energy sources may increase fungal abundance but they do not necessarily cause increases in nematode parasitism and the fungus appears to remain in its saprotrophic phase (Atkins et al. 2003).

In laboratory-based assays comparing the growth of different isolates of the fungus in different soils and their virulence, host preferences were demonstrated and there were indications of a significant fitness cost associated with parasitic activity; the most virulent isolates spread less rapidly in soil (Siddiqui et al. 2009). Such differences suggest that careful selection of isolates is necessary to identify those with potential as biological control agents. These authors also suggested that there was a positive correlation between the abundance of the fungus in soil and in the rhizosphere. However, there was no significant relationship between the abundance of the fungus on roots and the numbers of nematode eggs infected, which also depended on fungal biotype host-preference, host-plant species and nematode abundance. Clearly the interactions between the fungus and its nematode hosts in

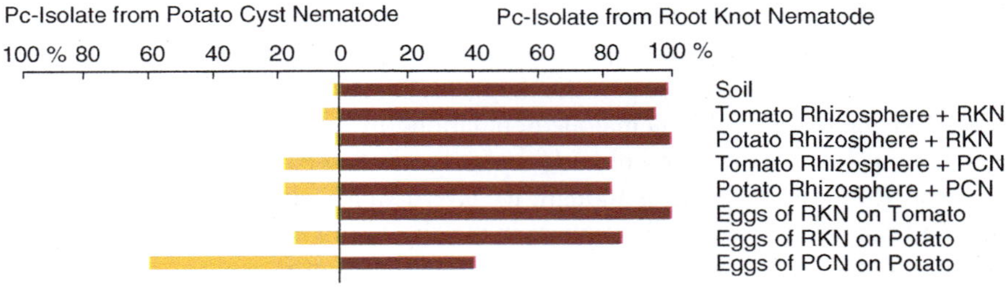

Fig. 7.2 Changes in the structure of populations of two isolates of *P. chlamydosporia* added to soil as a combined application around the roots of tomato or potato plants infected with *M. incognita* or *G. pallida* respectively (After Mauchline et al. 2004)

the rhizosphere are complex and an understanding will depend on the exploitation of more discriminating techniques than dilution plates. For example, Mauchline et al. (2004) added equal numbers of chlamydospores of an isolate of *P. chlamydosporia* from cyst nematodes and from root-knot nematodes which were mixed in a sterilised soil planted with either a potato tuber piece or a tomato seedling. Second-stage juveniles of *Globodera pallida* or *M. incognita* were added around the roots of the potato or tomato plants, respectively, and the numbers of eggs infected and the proportion of colonies of each fungal isolate in the rhizosphere was determined using molecular diagnostic methods. The isolate collected from root-knot nematode eggs was much more abundant in the soil and in the rhizosphere of both plant species than the isolate from cyst nematodes but the latter, despite forming <25% of the colonies from roots, was present in >60% of the *G. pallida* eggs infected by the fungus (Fig. 7.2). Clearly, saprotrophic competitiveness may not relate to parasitic activity and host-preference has an important role to play in the dynamics of *P. chlamydosporia* in the rhizosphere of nematode-infected plants. However, careful selection of the appropriate isolate for use against a specific test can result in large numbers of parasitized eggs and effective biological control (Atkins et al. 2003).

7.5 Infection Processes and Their Regulation

Although the abundance of *P. chlamydosporia* was reported to be greater overall in nematode-infected roots than non-infected roots (Bourne et al. 1996), there was no evidence of increased abundance on roots adjacent to galls (Atkins et al. 2009), providing no evidence for tropism to galls. It is, therefore, more likely that the proliferation of the fungus observed in infected eggs explains this increase in abundance. The eggs of root-knot and cyst nematodes are surrounded by a gelatinous matrix, which fungal hyphae must traverse prior to infection. This gelatinous matrix is reported to contain antimicrobial compounds (Orion et al. 2001) but also to act as a substrate for invading microorganisms (Sharon et al. 2007). It is not known whether the gelatinous matrix is involved in the trophic switch in *P. chlamydosporia* but the fungus shows an exceptionally high level of resistance to antifungal agents, indicating that it possesses an active multidrug resistance mechanism (Atkins et al. 2004). The next stage in infection is the formation of appressoria on the egg surface, which may be the thigmotropic response of hyphal tips contacting a solid surface, increased by hydrophobicity of the surface (Lopez-Llorca et al. 2002b). Production of a relatively high amount of VCP1, required for subsequent dissolution of the vitelline membrane, has been visualised at the infection peg emanating from the appressoria (Segers et al. 1996). A similar observation had been made with the related species, *P. rubescens* (Lopez-Llorca and Robertson 1992). Production of VCP1 *in vitro* is repressed in rich media (catabolite repression), regained in nutrient-poor conditions, and further stimulated when nematode eggs are added. Collagen, an insoluble protein similar in structure to the vitelline membrane, and nitrate in solution, also stimulate VCP1 (Segers 1996). Analysis of the regulatory sequences upstream of the VCP1 gene detected motifs associated with catabolite repression and nitrate induction (Morton et al. 2003a). The pattern of repression and induction seen in VCP1 is similar to that of other serine proteinases of nematophgaous and entomopathogenic fungi (Morton et al. 2004).

Purified VCP1 alone can digest the eggshells of root-knot, but not cyst nematodes (Segers et al. 1996) which require both protease and chitinase (Tikhonov et al. 2002), possibly reflecting differences in the structure and thickness of the eggshell. In liquid culture, *P. chlamydosporia* protease activity reaches a peak 48 h before chitinase activity (Esteves et al. 2009b). However, the inducibility and relative activity of these enzymes produced in liquid culture does not correlate with the ability of different isolates to parasite nematode eggs. Further work will be needed to establish whether chitinase is produced concurrently with VCP1 during the infection process, or if it is induced subsequently. Similarly, it is as yet not clear whether the basis for fungal proliferation and chlamydospore production on nematode egg masses or cysts is a trophic response to increased nutrient availability, or the result of a specific stimulus.

7.6 Conclusions

Many questions remain unanswered but application of a range of new technologies has begun to improve our understanding of the relationship between *P. chlamydosporia* and its plant and nematode hosts, and the trophic switch to nematode egg parasitism followed by fungal spore formation. This will enhance our ability to manipulate cultivation regimes to favour egg parasitism and to optimise the performance of fungal inoculants. Also, knowledge in this area has a more general relevance to understanding the ecology of saprophytic fungi that have an additional ability to utilise a specialist niche, whether it is sedentary plant-parasitic nematode eggs, free-living nematodes or other soil invertebrates.

References

Arora DK, Hirsch PR, Kerry BR (1996) PCR-based molecular discrimination of *Verticillium chlamydosporium* isolates. Mycol Res 100:801–809

Atkins SD, Hildago-Diaz L, Kalisz H et al (2003) Development of a new management strategy for the control of root-knot nematodes (*Meloiogyne* spp.) in organic vegetable production. Pest Manag Sci 59:183–189

Atkins SD, Mauchline TH, Kerry BR et al (2004) Development of a transformation system for the nematophagous fungus *Pochonia chlamydosporia*. Mycol Res 108:654–661

Atkins SD, Peteira B, Clark IM et al (2009) Use of real-time quantitative PCR to investigate root and gall colonisation by co-inoculated isolates of the nematophagous fungus *Pochonia chlamydosporia*. Ann Appl Biol 155:143–152

Ayatollahy E, Fatemy S, Etebarian HR (2008) Potential for biological control of *Heterodera schachtii* by *Pochonia chlamydosporia* var. *chlamydosporia* on sugar beet. Biocontrol Sci Technol 18:157–167

Bailey DJ, Biran GL, Kerry BR et al (2008) Pathozone dynamics of *Meloidogyne incognita* in the rhizosphere of tomato plants in the presence and absence of the nematophagous fungus, *Pochonia chlamydosporia*. Plant Pathol 57:354–362

Bourne JM, Kerry BR (1999) Effect of the host plant on the efficacy of *Verticillium chlamydosporium* as a biological control agent of root-knot nematodes at different nematode densities and fungal application rates. Soil Biol Biochem 31:75–84

Bourne JM, Kerry BR, De Leij FAAM (1994) Methods for the study of *Verticillium chlamydosporium* in the rhizosphere. J Nematol 26(4S):587–591

Bourne JM, Kerry BR, De Leij FAAM (1996) The importance of the host plant on the interaction between root-knot nematodes (*Meloidogyne* spp.) and the nematophagous fungus, *Verticillium chlamydosporium*. Biocontrol Sci Technol 6:539–548

Ciancio A, Loffredo A, Paradies F et al (2005) Detection of *Meloidogyne incognita* and *Pochonia chlamydosporia* by fluorogenic molecular probes. Bulletin OEPP 35:157–164

Dackman C, Chet I, Nordbringhertz B (1989) Fungal parasitism of the cyst nematode *Heterodera schachtii*: infection and enzymatic activity. FEMS Microbiol Ecol 62:201–208

de Leij FAA, Kerry BR (1991) The nematophagous fungus *Verticillium chlamydosporium* as a potential biological control agent for *Meloidogyne arenaria*. Revue Nematol 14:157–164

Dupont A, Segers R, Coosemans J (1999) The effect of chitinase from *Verticillium chlamydosporium* on the egg of *Meloidogyne incognita*. Commun Agric Appl Biol 64:383–389

Esteves I, Peteira B, Powers S et al (2009a) Effects of osmotic and matric potential on radial growth and accumulation of endogenous reserves in three isolates of *Pochonia chlamydosporia*. Biocontrol Sci Technol 19:185–199

Esteves I, Peteira B, Atkins SD et al (2009b) Production of extracellular enzymes by different isolates of *Pochonia chlamydosporia*. Mycol Res 113:867–876

Flores-Camacho R, Manzanilla-Lopez RH, Cid del Pirado-Vera I et al (2007) Control of *Nacobbus aberrans* (Thorne) Thorne and Allen with *Pochonia chlamydosporia* (Goddard) Gams and Zare. Rev Mex Fitopatol 25:26–34

Flores-Camacho R, Manzanilla-Lopez RH, Cid del Pirado-Vera I et al (2008) Characterisation of Mexican isolates of *Pochonia chlamydosporia* var. *chlamydosporia* (Goddard) Gams and Zare for biological control of *Nacobbus aberrans* (Thorne) Thorne and Allen. Rev Mex Fitopatol 26:93–104

Gintis BO, Morgan-Jones G, Rodriguezkabana R (1983) Fungi associated with several developmental stages of *Heterodera glycines* from an Alabama soybean field soil. Nematropica 13:181–200

Heijbroek W (1983) Some effects of fungal parasites on the population development of the beet cyst nematode (*Heterodera schachtii* Schm.). Commun Agric Appl Biol Sci 48:433–439

Hernández MA, Hidalgo Díaz L (2008) KlamiC®: agricultural bionematicide from the fungus *Pochonia chlamydosporia* var. *catenulata*. Rev Protección Veg 23:131–134

Hidalgo-Díaz L, Bourne JM, Kerry BR et al (2000) Nematophagous *Verticillium* spp. in soils infested with *Meloidogyne* spp. in Cuba: isolation and screening. Int J Pest Manag 46:277–284

Hirsch PR, Atkins SD, Mauchline TH et al (2001) Methods for studying the nematophagous fungus *Verticillium chlamydosporium* in the root environment. Plant Soil 232:21–30

Jaffee B, Phillips R, Muldoon A et al (1992) Density-dependent host pathogen dynamics in soil microcosms. Ecology 73:495–506

Kerry B, Hidalgo-Diaz L (2004) Application of *Pochonia chlamydosporia* in the integrated control of root-knot nematodes on organically grown vegetable crops in Cuba. Bulletin OILB/SROP 27:123–126

Kerry BR (1975) Fungi and the decrease of cereal cyst-nematode populations in cereal monoculture. EPPO Bull 5:353–361

Kerry BR, Crump DH (1977) Observations on fungal parasites of females and eggs of the cereal cyst-nematode, *Heterodera avenae*, and other cyst nematodes. Nematologica 23:193–201

Kerry BR, Crump DH (1998) The dynamics of the decline of the cereal cyst nematode *Heterodera avenae* in four soils under intensive cereal production. Fundamental Appl Nematol 21:617–625

Kerry BR, Crump DH, Mullen LA (1982) Studies of the cereal cyst-nematode, *Heterodera avenae,* under continuous cereals, 1974–1978. II. Fungal parasitism of nematode females and eggs. Ann Appl Biol 100:489–499

Kerry BR, Kirkwood IA, de Leij FAA et al (1993) Growth and survival of *Verticillium chlamydosporium* Goddard, a parasite of nematodes, in soil. Biocontrol Sci Technol 3:355–365

Khambay BPS, Bourne JM, Cameron S et al (2000) A nematicidal metabolite from *Verticillium chlamydosporium*. Pest Manag Sci 56:1098–1099

Loffredo A, Bent E, McKenry MV et al (2007) Understanding a root-knot nematode suppressive soil. J Nematol 39:86

Lopez-Llorca LV, Robertson WM (1992) Immunocytochemical localization of a 32-kda protease from the nematophagous fungus *Verticillium suchlasporium* in infected nematode eggs. Exp Mycol 16:61–267

Lopez-Llorca LV, Bordallo JJ, Salinas J et al (2002a) Use of light and scanning electron microscopy to examine colonisation of barley rhizosphere by the nematophagous fungus *Verticillium chlamydosporium*. Micron 33:61–67

Lopez-Llorca LV, Olivares-Bernabeu C, Salinas J et al (2002b) Pre-penetration events in fungal parasitism of nematode eggs. Mycol Res 106:499–506

Maciá-Vicente JG, Jansson H-B, Talbot NJ et al (2009) Real-time PCR quantification and live-cell imaging of endophytic colonization of barley (*Hordeum vul*gare) roots by *Fusarium equiseti* and *Pochonia chlamydosporia*. New Phytol 182:213–228

Manzanilla-López RH, Atkins SD, Clark IM et al (2009) Measuring abundance, diversity and parasitic ability in two populations of the nematophagous fungus *Pochonia chlamydosporia* var. *chlamydosporia*. Biocon Sci Techn 19:391–406

Mauchline TH, Kerry BR, Hirsch PR (2002) Quantification in soil and the rhizosphere of the nematophagous fungus *Verticillium chlamydosporium* by competitive PCR and comparison with selective plating. Appl Environ Microbiol 68:1846–1853

Mauchline TH, Kerry BR, Hirsch PR (2004) The biocontrol fungus *Pochonia chlamydosporia* shows nematode host preference at the infraspecific level. Mycol Res 108:161–169

Meyer SLF, Huettel RN, Liu XZ et al (2004) Activity of fungal culture filtrates against soybean cyst nematode and root-knot nematode egg hatch and juvenile motility. Nematol 6:23–32

Morton CO, Hirsch PR, Peberdy JP et al (2003a) Cloning of and Genetic Variation in the Protease VCP1 from the nematophagous fungus *Pochonia chlamydosporia*. Mycol Res 107:38–46

Morton CO, Mauchline TH, Kerry BR et al (2003b) PCR-based DNA fingerprinting indicates host related genetic variation in the nematophagous fungus *Pochonia chlamydosporia*. Mycol Res 107:198–205

Morton CO, Hirsch PR, Kerry BR (2004) Infection of plant-parasitic nematodes by nematophagous fungi – a review of the application of molecular biology to understand infection processes and to improve biological control. Nematology 6:161–170

O'Flaherty SM, Hirsch PR, Kerry BR (2003) The influence of the root-knot nematode *Meloidogyne incognita*, the nematicide aldicarb and the nematophagous fungus *Pochonia chlamydosporia* on heterotrophic bacteria in soil and the rhizosphere. Eur J Soil Sci 54:759–766

Orion D, Kritzman G, SIF M et al (2001) A role of the gelatinous matrix in the resistance of root-knot nematode (*Meloidogyne* spp.) eggs to microorganisms. J Nematol 33:203–207

Rao MS, Kerry BR, Gowen SR et al (1997) Management of *Meloidogyne incognita* in tomato nurseries by integration of *Glomus deserticola* with *Verticillium chlamydosporium*. Z Pflanzenk Pflanzen – J Plant Dis Prot 104:419–422

Segers R (1996) The nematophagous fungus *Verticillium chlamydosporium*, aspects of pathogenicity. PhD thesis, University of Nottingham

Segers R, Butt TM, Kerry BR et al (1996) Title: The role of the proteinase VCP1 produced by the nematophagous *Verticillium chlamydosporium* in the infectionprocess of nematode eggs. Mycol Res 100:421–428

Sharon E, Chet I, Viterbo A et al (2007) Parasitism of *Trichoderma* on *Meloidogyne javanica* and role of the gelatinous matrix. Eur J Plant Pathol 118:247–258

Siddiqui IA, Atkins SD, Kerry BR (2009) Relationship between saprotrophic growth in soil of different biotypes of *Pochonia chlamydosporia* and the infection of nematode eggs. Ann Appl Biol 155:131–141

Sneh B, Humble SJ, Lockwood JL (1977) Parasitism of oospores of *Phytophthora megasperma* var *sojae*, *Phytophthora cactorum*, *Pythium* spp, and *Aphanomyces euteiches* in soil by oomycetes, chytridiomycetes, hyphomycetes, actinomycetes, and bacteria. Phytopathology 67:622–628

Stirling GR, Kerry BR (1983) Antagonists of the cereal cyst-nematode *Heterodera avenae* Woll. in Australian soils. Aust J Exp Agric Anim Husb 23:318–324

Sun M-H, Gao L, Shi Y-X, Li Bao-Jü, Liu X-Z (2006) Fungi and actinomycetes associated with *Meloidogyne* spp. eggs and females in China and their biocontrol potential. J Invert Pathol 93:22–28

Sutherland ED, Papavizas GC (1991) Evaluation of oospore hyperparasites for the control of *Phytophthora* crown rot of pepper. J Phytopathol 131:33–39

Tahseen Q, Clark IM, Atkins SD et al (2005) Impact of the nematophagous fungus *Pochonia chlamydosporia* on nematode and microbial populations. Commun Agric Appl Biol Sci 70:81–86

Tikhonov VE, Lopez-Llorca LV, Salinas J et al (2002) Purification and characterization of chitinases from the nematophagous fungi *Verticillium chlamydosporium* and *V. suchlasporium*. Fungal Genet Biol 35:67–78

Willcox J, Tribe HT (1974) Fungal parasitism in cysts of *Heterodera* I Preliminary investigations. Trans Br Mycol Soc 62:585–594

Zaki MJ, Maqbool MA (1993) Occurrence of *Verticillium chlamydosporium* egg parasite of root-knot nematodes *Meloidogyne javanica* and *M. incognita* in Pakistan. Pak J Nematol 11:37–40

Zare R, Gams W (2004) A monograph of *Verticillium* section *Prostrata*. Rostaniha (Suppl 3):194

Chapter 8
Trichoderma as a Biological Control Agent

Edna Sharon, Ilan Chet, and Yitzhak Spiegel

Abstract *Trichoderma* species are free-living fungi that are common in soil and root ecosystems. Some strains establish root colonization and enhance growth and development, crop productivity, resistance to abiotic stresses and uptake and use of nutrients. *Trichoderma* species can antagonize and control a wide range of economically important plant pathogenic fungi, viruses, bacteria and nematodes. Root-knot nematodes, *Meloidogyne* spp., are sedentary, obligatory root endoparasites of great economic importance, and polyphagous species, such as *M. javanica* and *M. incognita* are among the major limiting factors of crops production worldwide. Therefore, these nematodes have been the main target for nematode biocontrol by *Trichoderma*. Several *Trichoderma* species and isolates have been evaluated as biocontrol agents against the nematodes with various crops and experimental conditions. Significant results of nematode control and plants growth were achieved. Aiming to improve the biocontrol process, modes of action of the fungus against the root-knot nematodes have been investigated and are described in this chapter. Mechanisms such as parasitism, enzymatic lysis, antibiosis and induced resistance were studied. Understanding the fungus-nematode-plant interactions and the mechanisms of the biocontrol process might contribute to improve the implementation of this biocontrol agent.

E. Sharon* (✉) and Y. Spiegel
Division of Nematology, Institute of Plant Protection, ARO, The Volcani Center,
Bet Dagan 50250, Israel
(*Part of the Ph.D. thesis supervised by Spiegel & Chet)
e-mail: ednash2@gmail.com; spiegely@volcani.agri.gov.il

I. Chet
Department of Microbiology and Plant Pathology, Faculty of Agricultural,
Food and Environmental Quality Sciences, Hebrew University of Jerusalem,
Rehovot 76100, Israel
e-mail: chet@agri.huji.ac.il

K. Davies and Y. Spiegel (eds.), *Biological Control of Plant-Parasitic Nematodes: Building Coherence between Microbial Ecology and Molecular Mechanisms*,
Progress in Biological Control 11, DOI 10.1007/978-1-4020-9648-8_8,

Abbreviations

CF	culture filtrate
gm	gelatinous matrix
GFP	green fluorescent protein
J2	second-stage juvenile
MAb	monoclonal antibody
PAb	polyclonal antibody
RKN	root-knot nematode
SC	surface coat

8.1 Introduction

Trichoderma species are free-living fungi that are common in soil and root ecosystems; some strains are known as opportunistic, avirulent plant symbionts and can establish robust and long-lasting colonizations of root surfaces and penetrate into the epidermis and a few cells below this level. Root colonization by *Trichoderma* spp. frequently enhances root growth and development, crop productivity, resistance to abiotic stresses and uptake and use of nutrients (Yedidia et al. 2001; Harman et al. 2004). *Trichoderma* species can antagonize and control a wide range of economically important plant-pathogenic fungi and have been known as biocontrol agents against soil-borne, foliar and postharvest phytopathogenic fungal pathogens and can control also viruses and bacteria (Sivan and Chet 1992; Herrera-Estrella and Chet 1998; Yedidia et al. 2003; Harman 2006).

Various mechanisms have been suggested for the biocontrol activity of *Trichoderma* against phytopathogenic fungi: direct interactions such as parasitism, enzymatic lysis, antibiosis and competition. Indirect interactions involve the stimulation of plant self-defence mechanisms, i.e., plant systemic induced resistance (Harman et al. 2004; Harman 2006; Viterbo et al. 2007a). Most of these processes are probably caused by multi-gene complexes (Harman 2000), and it can be assumed that biocontrol is a result of multi-mechanism action of the antagonist. Synergism between different forms of antagonism may occur (Elad and Freeman 2002; Howell 2003).

Most mechanisms, apart from competition, could potentially be involved in the biocontrol of nematodes. Enzymes such as chitinases, glucanases and proteases seem to be very important in the mycoparasitic process (Haran et al. 1996; Viterbo et al. 2002b). Chitinases and proteases of *Trichoderma* spp. are much similar to those of nematophagous fungi, and have the potential to attack nematodes (Morton et al. 2004). The processes of *Trichoderma* parasitism and the effects of fungal enzymes and metabolites on nematodes may occur in the soil, within roots and on the root surfaces, and induced systemic resistance mechanisms may also affect the nematodes. Microorganisms are affected by environmental conditions in the rhizosphere, and since nematodes influence the quantity and quality of root exudates, they are likely to affect the physiology of such microorganisms in the rhizosphere (Kerry 2000).

Root-knot nematodes (RKNs), *Meloidogyne* spp. are sedentary, obligatory root endoparasites of great economic importance, and polyphagous species, such as *M. javanica* and *M. incognita* are among the major limiting factors in the production of field and plantation crops worldwide. RKNs are difficult to control because of their wide host range, short life-cycle, high reproductive rates and endoparasitic nature (Trudgill and Blok 2001; Manzanilla-Lopez et al. 2004). Therefore, these nematodes have been the main target for biocontrol by *Trichoderma*. The second-stage juveniles (J2s), which penetrate the roots and develop within them, induce a cascade of changes in the host plant, which lead to the formation of giant cells and galls. About 1 month after J2s penetration, the females lay out egg masses that contain nematode eggs enveloped in a gelatinous matrix (gm).

Several attempts have been made to use *Trichoderma* species to control plant-parasitic nematodes. Windham et al. (1989) reported reduced egg production in the root-knot nematode *M. arenaria*, following soil treatments with preparations of *T. harzianum* (T-12) and *T. koningii* (T-8). A combination of *T. harzianum* with neem cakes reduced the population of the citrus nematode *Tylenchulus semipenetrans* (Reddy et al. 1996). Among several other plant-based formulations of *T. harzianum* that were evaluated for the management of *M. incognita,* castor cake extracts showed the best biocontrol activity (Rao et al. 1998). Direct interactions between *T. harzianum* and the potato cyst nematode *Globodera rostochiensis* were demonstrated *in vitro* by Saifullah and Thomas (1996). The effect of *T. viride* metabolites on nematodes was demonstrated by implementing root-dip treatments with the fungal culture filtrate (Khan and Saxena 1997). *In vitro* assays with *T. virens* culture filtrates showed that low-molecular-weight, non-enzymatic factors inhibited egg hatching and impaired *M. incognita* second-stage juvenile mobility. The fungus, applied as seed treatment or root drenches, did not affect nematode *M. incognita* inoculation in greenhouse tests with tomato, but did achieve reductions in the nematode population on pepper roots (Meyer et al. 2000, 2001). *Trichoderma*-nematode interactions, has been studied by an Israeli group with main emphasis on RKNs, combining applied and fundamental research. Several *Trichoderma* species and isolates have been evaluated as biocontrol agents against *M. javanica* and *M. incognita* with various crops and experimental conditions. Significant results of nematode control and plants' growth improvements were achieved (Sharon et al. 2001; Spiegel et al. 2007).

Aiming to improve the biocontrol process, modes of action of the fungus against the root-knot nematodes have been investigated and are described in this chapter: Attachment and parasitic capabilities of *Trichoderma* on RKNs were demonstrated and the mechanisms were investigated (Sharon et al. 2007). Antibodies that bind to *M. javanica* surface served as a tool for further investigations of the fungal attachment to nematodes; antibodies were found to improve parasitism *in vitro* (Sharon et al. 2009b). Involvement of proteolytic and chitinolytic activities during parasitism has been investigated. *Trichoderma* metabolites affected the nematodes and differences were observed between the various isolates. Indirect effects of fungal root colonization on the nematodes were demonstrated using split-root systems, suggesting induced systemic resistance mechanisms in the host plants (Sharon et al. 2009a). Understanding the fungus-nematode-plant interactions and the mechanisms involved in the biocontrol process for various *Trichoderma* species and isolates might contribute to the development of optimal implementation methods to improve biocontrol agents.

8.2 *Trichoderma* Biocontrol Activity Against Root-Knot Nematodes

Trichoderma asperellum-203 and *T. atroviride* IMI 206040 (both fungi were previously defined as strains of *T. harzianum*) exhibited biocontrol activity against *M. javanica* in soil (Sharon et al. 2001). Several other *Trichoderma* species and isolates (3 isolates of *T. asperellum*: 44, GH11 and 34; *T. harzianum* 248; *T. hamatum* 382) have been also evaluated as biocontrol agents against *M. javanica* and *M. incognita*. Those *Trichoderma* isolates had shown biocontrol activity against plant pathogenic fungi. Significant biocontrol activities against the RKNs were obtained with several vegetable crops, such as: tomatoes, cucumbers, egg plants and lettuce, as well as with ornamentals. Experiments were conducted with pots, up to 50 L containers, in growth-chambers and in microplots. Peat-wheat bran *Trichoderma* preparations were applied to different soils (or potting mixes) 1–2 weeks before planting and/or to the potting mix of the growing seedlings. *Trichoderma*-treated plants exhibited reduced galling indices and egg production, while weights of shoots, fruits and roots were higher and flowering was improved. Fungal application to both seedlings and pots improved the results (Sharon et al. 2001, 2007; Spiegel et al. 2007). *Trichoderma* species and isolates were tested for compatibility on agar plates. Some isolates belonging to same species (*T. asperellum*) showed compatibility - no distinct barrier was observed in their meeting line. In pot experiments of dual isolates combinations, improved biocontrol was achieved with some combinations, while others did not show better results and sometimes were even worse than each isolate alone (Spiegel, Sharon, Chet unpubl.).

8.3 Attachment and Parasitism

Parasitism is probably an important mode of action and attachment is one of the initial steps of it. *Trichoderma asperellum*-203 and *T. atroviride* showed the ability to parasitize nematode eggs and J2s (Sharon et al. 2001). Mechanisms involved in the attachment and parasitism processes were investigated, with special attention to the role of the gelatinous matrix (gm) in direct nematode-fungus interactions. It was found that the gm enables fungal attachment and enhances parasitic abilities of most isolates, which could also utilize it as a nutrient source. Fungal conidia can attach to nematode egg masses and to eggs and J2s that had contact with the gm, whereas gm-free J2s and eggs are almost unattached by fungal conidia. However, differences were observed among the various *Trichoderma* species and isolates, in their attachment and parasitism capabilities, indicating the specificity of the processes (Sharon et al. 2007).

Conidia attachment and parasitism processes were microscopically monitored *in vitro*. A green fluorescent protein (GFP)-expressing *T. asperellum*-203 construct was used to observe the parasitism process (Fig. 8.1). Observations made by scanning electron microscopy (SEM) enabled a more detailed insight on this process, showing typical fungal parasitic behavior, including tight attachment of spores and hyphae, coiling of hypae around J2 and appressoria-like structures formation. Gelatinous matrix-free eggs and J2s were penetrated only by few fungal hyphae and colonized. The role of the gm in attachment was studied by using gm that had been separated from egg-masses. Conidia were agglutinated by a gm suspension (enhanced in presence of Ca^{2+}) and their germination was improved (Sharon et al. 2007). A biomimetic system based on nylon fibers, originally developed and used by Inbar and Chet (1994) for investigations of mycoparasitism, was modified and used with gm-coated fibers. It successfully expressed the specific triggering of fungal attachment and parasitic growth patterns by the gm, similar to the parasitism on the nematodes (Fig. 8.2) (Sharon et al. 2007). *Trichoderma* parasitic patterns on nematodes and nylon fibers, resembled mycoparasitc behavior (Viterbo et al. 2007a) and patterns induced by lectins derived from host fungi (Inbar and Chet 1994, 1997).

Hyphae of *T. atroviride*, which was the most effective parasite of the J2s, showed higher tendency to coil around the J2s than those of *T. asperellum*-203. Similar results with respect to the coiling process have been obtained in fungal-fungal biomimetic interactions using nylon fibers, especially after induction with a G-protein activator (Omero et al. 1999). The signal-transduction pathways downstream of the recognition event have recently been intensively investigated, with a focus on the role of G-protein α-subunit genes (Zeilinger et al. 2005). Further investigations may determine whether similar pathways are involved in gm induction of fungal parasitic behavior.

To understand *Trichoderma*–nematode direct interactions, the effects of *M. javanica* surface-binding antibodies on the parasitism was studied. The nematode's surface coat (SC) is considered to be important in recognition events involving plant hosts and microbial antagonists (Spiegel and McClure 1995; Kerry and Hominick 2001; Koltai et al. 2002; Morton et al. 2004). The nature of *Meloidogyne* species SCs has been studied (Spiegel et al. 1995, 1997; Lin and McClure 1996) and antibodies have been used to characterize surface antigens and the interactions with plant hosts (Gravato-Nobre and Evans 1998; Lopez de Mendoza et al. 1999) and microorganisms (Spiegel et al. 1996; Davies 2005).

Meloidogyne javanica (J2s or J2s and eggs) surface-binding monoclonal (MAb) and polyclonal (PAb) antibodies were tested for their effects on the

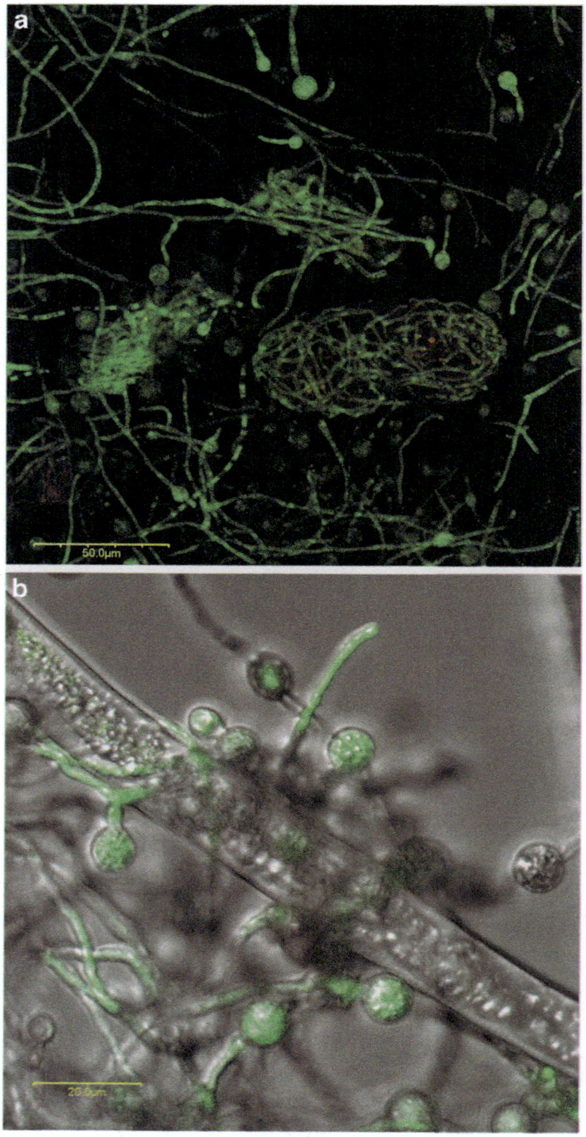

Fig. 8.1 Parasitism of *Trichoderma asperellum*-203 (constitutively expressing GFP construct) on *Meloidogyne javanica* on (**a**) eggs, bar = 50 μm, and on (**b**) second-stage juvenile, bar = 20 μm

nematode-*Trichoderma* interactions. Those antibodies inhibited J2s movement and therefore reduced root penetration (Sharon et al. 2002, 2009b). Parasitism of *T. asperellum*-203 and *T. atroviride* on nematode egg masses, eggs and juveniles was enhanced when antibodies were incorporated into *in-vitro* parasitism bioassays. Parasitism on gm-free and J2s was also improved, compared to controls without antibodies that almost did not attach fungal conidia. Improved parasitism could be

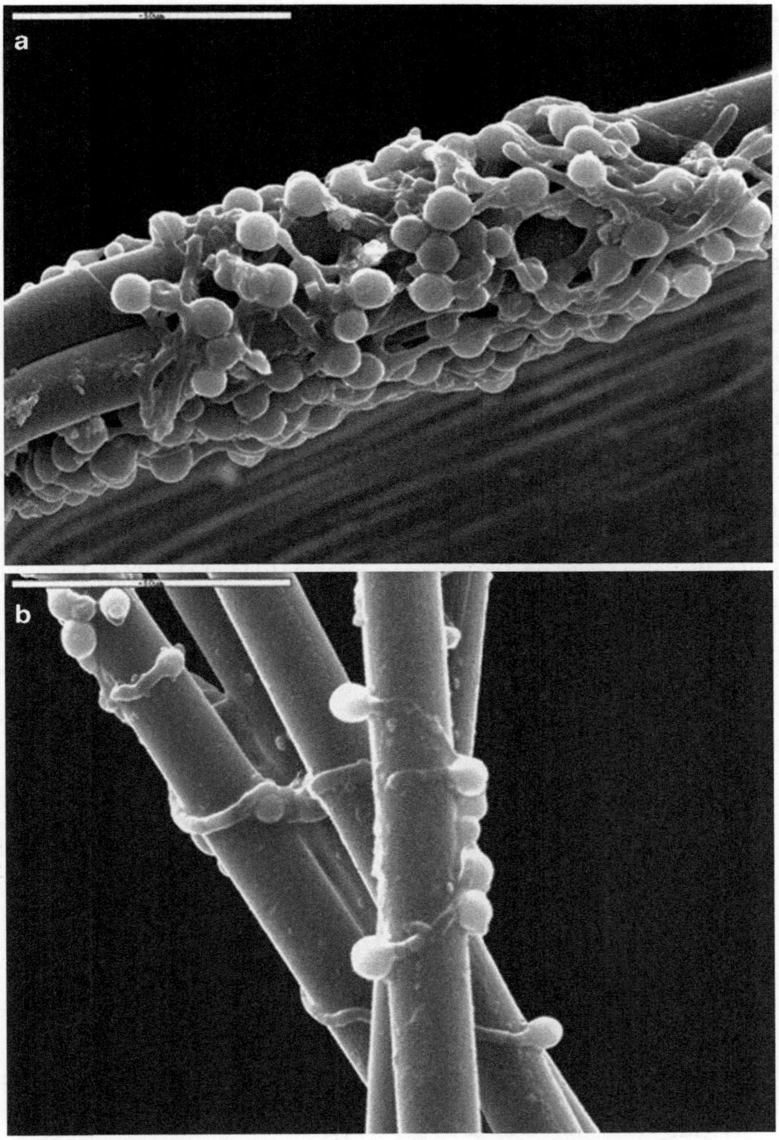

Fig. 8.2 Scanning electron micrographs of nylon fibers coated with gelatinous matrix (gm). (**a**) *Trichoderma asperellum*-203 conidium attachment, bar = 50 μm. (**b**) Fungal parasitic-like behavior of *T. atroviride*: tight adhesion of hyphae and coiling, bar = 50 μm

due to bilateral binding of the antibodies to the nematodes and conidia, enabling better conidial attachment to the nematodes. Enhanced germination of antibody-bound conidia further improved parasitism. Differences were observed among antibodies in their effects on fungal parasitism and their interaction with *Trichoderma* species.

Focus was made mainly on the egg- and juvenile-binding MAb MISC, which had been raised against *M. incognita* (race 3) and had exhibited specificity to fucosyl-bearing epitopes (Gravato-Nobre et al. 1999). Binding of MISC to *M. javanica* egg masses, eggs and J2s was inhibited by pretreatment of the MAb with fucose; therefore, the fucose-specific lectin, UEA-I, was used, and it also resulted in specific enhancement of conidial binding to nematodes and conidial agglutination, similar to the effect of the antibody. The labeling of gm and gm-originated eggs with UEA-I and its specific inhibition by the carbohydrate fucose indicate that the gm contains fucose residues.

A model for fungal conidia attachment to nematodes (Fig. 8.3) suggests that carbohydrate-lectin-like interactions might be involved in this process; such interactions are sometimes Ca^{2+}-dependent (Sharon et al. 2007, 2009b). This model addresses the roles of fucose and fucose-specific antibody and lectin; nevertheless, other carbohydrates/lectins interactions might be involved in these attachment processes. On the surface coat of gm-free J2s of *M. javanica* there are fucose-, mannose- and glucose-binding proteins (carbohydrate recognition domains CRDs) (Sharon and Spiegel 1996; Spiegel et al. 1995, 1997), and fucose residues. The gm contains fucose and fucose-binding domains (FBD) (Sharon and Spiegel 1993), molecules that also occur on *Trichoderma* conidia (Elad et al. 1983). Fucose inhibited conidia attachment to J2s, conidia agglutination by gm suspension and their attachment to nylon fibers; attachment was also inhibited after periodate treatment of nematodes. The model suggests that during J2's hatch from egg mass, gm, which

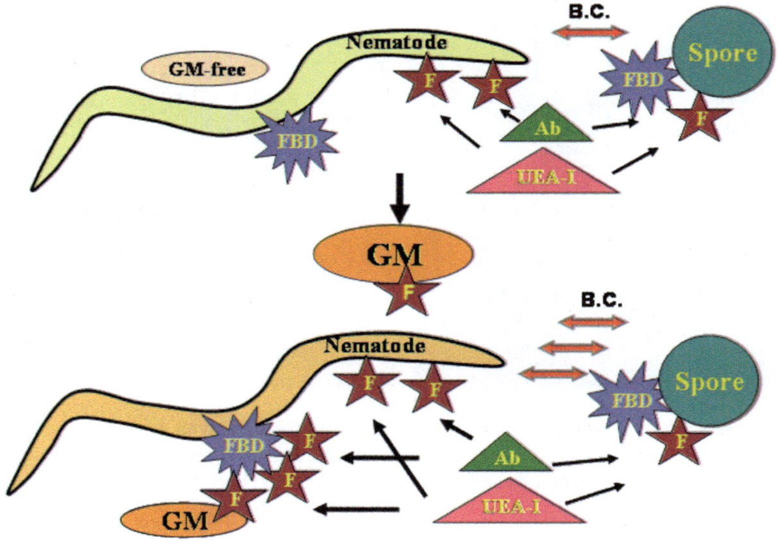

Fig. 8.3 Attachment model of *Trichoderma* spores to *Meloidogyne javanica* second-stage juveniles. *GM* gelatinous matrix, *F* fucose, *FDB* fucose-binding domain, *Ab* antibody, *Bc* biocontrol

contains carbohydrates such as fucose, binds to the J2s surface coat and this can alter their binding affinity to the fungal conidia that contain fucose-binding domains. As a result, gm-J2s are efficiently attached and parasitized by the fungus. Carbohydrate residues, such as fucose, on the surface of the nematode and fungal conidia can be involved in the antibody- and lectin-mediated improved attachment and parasitism.

One of the most interesting features of the nematode SC is its dynamic nature: there is a continuous turnover that involves shedding and replacement of the surface antigens (Lin and McClure 1996; Spiegel et al. 1997; Blaxter and Robertson 1998). However, surface-characterization studies of *Meloidogyne* species have been performed mainly on gm-free J2s and no attention has been paid to the role of the gm and its effect on interactions between nematodes and microorganisms. The MAb MISC has also been observed to label the gm of *M. incognita* (race 3) and the rectal glands, where the gm originates (Hu et al. 2000). Nevertheless, the fate of gm-originated components on the surface of *Meloidogyne* J2s and during the SC turnover process remains unclear.

The gm plays a key role in the process of *Trichoderma* conidia attachment to the nematode and in the ensuing parasitism. The gm is usually considered a defensive envelope that protects the eggs against microorganisms and enables the egg mass to survive in the soil (Sharon et al. 1993). Bacteria that were agglutinated by the gm could not reproduce in its presence, whereas others, which were not agglutinated, utilize the gm as a nutrition source and reproduce (Sharon et al. 1993). Thus, the ability of some *Trichoderma* species to be agglutinated by the gm and grow on it is unique, and partially accounts for their ability to attack RKNs.

Direct parasitism of *Trichoderma* on nematode life-stages on the roots might be important for a successful biocontrol process. The potential ability to parasitize nematode life-stages *in planta* was demonstrated with *T. asperellum*-203, which interacted with penetrating J2, and with females and egg masses on roots in soil, thereby interfering with the reproduction process (Sharon et al. 2007). The high affinity of this isolate as a root-surface colonizer (Yedidia et al. 1999) probably enhances these parasitic fungus-nematode interactions on the root surface. The ability of the fungus to colonize nematode penetration holes in the root might contribute to plant defense against secondary pathogens that usually exploit the penetration of the roots by the nematodes.

8.4 Lytic Enzymes and Metabolites

Following the attachment process, fungal lytic activities are induced in order to digest the host. For efficient parasitism, a biocontrol agent should overcome several barriers that protect the nematodes from the external environment. The eggshell forms an important barrier that is composed of three layers: an outer – vitelline (protein), middle – chitinous and an inner lipo-protein layer. The amino acid composition of the eggshell indicates that it probably contains collagen-like proteins, which provide the tough, resilient properties associated with eggshells (Morton et al. 2004). The chitinous layer provides strength to the eggshell and is the thickest and most obvious layer; that of *M. javanica* is thicker than those of other plant-parasitic species. This layer protects the lipid layer, which determines the permeability and protects from harmful chemicals (Wharton 1980). A combination chitinolytic and proteolytic enzymes is required to disrupt the eggshell (Tikhonov et al. 2002; Khan et al. 2004), although chitinolytic capacity is probably the most important activity on the eggshells (Morton et al. 2004).

Another barrier is the cuticle that is composed mainly of collagens (Blaxter and Robertson 1998).

The role of *Trichoderma* lytic enzymes in plant defense and in fungal biocontrol processes was reviewed by Viterbo et al. (2002b), Markovich and Kononova (2003) and by Steyaert et al. (2003). Synergistic actions of different hydrolytic enzymes have been reported (Elad and Freeman 2002). Proteolytic activities of *Trichoderma* have not been investigated extensively as those of other lytic enzymes such as chitinases, but they have recently begun to be explored.

8.4.1 Proteases

One of the most studied *Trichoderma* proteases is a 31-kDa basic proteinase (Prb1), produced by *T. atroviride* strain IMI 206040, which was identified and characterized as a serine protease and belongs to the S8 family. The gene encoding this proteinase was cloned by Geremia et al. (1993). The gene expression was repressed by glucose and induced by fungal cell wall preparations of *R. solani* or chitin (Flores et al. 1997). This enzyme was subjected to nitrogen catabolite repression (Olmedo-Monfil et al. 2002). Transgenic fungal lines, carrying multiple copies of *prb1,* revealed improved biocontrol activity against *R. solani* in cotton plants (Flores et al. 1997). Those lines were used also to study the role of this proteinase in fungus-nematode interactions. Line P-2 exhibited improved nematode biocontrol capacity in soil and on all nematode life- stages that were tested *in vitro*, indicating that this proteinase is involved in the nematode biocontrol process (Sharon et al. 2001). Involvement of the *prb1* gene in nematode parasitism was supported by microscope observations, using a GFP inducible reporter construct (Provided by Prof. A. Herrera-Estrella, Mexico), which showed that expression of this gene was induced during fungal parasitism on the various life stages of the root-knot nematode, especially those that involve gm (Sharon et al. 2007).

Other protease activities were detected in *T. atroviride* during nematode parasitism process. Amino acid sequencing of peptides from these proteases revealed peptides with similarity to some acid proteases. The proteolytic profile of *T. asperellum* strains differed from that of *T. atroviride*; in *T. asperellum*-203, the Prb1 seems not to be involved in nematode parasitism. Nevertheless, some proteases presented alleviated activities during parasitism. In some *Trichoderma* isolates, protease activities were very low or not detected (Spiegel, Sharon, Chet unpubl.). Differences in proteolytic capabilities of the various *Trichoderma* species and isolates might partially account for their different capabilities in parasitism on *Meloidogyne* life-stages.

PRA1, a serine-protease with trypsin-like activity was isolated from *T. harzianum* CECT 2413 (Suarez et al. 2004). This 28 kDa protease might be related to mycoparasitic interactions and exhibited nematicidal activity. PRA1 was found to be induced by conditions simulating antagonism, to be subject to nitrogen and carbon derepression, and to be affected by the pH of the culture medium, its optimal pH range being 7–8. Purified preparations of PRA1 reduced *M. incognita* egg hatch during *in vitro* assays and this nematicidal effect was enhanced by the use of fungal culture filtrates (CFs), suggesting that PRA1 has additive or synergistic interactions with other proteins produced during the antagonistic activity of the fungus (Suarez et al. 2004). Suarez et al. (2007) characterized the genes of six novel endopeptidases from *T. harzianum* CECT 2413, belonging to different families. Gens within a family are differently regulated in response to different culture conditions, suggesting that they have diverse functional roles.

8.4.2 Chitinases

The chitinolytic system of *Trichoderma* and its role in mycoparasitism have been intensively investigated (Kubicek et al. 2001). Several chitinases and their related genes have been isolated from *Trichoderma* spp. growing in media containing chitin as a sole carbon source. Generally, carbon starvation, products of chitin degradation, fungal cell-walls, and colloidal chitin are thought to induce chitinolytic enzyme expression, whereas glucose and other easily fermented carbon sources serve as repressors (Viterbo et al. 2002b).

N-acetylglucosaminidases. Two GlcNAcases, CHIT73 and CHIT102, were detected, isolated and identified in *Trichoderma asperellum*-203 growth medium (Ramot et al. 2004); the genes *exc1* and *exc2* encode for these enzymes, respectively. These enzymes were up-regulated by glucosamine and CHIT102 formed homodimers. CHIT102 was the first chitinase to appear upon contact with *S. rolfsii*, therefore, it has been speculated that it plays a unique role in triggering the expression of other chitinolytic enzymes (Haran et al. 1996; Viterbo et al. 2002b). The gene *nag1* in *T. atroviride* is a homologue of *exc1y* from *T. asperellum*. The Nag1 was extensively investigated by Brunner et al. (2003), who showed that it is essential for chitinase induction by chitin and, therefore, is of major relevance in biocontrol.

Exochitinases (Chitobiosidases). When grown on crab-shell chitin as the sole carbon source, a chitobiosidase of 40 kDa was secreted from *T. atroviride* P1 (Harman et al. 1993).

Endochitinases. An endochitinase of 42 kDa has been isolated from several different strains of *Trichoderma*; it is believed also to be a key enzyme in the mycoparasitic interaction (Carsolio et al. 1999; Zeilinger et al. 1999). Two more endochitinases – of

37 and 33 kDa – were reported in *T. harzianum* (Viterbo et al. 2002b), and a new endochitinase termed CHIT36 (previously designated CHIT33 by Haran et al. (1996)), was isolated from *T. harzianum* isolate TM (Viterbo et al. 2001). The CHIT36 from *T. asperellum*- 203 is very similar (Viterbo et al. 2002a). A 37-kDa endochitinase has been isolated from *T. harzianum* 109 (De Marco et al. 2000).

Induction of chitinolytic activities during fungal parasitism on nematodes was demonstrated using GFP reporter constructs: the endochitinases CHIT36 and CHIT42 in *T. asperellum*-203 and *T. atroviride* P1, respectively and the *β-N*-acetyl-D-glucoseaminidases (Hexoaminidases) CHIT102 and Nag1 in those species, respectively. As in the case of proteinase Prb1, the presence of gm enhanced the production of chitinilytic enzymes (Sharon et al. 2009a).

Steyaert et al. (2003) suggested that there was co-regulation of the genes *prb1* and *chit42* in *T. hamatum*. The genes *prb1* and *chit42* were not induced by lectins in the fungal biomimetic system but were induced by diffusible factors from the host fungus (Cortes et al. 1998), a process that involves regulation pathways other than coiling and conidiation processes (Rocha-Ramirez et al. 2002).

8.4.3 Effects of Environmental Conditions on Enzymes

There is growing evidence for the effect of ambient pH on the expression and activity of fungal extracellular enzymes. In the regulation of many proteases, pH plays prominent role (St. Leger et al. 1998). In response to environmental signals, enzymes production during pathogenesis is probably regulated by the structural elements of the host, nutrient limitation, and the ambient pH. Fungi may be able to adjust the pH of micro-environments to facilitate optimal enzyme activity (Morton et al. 2004) and ecological niches may be actively improved and protected by the fungi during plant-*Trichoderma* interactions (Suarez et al. 2004).

Environmental different pH and nutritional conditions can be crucial for the production and activity of fungal enzymes and metabolites that affect the nematodes. Proteases that are apparently involved in nematode biocontrol required acidic pH conditions for activity, similar to the optimal conditions for chitinases activities (Schickler et al. 1998). Some of the parasitism-related proteolytic and chitinolytic enzymes presented nitrogen catabolite repression. Enzymes such as Prb1 are induced in presence of nitrate and repressed by ammonium and are capable of responding to different environmental conditions that may reflect stress conditions (Olmedo-Monfil et al. 2002). Micro-environments, in soil and rhizosphere, supporting preferable conditions for *Trichoderma* biocontrol activity may improve the process.

8.5 Antibiotics Production

Trichoderma species can produce a variety of secondary metabolites, including antibiotic compounds, which may contribute to the biocontrol processes (Howell 1998). The nature and roles of antibiotic peptides that belong to the peptaibol group have been intensively studied (Szekeres et al. 2005). Peptaibols generally exhibit antimicrobial activity, which is thought to arise from their membrane activity and their ability to form pores in lipid membranes. A peptaibol synthetase gene has been cloned (Wiest et al. 2002) and further studies suggested that peptaibols are critical in the chemical communication between *Trichoderma* and plants as triggers of non-cultivar-specific defense responses (Viterbo et al. (2007b)). *Trichoderma virens* produces gliotoxin and gliovirin and also peptaibols. Parallel formation and synergism of hydrolytic enzymes and peptaibol antibiotic action of *Trichoderma* against phytopathogenic fungi has been reported (Schirmböck et al. 1994). The antifungal action of enzymes reinforced by synergism with antibiotics was comprehensively reviewed by Kubicek et al. (2001).

Nematicidal activity against *M. javanica* J2s was detected mainly in *T. atroviride* culture filtrates (CFs); the active component/s showed low molecular weight (MW) and heat sensitivity. The nematicidal activity was specifically increased during parasitism on egg-masses (Sharon et al. 2007). Immature eggs exhibited reduced hatching rates in presence of CF, whereas hatching of mature eggs was enhanced. The effect of CF on eggs was contributed by both the enzymatic fraction, which contained proteases and chitinases, and by the low-MW component/s. (Spiegel, Sharon, Chet unpubl.). Appropriate candidates responsible for such nematicidal activity might be antibiotic peptides, such as peptaibols (Sharon et al. 2007). Such molecules have been identified and sequenced in *T. atroviride* by Oh et al. (2000).

8.6 Induced-Resistance

Trichoderma strains that are capable of root interaction induce metabolic changes in plants that increase resistance to a wide range of plant-pathogenic microorganisms and viruses. This response seems to be broadly effective for many plants, which indicates that there is little or no plant specificity (Harman et al. 2004). *Trichoderma* strains produce or release a variety of compounds that induce localized or systemic responses, and this explains their lack of pathogenicity to plants. These elicitors include peptides, proteins and low-molecular weight compounds (Yedidia et al. 2000; Harman et al. 2004; Viterbo et al. 2004). These root-microorganism associations cause substantial changes to the plant proteome and metabolism. Plants are protected from numerous classes of plant pathogens by responses that are similar to systemic acquired resistance (SAR) and rhizobacteria-induced systemic resistance (RISR). In the SAR pathways there is direct production of pathogenesis-related (PR) proteins by the plant, mediated by either salicylic acid or jasmonic acid as signaling molecules. However, in RISR the PR proteins and phytoalexins are not induced in the absence of attack by plant pathogens (Harman et al. 2004).

Analysis of signal molecules involved in defense mechanisms, and application of specific inhibitors, indicated the involvement of jasmonic acid and ethylene in the protective effect conferred by *Trichoderma* spp. against the leaf pathogen *Pseudomonas syringae* pv. *lachrymans*. Moreover, examination of local and systemic gene expression revealed that *T. asperellum*-203 modulated the expression of genes involved in the jasmonate/ethylene signalling pathways of ISR in cucumber plants. Subsequent challenge of *Trichoderma*-preinoculated plants with the leaf pathogen resulted in higher systemic expression of the pathogenesis-related genes encoding for chitinase 1, β-1, 3-glucanase, and peroxidase relative to non-inoculated, challenged plants (Shoresh et al. 2005). The MAPK signal transduction pathways, both of the plant and *Trichoderma,* are important for the induction of systemic resistance (Viterbo et al. 2007a). Alfano et al. (2007) showed that *T. hamatum* 382 induced resistance response in tomato against bacterial spot of tomato and its pathogen *Xanthomonas euvesicatoria*. Fungal actively induced systemic changes in plant physiology and disease resistance through systemic modulation of the expression of stress and metabolism genes.

The indirect effects of fungal root colonization on *M. javanica* in pot experiments, using split-roots systems, was examined. Nematode infection was reduced, and inhibition of nematode development and egg production were recorded in root-halves that had not been exposed to the fungus, indicating potential involvement of systemic induced resistance mechanisms in the nematode biocontrol process (Sharon et al. 2009a). The effect on egg production might be due to the higher systemic expression of pathogenesis-related (PR) proteins, such as chitinases and peroxidases (Yedidia et al. 2000; Shoresh et al. 2005) that are produced in *Trichoderma*-preinoculated plants during the systemic response.

Jasmonic acid suppresses nematode infestation on tomato roots (Cooper et al. 2005). It is known to be transported from foliage to roots, where it can have a wide range of effects on plants development and metabolism. Jasmonates influence root growth and nutrient partitioning, which could potentially affect nematode parasitism (Cooper et al. 2005). Induced resistance seems to be an important indirect mechanism of the nematode biocontrol process. Further investigations are required to elucidate the pathways that mediate these systemic responses that affect the nematodes.

8.7 Interactions of *Trichoderma* with Other Nematodes and Microorganisms

The potential of *Trichoderma* to control other phytophagous nematodes is most promising, as was demonstrated *in vitro* with several nematode species. Culture filtrates (CFs) immobilized different plant-parasitic nematodes; nevertheless, non-target and beneficial nematode species were not harmed by direct parasitism with *T. asperellum*-203, nor by *T. atroviride* CFs (Spiegel, Sharon, Chet unpubl.).

The differing responses of nematode groups to fungal metabolites might be due to wide differences among the structures and compositions of their cuticles, which affect the permeability. The epicuticle is made up of lipids, and it appears to act as a hydrophobic barrier to diffusion. Lipids and glycolipids are presented on the surfaces of free-living nematodes and animal parasites, but there is very limited knowledge of their presence in plant-parasitic nematodes (Blaxter and Robertson 1998).

Interactions of different *Trichoderma* isolates with other organisms and microorganisms in the ecological systems and the possible effects on nematode biocontrol processes have not been deeply investigated. Competition between *Trichoderma* and other microorganisms might interfere with root colonization and biocontrol processes. *Trichoderma* spp. in general have been found to be highly resistant to a variety of toxins and other compounds, including antibiotics produced by other microorganisms, plant antimicrobial compounds and chemical fungicides (Harman et al. 2004). Synergy between mycorrhizal fungi and *Trichoderma* has been shown, as well as synergy between *Trichoderma* enzymes and bacterial antibiotics (Harman et al. 2004). Mixtures of different root-colonizing biocontrol agents can provide better results than any one agent used on its own (Whipps 2001). However, the abilities of combinations of beneficial root-colonizing microorganisms to improve plant performance have been inadequately examined in either managed or natural ecosystems (Harman et al. 2004).

Kok et al. (2001) reported that egg masses of *Meloidogyne* species from soils contained a bacterial community significantly greater than that of the rhizosphere. They suggested that the egg masses microflora may be an important factor in determining the success of nematode biocontrol agents. Interestingly, a strain of *Trichoderma* that strongly reacted against the biocontrol agent *V. chlamydosporium* was found among *M. fallax* egg masses microflora.

8.8 Concluding Remarks and Future Prospects

Trichoderma isolates are very unique biocontrol agents as they present a wide range of activities and interactions in their ecosystems as free-living, plant symbionts or parasites of plant pathogens such as fungi and nematodes. Some isolates have the potential to serve as broad-spectrum plant protection agents and growth promoters.

Several modes of action are involved in the activity of *Trichoderma* against nematodes: direct parasitism, which involves attachment and enzymatic digestion by proteolytic and chitinolytic enzymes, production of nematicidal metabolites, as well as indirect effects of induced systemic resistance in the host plants. Different species and isolates can specialize in distinct modes, so that the combined application of some compatible isolates may result in a synergistic effect. Understanding the main mode of action of different isolates may improve their application methods, implementation sites (i.e. soil, roots) rates and timing. Formulations can be improved and designed to support and enhance production of enzymes and metabolites. Better understanding of the processes involved in biocontrol could lead to developments in selection of active biocontrol isolates.

Chen and Dickson (2004) described different groups of fungal antagonists of nematodes: predacious fungi, endoparasites of vermiform nematodes, parasites of sedentary females and eggs, fungi that produce antibiotic substances, and vesicular-arbuscular mycorrhizal fungi (VAM) or VAM-like fungi. There is no clear-cut distinction between the categories, and some fungi can belong to more than one category. It is evident that in some *Trichoderma* species we can find all of these features.

The mechanisms potentially involved in the *Trichoderma* parasitism process on the nematode were studies mainly in the nematode-fungus interaction level. The presence of host plant triggers the *Trichoderma* to produce specific enzymes and metabolites (Harman et al. 2004; Viterbo et al. 2007b), that may also affect the nematodes. The tritrophic direct and indirect interactions should be further studied from ecology to molecular levels. Further applied and fundamental studies will enable the development of *Trichoderma* biocontrol potential also against other plant-parasitic nematodes.

References

Alfano G, Lewis Ivey ML, Cakir C et al (2007) Systemic modulation of gene expression in tomato by *Trichoderma hamatum* 382. Phytopathology 97:429–437

Blaxter ML, Robertson WM (1998) The cuticle. In: Perry RN, Wright DJ (eds) Free-living and plant-parasitic nematodes. CAB International, Wallingford

Brunner K, Peterbauer CK, Mach RL et al (2003) The Nag1 *N*-acetylglucosaminidase of *Trichoderma atroviride* is essential for chitinase induction by chitin and of major relevance to biocontrol. Curr Genet 43:289–295

Carsolio C, Benhamou N, Haran S et al (1999) Role of the *Trichoderma harzianum* endochitinase gene, *ech42*, in mycoparasitism. Appl Environ Microbiol 65:929–935

Chen S, Dickson DW (2004) Biological control of nematodes by fungal antagonists. In: Chen ZX, Chen SY, Dickson DW (eds) Nematology – advances and prospectives, vol II, Nematode management and utilization. CABI Publishing, Cambridge

Cooper WR, Jia L, Goggin L (2005) Effects of jasmonate-induced defences on root-knot nematode infection of resistant and susceptible tomato cultivars. J Chem Ecol 31:1953–1967

Cortes C, Gutierrez A, Olmedo V et al (1998) The expression of genes involved in parasitism by *Trichoderma harzianum* is triggered by a diffusible factor. Mol Gen Genet 260:218–225

Davies K (2005) Interactions between nematodes and microorganisms: bridging ecological and molecular approaches. Adv Appl Microbiol 37:53–78

De Marco JL, Lima LHC, valle de Sousa M et al (2000) A *Trichoderma harzianum* chitinase destroys the cell wall of the phytopathogen *Crinipellis perniciosa*, the casual agent of witches' broom disease of cocoa. World J Microbiol Biotechnol 16:383–386

Elad Y, Freeman S (2002) Biocontrol of fungal plant pathogens. In: Kempken F (ed) The Mycota, vol XI, Agricultural applications. Springer, Berlin/Heidelberg

Elad Y, Barak R, Chet I (1983) Possible role of lectins in mycoparasitism. J Bacteriol 154:1431–1435

Flores A, Chet I, Herrera-Estrella A (1997) Improved biocontrol activity of *Trichoderma harzianum* by over-expression of the proteinase-encoding gene *prb1*. Curr Genet 31:30–37

Geremia R, Goldman G, Jacobs D et al (1993) Molecular characterization of the proteinase-encoding gene *prb1* related to mycoparasitism by *Trichoderma harzianum*. Mol Microbiol 8:603–613

Gravato-Nobre MJ, Evans K (1998) Plant and nematode surfaces: their structure and importance in host-parasite interactions. Nematologica 44:103–124

Gravato-Nobre MJ, McClure MA, Dolan L et al (1999) *Meloidogyne incognita* surface antigen epitopes in infected *Arabidopsis* roots. J Nematol 31:212–223

Haran S, Schickler H, Chet I (1996) Molecular mechanisms of lytic enzymes involved in the biocontrol activity of *Trichoderma harzianum*. Microbiology 142:2321–2331

Harman GE (2000) Myths and dogmas of biocontrol. Plant Dis 84:377–391

Harman GE (2006) Overview of mechanisms and uses of *Trichoderma* spp. Phytopathology 96:190–194

Harman GE, Hayes CK, Lorito M et al (1993) Chitinolytic enzymes of *Trichoderma harzianum*: purification of chitobiosidase and endochitinase. Phytopathology 83:313–318

Harman GE, Howell CR, Viterbo A et al (2004) *Trichoderma* spp. – opportunistic avirulent plant symbionts. Nat Microbiol Rev 2:43–56

Herrera-Estrella A, Chet I (1998) Biocontrol of bacteria and phytopathogenic fungi. In: Altman A (ed) Agricultural biotechnology. Marcel Dekker, New York/Basel/Hong-Kong

Howell CR (1998) The role of antibiosis in biocontrol. In: Harman GE, Kubicek CP (eds) *Trichoderma* and *Gliocladium*, vol II. Taylor & Francis, Padstow

Howell CR (2003) Mechanisms employed by *Trichoderma* species in the biological control of plant diseases: the history and evolution of current concepts. Plant Dis 87:4–10

Hu GG, McClure MA, Schmitt ME (2000) Origin of a *Meloidogyne incognita* surface coat antigen. J Nematol 32:174–182

Inbar J, Chet I (1994) A newly isolated lectin from the plant pathogenic fungus *Sclerotium rolfsii*: purification, characterization, and its role in mycoparasitism. Microbiology 140:651–657

Inbar J, Chet I (1997) Lectins in biocontrol. Crit Rev Biotechnol 17:1–20

Kerry BR (2000) Rhizosphere interactions and exploitation of microbial agents for the biological control of plant-parasitic nematodes. Annu Rev Phytopathol 38:423–441

Kerry BR, Hominick WM (2001) Biological control. In: Lee DL (ed) Biology of nematodes. Taylor & Francis, London

Khan TA, Saxena SK (1997) Effect of root-dip treatment with fungal filtrates on root penetration, development and reproduction of *Meloidogyne javanica* on tomato. Int J Nematol 7:85–88

Khan A, Williams KL, Nevalainen HKM (2004) Effects of *Pacecilimyces lilacinus* protease and chitinase on the eggshell structures and hatching of *Meloidogyne javanica* juveniles. Biol Control 31:346–352

Kok CJ, Papert A, Hok-A-Hin CH (2001) Microflora of *Meloidogyne* egg masses: species composition, population density and effect on the biocontrol agent *Verticillium chlamydosporium* (Goddard). Nematology 3:729–734

Koltai H, Sharon E, Spiegel Y (2002) Root-nematode interactions: recognition and pathogenicity. In: Waisel Y, Eshel AA, Kafkafi U (eds) Plant roots: the hidden half, 3rd edn. Marcel Dekker, New York

Kubicek CP, Mach RL, Peterbauer CK et al (2001) *Trichoderma*: from genes to biocontrol. J Plant Pathol 83:11–23

Lin H, McClure MA (1996) Surface coat of *Meloidogyne incognita*. J Nematol 28:216–224

Lopez de Mendoza ME, Curtis R, Gowen S (1999) Identification and characterization of excreted-secreted products and surface coat antigens of animal- and plant-parasitic nematodes. Parasitology 118:397–405

Manzanilla-Lopez RH, Kenneth E, Bridge J (2004) Plant diseases caused by nematodes. In: Chen ZX, Chen SY, Dickson DW (eds) Nematology – advances and prospectives, vol II, Nematode management and utilization. CABI Publishing, Cambridge

Markovich NA, Kononova GL (2003) Lytic enzymes of *Trichoderma* and their role in plant defense from fungal diseases: a review. Appl Biochem Microbiol 39:389–400

Meyer SLF, Massoud SI, Chitwood DJ et al (2000) Evaluation of *Trichoderma virens* and *Burkholderia cepacia* for antagonistic activity against root-knot nematode, *Meloidogyne incognita*. Nematology 2:871–879

Meyer SLF, Roberts DP, Chitwood DJ et al (2001) Application of *Burkholderia cepacia* and *Trichoderma virens*, alone and in combinations, against *Meloidogyne incognita* on bell pepper. Nematropica 31:75–86

Morton CO, Hirsch PR, Kerry B (2004) Infection of plant-parasitic nematodes by nematophagous fungi – a review of application of molecular biology to understand infection processes and to improve biological control. Nematology 6:161–170

Oh S-U, Lee S-J, Kim J-H et al (2000) Structural elucidation of new antibiotic peptides, atroviridins A, B and C from *Trichoderma atroviridae*. Tetrahedron Lett 41:61–64

Olmedo-Monfil V, Mendoza-Mendoza A, Gómez I et al (2002) Multiple environmental signals determine the transcriptional activation of the mycoparasitism related gene prb1 in *Trichoderma atroviridae*. Mol Genet Genomics 267:703–712

Omero C, Inbar J, Rocha Ramirez V et al (1999) G protein activators and cAMP promote mycoparasitic behavior in *Trichoderma harzianum*. Mycol Res 103:1637–1642

Ramot O, Viterbo A, Friesem D et al (2004) Regulation of two homodimer hexosaminidases in the mycoparasitic fungus *Trichoderma asperellum* by glucosamine. Curr Genet 45:205–213

Rao MS, Reddy PP, Nagesh M (1998) Evaluation of plant based formulations of *Trichoderma harzianum* for the management of *Meloidogyne incognita* on egg plant. Nematol Medit 26:59–62

Reddy PP, Rao MS, Nagesh M (1996) Management of citrus nematode, *Tylenchulus semipenetrans*, by integration of *Trichoderma harzianum* with oil cakes. Nematol Medit 24:265–267

Rocha-Ramirez V, Omero C, Chet I et al (2002) *Trichoderma atroviride* G-protein α-subunit gene *tga1* is involved in mycoparasitic coiling and conidiation Eukar Cell 1:594–605

Saifullah K, Thomas BJ (1996) Studies on the parasitism of *Globodera rostochiensis* by *Trichoderma harzianum* using low temperature scanning electron microscopy. Afro-Asian J Nematol 6:117–122

Schickler H, Danin-Gehali B, Haran S (1998) Electrophoretic characterization of chitinases as a tool for the identification of *Trichoderma harzianum* strains. Mycol Res 102:373–377

Schirmböck M, Lorito M, Wong Y-L et al (1994) Parallel formation and synergism of hydrolytic enzymes and peptaibol antibiotic action of *Trichoderma harzianum* against phytopathogenic fungi. Appl Environ Microbiol 60:4364–4370

Sharon E, Spiegel Y (1993) Glycoprotein characterization of the gelatinous matrix in the root-knot nematode *Meloidogyne javanica*. J Nematol 25:585–589

Sharon E, Spiegel Y (1996) Gold-conjugated reagents for the labelling of carbohydrate-recognition domains and glycoconjugates on nematodes surfaces. J Nematol 28:124–127

Sharon E, Orion D, Spiegel Y (1993) Binding of soil microorganisms and red blood cells by the gelatinous matrix and eggs of *Meloidogyne javanica* and *Rotylenchulus reniformis*. Fundam Appl Nematol 16:5–9

Sharon E, Bar-Eyal M, Chet I (2001) Biocontrol of the root-knot nematode *Meloidogyne javanica* by *Trichoderma harzianum*. Phytopathology 91:687–693

Sharon E, Spiegel Y, Solomon R et al (2002) Characterization of *Meloidogyne javanica* surface coat using antibodies and their effect on nematode behaviour. Parasitology 125:177–185

Sharon E, Chet I, Viterbo A et al (2007) Parasitism of *Trichoderma* on *Meloidogyne javanica* and role of the gelatinous matrix. Eur J Plant Pathol 118:247–258

Sharon E, Chet I, Bar-Eyal M et al (2009a) Biocontrol of root-knot nematodes by *Trichoderma* – modes of action. Proceedings of IOBC Meeting on Multitrophic Interactions in Soil, Dijon, France. IOBC/WPRS Bulletin 42:159–163

Sharon E, Chet I, Spiegel Y (2009b) Improved attachment and parasitism of *Trichoderma* on *Meloidogyne javanica in vitro*. Eur J Plant Pathol 123:291–299

Shoresh M, Yedidia I, Chet I (2005) Involvement of jasmonic acid/ethylene signaling pathway in the systemic resistance induced in cucumber by *Trichoderma asperellum* T203. Phytopathology 95:76–84

Sivan A, Chet I (1992) Microbial control of plant diseases. In: Mitchell R (ed) New concepts in environmental microbiology. Wiley-Liss, New York

Spiegel Y, McClure MA (1995) The surface coat of plant-parasitic nematodes: chemical composition, origin and biological role: a review. J Nematol 27:127–134

Spiegel Y, Inbar J, Kahane I et al (1995) Carbohydrate-recognition domains on the surface of phytophagous nematodes. Exp Parasitol 80:220–227

Spiegel Y, Mor M, Sharon E (1996) Attachment of *Pasteuria penetrans* endospores to the surface of *Meloidogyne javanica* second-stage juveniles. J Nematol 28:328–334

Spiegel Y, Kahane I, Cohen L et al (1997) *Meloidogyne javanica* surface proteins: characterization and lability. Parasitology 115:513–519

Spiegel Y, Sharon E, Bar-Eyal M (2007) Evaluation and mode of action of *Trichoderma* isolates as biocontrol agents against plant-parasitic nematodes. IOBC WPRS Bull 30:129–133

St. Leger RJ, Joshi L, Roberts D (1998) Ambient pH is a major determinant in the expression of cuticle-degrading enzymes and hydrophobin by *Metarhizium anisopliae*. Appl Environ Microbiol 64:709–713

Steyaert JM, Ridgway HJ, Elad Y et al (2003) Genetic basis of mycoparasitism: a mechanism of biological control by species of *Trichoderma*. N Z J Crop Hort Sci 31:281–291

Suarez B, Rey M, Castillo P et al (2004) Isolation and characterization of PRA1, a trypsin-like protease from the biocontrol agent *Trichoderma harzianum* CECT 2413 displaying nematicidal activity. Appl Microbiol Biotechnol 65:46–55

Suarez MB, Vizcaino JA, Llobell A (2007) Characterization of genes encoding novel peptidases in the biocontrol fungus *Trichoderma harzianum* CECT 2413 using the TrichoEST functional genomics approach. Curr Genet 51:331–342

Szekeres A, Leitgeb B, Kredics L et al (2005) Peptaibols and related peptaibiotics of *Trichoderma:* A review. Acta Microbiol Immunol Hung 52:137–168

Tikhonov VE, Lopez-Llorca LV, Salinas J et al (2002) Purification and characterization of chitinases from the nematophagous fungi *Verticillium chlamydosporium* and *V Suchlasporium*. Fungal Gen Biol 35:67–78

Trudgill DL, Blok VC (2001) Apomictic, polyphagous root-knot nematodes: exceptionally successful and damaging biotrophic root pathogens. Ann Rev Phytopathol 39:53–77

Viterbo A, Haran S, Friesem D et al (2001) Antifungal activity of a novel endochitinase gene (chit36) from *Trichoderma harzianum* Rifai TM. FEMS Microbiol Lett 200:169–174

Viterbo A, Montero M, Ramot O et al (2002a) Expression regulation of the endochitinase chit36 from *Trichoderma asperellum* (*T. harzianum* T-203). Curr Genet 42:114–122

Viterbo A, Ramot O, Chernin L et al (2002b) Significance of lytic enzymes from *Trichoderma* spp. in the biocontrol of fungal plant pathogens. Antonie Leeuwenhoek 81:549–556

Viterbo A, Harel M, Chet I (2004) Isolation of two asparyl proteases from *Trichoderma asperellum* expressed during colonization of cucumber roots. FEMS Microbiol Lett 238:151–158

Viterbo A, Inbar J, Hadar Y et al (2007a) Plant disease biocontrol and induced resistance via fungal mycoparasites. In: Kubicek CP, Deruzhinina IS (eds) The mycota IV: environmental and microbial relationships, 2nd edn. Springer-Verlag, Berlin/Heidelberg

Viterbo A, Wiest A, Brotman Y et al (2007b) The 18mer peptaibols from *Trichoderma virens* elicit plant defence responses. Mol Plant Pathol 8:737–746

Wharton D (1980) Nematode eggshells. Parasitology 81:447–463

Whipps JM (2001) Microbial interactions and biocontrol in the rhizosphere. J Exp Bot 52:487–511

Wiest A, Grzegorski D, Xu B-W et al (2002) Identification of peptaibols from *Trichoderma virens* and cloning of a peptaibol synthetase. J Biol Chem 277:20862–20868

Windham GL, Windham MT, Williams WP (1989) Effects of *Trichoderma* spp. on maize growth and *Meloidogyne arenaria* reproduction. Plant Dis 73:493–494

Yedidia I, Benhamou N, Chet I (1999) Induction of defense response in cucumber plants (*Cucumis sativus* L.) by the biocontrol agent *Trichoderma harzianum*. Appl Environ Microbiol 65:1061–1070

Yedidia I, Benhamou N, Kapulnik Y et al (2000) Induction and accumulation of PR proteins activity during early stages of root colonization by the mycoparasite *T. harzianum* strain T-203. Plant Physiol Biochem 38:863–873

Yedidia I, Srivastva AK, Kapulnik Y et al (2001) Effect of *Trichoderma harzianum* on microelement concentrations and increased growth of cucumber plants. Plant Soil 235:235–242

Yedidia I, Shoresh M, Kerem Z et al (2003) Concomitant induction of systemic resistance to *Pseudomonas syringae* pv. *lachrymans* in cucumber by *Trichoderma asperellum* (T-203) and accumulation of phytoalexins. Appl Environ Microbiol 69:7343–7353

Zeilinger S, Galhaup C, Payer K et al (1999) Chitinase gene expression during mycoparasitic interaction of *Trichoderma harzianum* with its host. Fungal Genet Biol 26:131–140

Zeilinger S, Reithner B, Scala V et al (2005) Signal transduction by Tga3, a novel G protein α subunit of *Trichoderma atroviride*. Appl Environ Microbiol 71:1591–1597

Chapter 9
New Insights on the Mode of Action of Fungal Pathogens of Invertebrates for Improving Their Biocontrol Performance

Jose G. Maciá-Vicente, Javier Palma-Guerrero, Sonia Gómez-Vidal, and Luis V. Lopez-Llorca

Abstract The main fungal pathogens of invertebrates (FPI), nematophagous fungi and entomopathogenic fungi, have an important lifestyle overlapping. This is mainly due to the characteristics in common that their host share. Both groups of biocontrol agents share pathogenic determinants because the barriers of nematodes and insects are evolutionary conserved. Recently endophytism has been found a new aspect of the mode of action of FPI which has a potential relevance in biocontrol performance. The rationale is because they can modulate plant defences and because they act where their pest targets live and act. Natural vegetation is a reservoir root endophytes and subsequently a microbe group to screen for new biocontrol agents of plant-parasitic nematodes and root dwelling insect pests. We have found that FPI are compatible with chitosan natural compounds which may enhance their biocontrol potential.

9.1 Introduction

Fungal parasites of invertebrates (FPI) include nematophagous and entomopathogenic fungi (NEF). Their action versus plant- and animal-parasitic nematodes or insects is well known. Among their targets there are important threats to our crops (e.g. the root rot nematodes, or sap sucking insects) or to animal or human health (e.g. vectors of diseases). In spite of their main handicap, which is usually a slow action versus their targets (entomopathogenic fungi) or poor performance due to competence with existing microbiota (nematophagous fungi), they are a useful potential tool for sustainable pest management. For nematode control, the withdrawal of the fumigant

J.G. Maciá-Vicente, J. Palma-Guerrero, S. Gómez-Vidal, and L.V. Lopez-Llorca (✉)
Laboratory of Plant Pathology, Department of Marine Sciences and Applied Biology, Multidisciplinary Institute for Environmental Studies (MIES) Ramón Margalef, University of Alicante, Apto. 99, Alicante 03080, Spain
e-mail: jgmv@ua.es; jpalma@ua.es; sonia.gomez@ua.es; luisvilopezllorca@gmail.com

K. Davies and Y. Spiegel (eds.), *Biological Control of Plant-Parasitic Nematodes: Building Coherence between Microbial Ecology and Molecular Mechanisms*, Progress in Biological Control 11, DOI 10.1007/978-1-4020-9648-8_9,

methyl bromide has put biocontrol as a necessary component for integrated pest management of nematodes.

In this chapter we give a new insight to the mode of action of FPI. We analyse together nematophagous and entomopathogenic fungi. This is because their targets share similarities and therefore the two groups of fungal antagonists have coevolved common pathogenicity traits. Leaving aside their canonical hosts (nematodes and insects) we analyse their behaviour in the host of their hosts: the crop plant (plant-parasitic nematodes and insects).

Endophytism of FPI, which had been largely neglected, is a key factor in their biocontrol performance. The FPI can improve plant growth, which could then "escape" (at least partially) pest or disease. This could be combined with the genetic resources selected by plant breeding along the history of agriculture. Apart from this, endophytism would help FPI by providing them shelter for competence by rhizosphere or phylloplane microbials. Other factor, which may increase the performance of the biocontrol activity of NEF, is their combination with natural, active compounds. In the last section of this chapter we bring chitosan as a case study of such compounds, which is a potential enhancer of the nematophagous and/or entomopathogenic effect of FPI due to its interesting biological properties.

9.2 Fungal Pathogens of Invertebrates

Nematophagous and entomopathogenic fungi constitute an important group of fungal parasites of invertebrates, representing a wide range of fungal taxa. Most of them are facultative parasites (with some exceptions: e.g. most species within the Entomopthorales), with low host specificity (i.e. broad host-range). They maintain a capacity to grow saprophytically and independently of their canonical hosts. Consequently most important species of NEF have been described as soil inhabitants, which is the habitat of plant-parasitic nematodes, root-parasitic insects and many pupal stages of aerial insect pests.

In the Ascomycota, the family Clavicipitaceae, within the order Hypocreales, includes most fungal species pathogenic to invertebrates (75% out of ca. 500 species) (Artjariyasripong et al. 2001). The majority of these FPI are comprised within the genus *Cordyceps*, mostly consisting of fungal pathogens of insects and spiders, and to a lesser extent to nematodes and other fungi (Artjariyasripong et al. 2001). *Cordyceps* is the teleomorph of most mitosporic (=Deuteromycetes) NEF.

Following Lopez-Llorca and Jansson (2006), we will consider nematodes and insects the canonical hosts of nematophagous and entomopathogenic fungi, respectively. Their multimodal behaviour is reflected in their capacity to infect non-canonical hosts, e.g. other than nematodes or insects, respectively.

Development of both NEF groups as biological control agents for management of either plant pathogens or insect pests requires a better understanding of physiological aspects of growth, metabolism or genetic basis of virulence as well as their ecological performance (Lopez-Llorca and Jansson 2006).

9.2.1 Nematophagous Fungi

The so-called nematophagous fungi are a phylogenetically diverse (non-related) group of fungi which is able to infect and kill nematodes, at any developmental stage (e.g. eggs, juveniles or sessile females). The biology and infective processes of the different groups of nematophagous fungi have been extensively reviewed (Lopez-Llorca et al. 2006, 2008; Lopez-Llorca and Jansson 2006), and therefore we will only give here a brief description. According to their mode of infecting nematodes, nematophagous fungi can be divided into four groups (Lopez-Llorca et al. 2006, 2008). The nematode-trapping fungi (e.g. *Arthrobotrys oligospora*, *Drechslerella dactyloides*, *Nematoctonus robustus*), as the name implies, capture nematodes by means of hyphal trapping devices of various shapes and sizes, e.g. adhesive three-dimensional nets, adhesive knobs, or non-adhesive constricting rings. These have a low host specificity and parasitic ability. The endoparasitic fungi (e.g. *Drechmeria coniospora*, *Hirsutella rhossiliensis*, *Nematoctonus pachysporus*) use their spores (conidia or zoospores) to infect nematodes, being most of these obligate parasites (i.e. high host specificity) of nematodes which spend their entire vegetative live inside infected nematodes. The egg- and female-parasitic fungi (e.g. *Lecanicillium lecanii*, *Pochonia chlamydosporia*, *P. rubescens*) infect nematode females and the eggs they contain, using appressoria or zoospores. Finally, the toxin-producing fungi (e.g. *Coprinus* spp., *Pleurotus* spp.) immobilize the nematodes by means of toxins, prior to hyphal penetration through the nematode cuticle. In every case the final result of the nematode infection is the complete digestion of the host by the fungus.

Although these fungi have been proposed as promising candidates for biological control of plant-parasitic nematodes, experimental trials for this purpose have met little or no success (Stirling 1991). This is partly due to our still scarce understanding on the ecological performance of this group of fungi under non-controlled field conditions.

9.2.2 Entomopathogenic Fungi

The entomopathogenic fungi, also called entomophagous or entomogenous fungi, is another heterogeneous group of FPI which infect insects being the fungal group most exploited commercially for biological control purposes.

Within the broad concept of entomopathogenic fungi, a monophyletic group is mostly composed by highly specialised parasites of insects: the Entomophtorales (Zygomycota). Although these are known to cause epizootics under natural conditions, their production and application for biological control of insect pests is difficult due to their dependency on the canonical host (i.e. the insect). Therefore, a non so-highly specialised group of entomopathogens, comprising mostly anamorphic (Ascomycota and Basidiomycota) fungal species (e.g. *Beauveria bassiana* and *Metarhizium anisopliae*), are developed for biological control, due to their ability to grow on growth media and assorted substrates (Butt 2002). In spite of such facultative lifestyle, these fungi have been found causing epizootics in natural insect populations (Hajek and Leger 1994; Charnley 1997; Butt 2002).

The infection process of an insect by most entomopathogenic fungi takes place in a series of common steps beginning with the attachment of the fungal conidia to the insect's cuticle with the subsequent germination. After conidial germination, most entomopathogenic fungi belonging to Ascomycota penetrate the insect cuticle, mainly through generation of appressoria, with exertion of mechanical pressure together with secretion of extracellular cuticle degrading enzymes. Other fungi, however, are only able to penetrate the insect host through natural openings such as the mouth or the anus. Penetration of the insect host is followed by fungal multiplication within the host's haemocoel, usually with generation in some cases of budding cells (e.g. blastospores or hyphal bodies). This step may be accompanied by production of an array of toxins in certain species or strains of the entomopathogenic fungi such as *M. anisopliae* or *B. bassiana*. All this sequence finishes with the death of the insect, entailing a subsequent saprophitic growth of the fungus to completely colonise the carcass. Emergence of the fungus from the insect's body surface, and sporulation for further dissemination of propagules finally take place (Hajek and St Leger 1994).

9.2.3 Lifestyle Overlapping

Several fungal taxa include both nematophagous and entomopathogenic species. Within the Ascomycota, the teleomorphic genus *Cordyceps* (a polyphyletic group recently split in *Ophiocordyceps*, *Eucordyceps*, *Neocordyceps*) includes a high diversity of NEF species (e.g. nematophagous *C. chlamydosporia* (=*P. chlamydosporia*); entomopathogenic *C. bassiana* (=*B. bassiana*)). The genus *Hirsutella* (*Ophiocordyceps*) comprises both representatives of nematophagous, such as the specialised endoparasite *H. rhossiliensis*, as well as entomopathogenic species, such as *H. thompsonii*. The genus *Paecilomyces* also comprises representatives of the nematophagous (e.g. *P. lilacinus*) and entomogenous (e.g. *P. fumosoroseus*) fungi. The genus *Verticillium*, which included several species of NEF, was divided into *Pochonia* (nematophagous, egg- and cyst-parasitic), *Haptocillium* (nematophagous, with adhesive conidia and sometimes dictyochlamydospores), and *Lecanicillium* (entomopathogenic and fungicolous), to differentiate them from the plant pathogens which remained within the old genus (Gams and Zare 2001).

Nematode and insect cuticles share similar features for protection against environment. In both cases, the outer cuticle is composed of a matrix, mostly formed by protein, embedding structural microfibrils consisting of common or different materials: collagen in nematode cuticles; chitin in nematode eggshell and insect cuticle. These structures are also similar to those forming cell walls of potential non-canonical hosts of NEF: fungal cell walls, whose major microfibrils component is also chitin, and plant cell walls, mostly formed by cellulose and hemicelluloses (Agrios 1997). To further strengthen the structure of nematodes and insect cuticles, the matrix proteins are often cross-linked and contain phenolic and other compounds, such as melanin and dihydroxy-phenylalanine (DOPA), which in the case of insects can act as antimicrobial agents, among other functions (Lopez-Llorca and Jansson 2006).

The nematode cuticle consists of several layers of proteins (mainly collagen), lipids and carbohydrates (Bird and Bird 1991). Externally to the cuticle a surface coat (or glycocalix) consisting of glycoproteins can be found (Bird and Bird 1991). The surface coat is probably the part of the nematode surface most relevant to recognition and adhesion of nematophagous fungi, since proteolytic removal of this structure results in reduced adhesion of microorganisms (Bird 1985; Jansson 1993). The egg-shell of plant-parasitic nematodes is also a multilayered structure (Wharton 1980; Bird and Bird 1991). The thickest and most important is the chitinous layer which, like the insect cuticle, is also made of chitin and proteins. Regarding nematodes, chitin is, for instance, more abundant in *Meloidogyne* spp. eggshells, whereas protein is present to a higher degree in *Globodera* spp. Structurally, *Globodera* spp. have thicker egg-shells than those of *Heterodera* spp. (Lopez-Llorca and Robertson 1992).

Regarding insects, the cuticle is formed by several layers (Andersen 1979). The epicuticle is the outer of these, mainly made of lipoproteins and waxes, whereas the procuticle is mainly made of proteins and chitin. Other compounds such as lipids, pigments and other small organic molecules and inorganic materials are minoritary, although they may affect the performance of entomopathogenic fungi. Chitin microfibrils (20–50% of procuticle composition) are embedded in a protein matrix (resilin) in different amounts, depending on the insect groups (Lopez-Llorca and Jansson 2006).

Molecular components of insect and nematode barriers undergo several types of modifications, e.g. protein cross-linking and melanization (Andersen 1979; Lopez-Llorca and Fry 1989). These processes can be highly relevant in the susceptibility of both nematodes and insects to FPI (Lopez-Llorca and Jansson 2006).

The similarities in structure and molecular composition of protection barriers between nematodes and insects can also be largely responsible for deep similarities between NEF lifestyles. These are specially apparent during the first steps of host infection such as spore adhesion and cuticle penetration. One of the most important determinants of fungal virulence is the secretion of extracellular enzymes that degrade the outer layers of the host cuticle. They are present in the extracellular matrix which surrounds appressoria or hyphae at the penetration site. These enzymes are an essential factor to facilitate cuticle penetration and subsequent degradation of structural components of this host barrier. Penetration of the host cuticle is a complex process which may depend on several mechanisms of the fungal pathogen, mainly mechanical pressure plus enzymatic digestion of cuticle components. Recent studies have been performed to evaluate the independent role of each factor. In particular, a large number of cuticle degrading enzymes from NEF have been identified, characterised and cloned.

Proteases and chitinolytic enzymes are perhaps the major contributors to insect and nematode cuticle penetration by fungi (St Leger et al. 1996b). Serine proteases (EC3.4.21.-), specially those from the subtilisin-like family (EC3.4.21.14), seem to play an important role in cuticle penetration by NEF, since they are frequently found in the extracellular matrix surrounding appressoria or hyphae at the penetration sites. Subtilisin-like proteases are a large family of endopeptidases present only in fungi and Gram-negative bacteria, in contrast to the trypsin-like family (EC3.4.21.4). Broad-spectrum subtilisin-like proteases are the major proteins produced by the insect pathogens *M. anisopliae* and *B. bassiana* during infection processes, and have much greater ability than the trypsin-like enzymes to degrade insect cuticle (Bidochka et al. 1999). Collagenase secretion upon penetration of nematode cuticles by *Arthrobotrys* spp. has also been recently studied (Yang et al. 2007).

Since the first isolation of a serine protease, P32 from the nematode egg-parasite *P. rubescens* (Lopez-Llorca 1990), a large array of other serine-proteases have been identified in many species of NEF. Examples of these are VCP1 from *P. chlamydosporia* (Segers et al. 1994), a close relative species to *P. rubescens*, PII and Aoz1 from the nematode-trapping *A. oligospora* (Tunlid et al. 1994; Zhao et al. 2004), Mlx from nematophagous *Monacrosporium microscaphoides* (Wang et al. 2006a), Ds1 from *Dactylella shizishanna* (Wang et al. 2006b), pSP-3 from *Paecilomyces lilacinus* (Bonants et al. 1995), Ver112 from *Lecanicillium psalliotae* (Yang et al. 2005), PrC from *Clonostachys rosea* (Li et al. 2006), CDEP-1 from *B. bassiana* (Zhang et al. 2008b), or Csp1 and Csp2 from the entomopathogenic *Hirsutella sinensis* (=*Cordyceps sinensis*) (Zhang et al. 2008a).

Morton et al. (2003) detected homology among the nematophagous fungal serine protease VCP1 and the protease PRI from the entomopathogen *M. anisopliae*. Furthermore, in a recent structural comparative study among the serine proteases PRI, Ver112 and VCP1, Liu et al. (2007) identified a virtually identical backbone topology and similar structural properties. The three enzymes however differed in the electrostatic surface potential, hydrophobicity and size of one of the substrate binding pockets, which may influence substrate specificity and catalytic efficiency. Serine proteases from both nematophagous and entomopathogenic fungi seem to have a common ancestor, different from that of plant-parasitic fungi (Yang et al. 2007). Production of subtilisins may be a retained character from a saprophytic ancestor, since subtilisin-like proteases are the principal broad-spectrum proteases produced by many saprophytes (Bidochka et al. 1999).

Chitinases (EC3.2.1.14) have also been shown to be an important factor for nematode and insect cuticle degradation, and have also been found in important species of NEF. Examples are endochitinase CHI43 from *P. chlamydosporia* and *P. rubescens* (Tikhonov et al. 2002), Bbchit1 from *B. bassiana* (Fang et al. 2005), Lpchi1 from *L. psalliotae* (Gan et al. 2007), endochitinases CH1, CH2 and CH3 and an exochitinase from *M. anisopliae* (St Leger et al. 1993), or several chitinases from *P. lilacinus* (Dong et al. 2007).

As shown for subtilisin proteases, some chitinases are a conserved trait of NEF for host cuticle degradation. Diversity of chitinases produced by NEF seems to be greater than that of proteases, and probably different chitinases have appeared over time. Therefore, whereas in a phylogenetic study, chitinase Lpchi1 from the nematophagous fungus *L. psalliotae* clustered together with chitinases from entomopathogenic fungi (Gan et al. 2007), other chitinases such as Bbchit1 from *B. bassiana* had low homology to other chitinase genes from entomopathogenic fungi (Fang et al. 2005).

In some cases, proteases of NEF have been shown to be directly related to canonical host infection. Lopez-Llorca and Robertson (1992) confirmed the role of P32 in pathogenicity by immuno-localisation in infected nematode eggs. For nematode-trapping fungi, Ahman et al. (2002) found that transformants of *A. oligospora* with multiple copies of the protease PII gene showed increased adherence and immobilisation of free-living nematodes in their traps. Mutants lacking the PII gene were less pathogenic than the wild type. In entomopathogenic fungi, CDEP-1 from *B. bassiana* has been shown to increase virulence of the fungus towards the apterous green peach aphid *Myzus persicae* (Zhang et al. 2008b). In some studies, NEF chitinases have also been related to fungal virulence towards the canonical host (Fan et al. 2007).

There are evidences of the synergistic action of cuticle-degrading enzymes of NEF on their hosts. An example of this is the experimental effect of enzyme combinations on cuticle degradation and the dynamics of secretion of the responsible enzymes by NEF. Eggs of the plant-parasitic nematode *Globodera pallida* treated with the protease P32 and the endochitinase CHI43, alone or in combination, showed surface damage as compared with untreated controls. Both enzymes in combination caused greater eggshell damage than when they were applied separately (Tikhonov et al. 2002). Similar results were obtained when treating *M. incognita* eggs with serine protease Ver112 and chitinase Lpchi1 from *L. psalliotae*, respectively (Gan et al. 2007).

A further example of the synergistic cooperation of cuticle-degrading enzymes of NEF is the dynamics of their production in host infection. In entomopathogenic fungi, proteases are produced earlier and more abundantly than chitinases (St Leger et al. 1996a). This has been explained by the abundance of the amorphous protein matrix in the infected cuticle which would shield chitin microfibrils from enzymatic degradation. Secreted in this order, both enzymes would finally structurally degrade the host cuticle. No such studies have been performed for nematophagous fungi, but substrate degradation assays on Petri dishes always show easier induction of proteolytic than chitinolytic enzymes (Olivares-Bernabeu and Lopez-Llorca 2002).

Taxonomic and physiological similarities between NEF are reflected in the multimodal capacity to infect hosts by these biocontrol agents. This sustains the ability of NEF to parasitize hosts other than their canonical ones (Lopez-Llorca and Jansson 2006). Several examples of NEF illustrate the elusive borders between lifestyles. An illustrative case is the fungal species *L. lecanii*, which can be found occurring naturally as either nematophagous or entomopathogen. In addition *L. lecanii* has also been described as a mycoparasite and as root endophyte (Lopez-Llorca and Jansson 2006, and references therein). *P. chlamydosporia*, a nematode egg-parasitic fungus, is also an example of a facultative fungus with a wide array of alternative hosts. The fungus has been found parasitizing rust uredospores (Leinhos and Buchenauer 1992), *Phytophthora* sp. oospores or snail eggs (Domsch et al. 1980). In any case it is most often found as a causal agent of nematode suppressive soils worldwide infecting females and eggs.

9.3 Endophytism: Relevance in Biocontrol Performance

Both nematophagous and entomopathogenic fungi have demonstrated their ability to endophytically colonise plant tissues (Lopez-Llorca et al. 2006; Vega et al. 2008). It has been suggested that FPI may have arisen from plant pathogens to escape competition via further specialisation in alternative hosts (Barron 1992; Bidochka et al. 1999). FPI capabilities of changing hosts between invertebrates (insects or nematodes), endophytism and mycoparasitism could then be the consequences of this evolutionary process (Lopez-Llorca and Jansson 2006).

9.3.1 The Fungal Endophytes

The largest component of the fungal diversity of our ecosystems is somehow associated with plants. Fungi which colonise internally plant tissues without causing disease symptoms are usually known as endophytes. The term endophyte means literally "within the plant" (*endon* Gr.: within, *phyton*: plant), and was first introduced by de Bary (1866) to define those organisms which colonise and dwell within tissues or cells of a host plant. This broad definition would include not only several organisms (e.g. bacteria, fungi, algae and animals), but also different interactions with the host plant such as parasitism, commensalism or mutualism. The different interactions between the endophyte and its host-plant constitute a continuum from antagonism to mutualism, and this relationship may change over time and space (Sieber 2002). Probably all possible interactions between the endophyte and its host-plant can occur in nature. Many endophytes represent latent infections of plant-pathogenic fungi which remain inactive within the host-plant until it is subjected to stress conditions or reaches senescence. Certain strains may colonise asymptomatically some plants species while act as pathogens in others. In fact, it is thought that some pathogens may appear from non-pathogenic endophytes after the introduction in its habitat of potentially new host-plants, such as crop species (Summerell and Leslie 2004). A more practical definition was therefore given by Sieber (2002), who defined root endophytes as those inhabiting apparently healthy and functional roots at the moment of sampling.

The capacity of fungi to colonise endophytically plants in nature is extremely common. Endophytes have been isolated from nearly all plant species studied. Diversity of species, frequency and abundance of endophytes, depend on climatic and edaphic conditions and on the heterogeneity of habitats and niches occupied by their hosts (Sieber 2002). Many endophytes are not considered to be organ-specific, while others are solely found in roots (Jumpponen 2001; Sieber 2002) or aboveground organs (Schulz and Boyle 2005); in any case, every organ of the host can be colonised by fungal endophytes.

Endophytic populations colonising plant tissues may represent a wide range of interactions with the host, and these biological strategies may overlap. These interactions include fungi causing only localised infections in individual cells until plant senescence, low virulent pathogens, causing silent infections, or in latency periods, hemibiotrophs, or incompatible pathogens. Therefore, isolating an organism as an endophyte does not exclude the possibility that it may become aggressive or pathogenic in the future or under different conditions. The ability of a fungal isolate to cause disease does not only depends on its pathogenicity, but also on the status of the host-plant, constituting an equilibrium between fungal virulence and plant defences (Schulz et al. 1998, 2002). If this equilibrium is altered by a decrease of the plant defences or by an increase of the fungal virulence, disease develops. The endophyte has first to synthesize metabolites to compete with epiphytes and with pathogens in order to colonise its host, but then also to regulate the host's metabolism in a slightly balanced association.

Fungal endophytes may provide several benefits to plants such as protection against diseases (Carroll 1988; Vilich et al. 1998; Redman et al. 1999, 2001), production of secondary metabolites effective versus host pathogens (Noble et al. 1991; Calhoun et al. 1992; Schulz et al. 1995; Liu et al. 2001a), protection against insect pests (Latch 1993; Azevedo et al. 2000; Anke and Sterner 2002), herbivore resistance (Latch 1993), or growth promotion (Allen 1992; Varma et al. 1999, 2000; Jumpponen 2001; Ernst et al. 2003). Some endophytes may also improve the phosphorus uptake by the plant (Sieber 2002), its photosynthetic efficiency (Obledo et al. 2003), or increase the host's tolerance to abiotic stresses, such as drought (Read and Camp 1986; Bacon et al. 1996; Bacon and Hill 1996), metals (Read 1999), salts (Rodriguez and Redman 2008), or high temperatures (Redman et al. 2002). Nevertheless, no growth improvement has been recorded when the endophytes are inoculated in aerial organs, probably because these infections are localised, in contrast with those of the roots usually systemic (Boyle et al. 2001; Schulz et al. 2002).

9.3.2 Endophytic Behaviour and Biological Control

As stated above, colonisation of a plant tissue by endophytes may increase host tolerance to stress caused by pests and pathogens through a wide array of mechanisms. These would include induction of host's systemic resistance, antibiosis, reduction of palatability, or direct parasitism. Among these, parasitism of plant parasitic nematodes or pest insects are the best known effects of NEF. Many different organisms have been shown to produce a systemic activation of the plant defence mechanisms upon colonisation of the host plant tissues, either endophytically or not. In the specific case of fungi, several taxa are well known for producing secondary metabolites biologically active against bacteria, fungi, algae, plants or animals. Besides, colonisation of plant organs by these fungi may activate the synthesis of toxins not expressed during saprophytic growth. Alternatively or simultaneously endophytes may induce the synthesis of antibiotic compounds by the host-plant, or reduce its palatability for pests causing agents.

Regarding the canonical hosts of NEF, nematodes and insects, it is important to consider the predominant target where each organism is going to perform its parasitism or predatory attack. While most phytopathogenic nematodes, specially those with a sedentary lifestyle (e.g. root-knot and cyst nematodes) infect roots, the majority of insect pests, apart of some soil-dwelling insects, have a deleterious effect on aerial parts of the plants. Root colonisation patterns by fungal endophytes substantially differ from those of aerial organs. Endophytic growth of fungi within roots have shown to be mostly extensive and systemic, both inter- and/or intracellularly and occasionally with formation of specific structures (Capellano et al. 1987; Allen 1992; Jumpponen and Trappe 1998; Varma et al. 2000; Kuldau and Yates 2000; Schulz et al. 2002; Schulz and Boyle 2006). This is in contrast with the local colonisation which basically occurs in aerial tissues (Stone et al. 1994; Carroll 1995; Boyle et al. 2001; Schulz et al. 2002; Schulz and Boyle 2005).

9.3.2.1 Nematophagous Fungi as Endophytes

Nematode-trapping fungi have been shown to colonise the rhizosphere of different plant species, with a special occurrence in plants from family Leguminosae (Peterson and Katznelson 1964; Gaspard and Mankau 1986; Persmark and Jansson 1997; Persson and Jansson 1999). Riekert and Tiedt (1994) found *M. incognita* larvae captured in *D. dactyloides* trapping devices formed in the surface of infested maize roots. Nevertheless, fungal traps were never observed inside the roots, and these appeared mostly in the vicinity of secondary root bases. Rhizosphere of several plant species is also colonised by the nematode egg-parasite *P. chlamydosporia*, being its rhizospheric population increased by egg production of *M. incognita* females infecting the roots (Bourne et al. 1996). The endophytic behaviour of nematophagous fungi has been little studied. Both nematode-trapping (*A. oligospora*) and egg-parasitic (*P. chlamydosporia*) fungi can behave as root endophytes under axenic conditions (Bordallo et al. 2002). We have recently extended these experiments to representatives of the rest of the ecological groups of nematophagous fungi: nematode-trapping (*D. dactyloides*, *N. robustus*), toxin-producing (*P. djamor*) and endoparasitic fungi (*H. rhossiliensis*, *N. pachysporus*) (Lopez-Llorca et al. 2006). These experiments showed that all fungi tested colonised abundantly the rhizoplane of barley. Their presence was stable over time and root depth. *N. pachysporus* was the slowest coloniser of the rhizoplane for all species tested. *D. dactyloides*, *N. robustus* and *P. djamor* colonised endophytically barley roots, as shown by root plating and light microscopy. Similarly to that observed for *P. chlamydosporia* (Bordallo et al. 2002; Lopez-Llorca et al. 2002), *D. dactyloides* also formed coiling structures in barley root cells. Such structures are also formed by other root endophytes, e.g. *Piriformospora indica* (Varma et al. 1999), and presumably improve nutrient exchange. This would indicate a certain degree of specialisation for the endophytic behaviour. The endoparasitic fungi (*H. rhossiliensis* and *N. pachysporus*) tested failed to cross the root epidermis. *H. rhossiliensis* was however present abundantly all along the root, producing viable conidiophores. Endoparasites include mostly obligate parasites, strictly adapted to their nematode hosts, with low growth rates on culture media. This high adaptation to their canonical hosts may make it difficult to switch to non-canonical ones such as the rhizosphere. These results, together with those obtained by Bordallo et al. (2002) and Monfort et al. (2005), suggest that both the ecological lifestyle of the nematophagous fungi and the host-plant influence their endophytic behaviour.

Once the capacity of the nematophagous fungi to behave as root endophytes is evaluated in the laboratory, it is to be evaluated under more natural conditions. The final goal would be to optimise endophytism of nematophagous fungi for control of plant parasitic nematodes in the glasshouse or under field conditions.

Most investigations on biological control of plant-parasitic nematodes by mutualistic endophytic fungi have been performed with fungi which are non strictly nematophagous. These endophytes are ubiquitous fungal species which may be found in other substrates such as soil or plant material. The majority of this work has been carried out by inoculations of different host-plants with non-pathogenic *Fusarium oxysporum* strains (Sikora et al. 2008, and references therein). Direct negative effect of grass non-strict nematophagous fungal endophytes on nematodes have also been reported (West et al. 1988; Kimmons et al. 1990).

Few efforts have been invested, contrarily, to study the biocontrol capacity of strict nematophagous fungi colonising endophytically plant roots on parasitic nematodes. Nematode-trapping and egg-parasitic fungi can be observed more frequently in the rhizosphere than in the bulk soil (Persmark and Nordbring-Hertz 1997; Kerry 2000; Lopez-Llorca et al. 2006, 2008). Chemotropic growth towards roots of different plants by the nematode-trapping fungus *Arthrobotrys oligospora* has been described (Bordallo et al. 2002). Furthermore, induction of conidial-traps in this group of fungi by presence of roots or root exudates has also been reported (Persmark and Nordbring-Hertz 1997). Plant growth promotion has been observed after inoculation of *P. chlamydosporia* in the rhizosphere of several plant species (Monfort et al. 2005; Siddiqui and Akhtar 2008; Maciá-Vicente et al. 2009b). These observations would indicate a specific adaptation of certain species of nematophagous fungi to rhizosphere colonisation. This ability may provide the fungus with an opportunity to infect eggs inside the roots, while formation of trapping organs by nematode-trapping fungi (e.g. *A. oligospora*) in the root epidermis may serve the purpose of trapping newly hatched juveniles leaving the root. Root colonisation may be also a long-term survival strategy for these fungi and could explain suppressiveness to plant-parasitic nematodes in nature (Lopez-Llorca et al. 2008).

9.3.2.2 Entomopathogenic Fungi as Endophytes

Several species of entomopathogenic fungi have been reported as capable to colonise endophytically a broad array of plant species, in many cases occurring naturally within the plant host (Vega et al. 2008; Vega 2008, and references therein).

The endophytic ability of this group of fungi, and its importance in pest management has been more extensively studied than that of nematophagous fungi. Pioneer work on the endophytic behaviour of entomopathogenic fungi was carried out for biological control of the European corn borer (*Ostrinia nubilalis* (Hübner)) with *B. bassiana* endophytically colonising maize (*Zea mays* L.) (Lewis and Cossentine 1986; Lewis and Bing 1991; Bing and Lewis 1993). These authors illustrated the capacity of *B. bassiana* to control *O. nubilalis* after establishing endophytically within the maize aerial tissues. Although suppression of the pest caused by the insect was effective, e.g. a significant reduction of tunnelling of maize leaves was achieved, comparatively few insects appeared mycosed. This seemed to indicate that suppression of the pest was achieved by a feeding deterrence or antibiosis caused in the plant upon endophytic colonisation of *B. bassiana*. The endophytic growth seems not to be adequate for a direct parasitism of the fungus on the pest insect, since there are almost no reports on conidial formation within plant tissues by endophytes. On the other hand, as has been discussed earlier, endophytic colonisation of above-ground plant organs is usually restricted to few cells, contrarily to that (e.g. systemic colonisation) found in roots.

Other entomopathogens, including *L. lecanii*, *P. farinosus*, *P. varioti*, or *Cladosporium* spp., have subsequently been reported as naturally occurring endophytes of several plant species (Vega et al. 2008; Vega 2008, and references therein).

Laboratory and field assays in our laboratory have demonstrated the capability of the entomopathogenic fungi *B. bassiana*, *L. dimorphum* and *L.* c.f. *psalliotae* to endophytically colonise leaf petioles of *Phoenix dactylifera* L., without causing evident harm to the host-plant. The fungi remained inter- and intra-cellularly within the host tissues up to 30 days after the inoculation, and were able to move away from the inoculation site (Gómez-Vidal et al. 2006). Recent studies on the proteome of endophytically colonised date palms indicate that entomopathogenic fungi growing endophytically can up-regulate proteins related with stress and plant defence (Gómez-Vidal et al. 2009).

Current outbreaks of palm pests such as the red scale insect (*Phoenicococcus marlatti*) or the present epidemics of the red palm weevil (*Rhyncophorus ferrugineus*) are examples for practical application of endophytism for pest control. Besides, the multimodal action of FPI and their common endophytic behaviour opens up the use of entomopathogenic fungi for plant-parasitic nematode control.

9.4 Screening for New Biocontrol Agents: Root Endophytes of Natural Vegetation

Fungal endophytes are a valuable source of secondary metabolites with potential pharmacological and agricultural uses (Dimock et al. 1993; Lingham et al. 1993; Bills et al. 1994; Dreyfuss and Chapela 1994; Peláez et al. 1998). The broad diversity and taxonomic spectra of this group of fungi, and its relatively easy isolation, makes them specially interesting in search programs for natural active products

(Peláez et al. 1998). Positive correlations between the biological activity of metabolites and the biotope where the producing microorganisms occur have been found. Schulz et al. (2002) observed that a larger proportion of fungal endophytes, in comparison with fungi isolated from soil, inhibited at least one of the test organisms when searching for antialgal and herbicidal activities. This comparison between the production of metabolites by endophytes and soil isolates shows also a considerably higher production of novel substances by the endophytes, due to less efforts invested in the study of the former. Consequently, metabolites purified from endophytes usually have unknown chemical structures. On the other side, metabolic interactions of endophytes with their hosts favour the synthesis of biologically active secondary metabolites (Schulz et al. 2002). A fungus synthesizes of the secondary metabolites may correspond to its respective ecological niche, for instance mycotoxins in plant pathogens (Gloer 1997). Metabolic interactions may increase the production of secondary metabolites. Surveys for fungal isolation should therefore be performed on biotopes not studied previously, and where metabolic interactions with the environment are likely to occur (Schulz et al. 2002). Fungal endophytes fit both criteria: they grow within their host plants without apparently causing disease symptoms (Petrini 1991; Wilson 1995), and this implies a stable metabolic relationship between the fungus and its host (Schulz et al. 2002).

Protocols for screening large collections for biocontrol activities usually include an initial step where simple bioassays are aimed to detect bioactive compounds (Knudsen et al. 1997). These have unveiled active antibiotics versus bacterial and fungal plant pathogens. In any case the production in growth media tests of an inhibitory substance does not ensure its action in the field as has often been found with antibiotics of medical significance.

A working hypothesis is that natural vegetation which have evolved in undisturbed soils have been selected over time for its tolerance against abiotical or biotical stresses (e.g. plant pathogens). Recent studies seem to indicate that mutualistic associations with fungi, not necessarily mycorrhizal, play an important (and probably necessary) role in tolerance against environmental stresses. Furthermore, root colonisation, particularly endophytical, may help for a stable establishment of a biocontrol agent where it should perform its function, avoiding rhizospheric competence with other microorganisms. On the other hand, the establishment of the fungi within the plant tissues supposes a "metabolic fight" between the endophyte and the host which may result in antagonism vs. an arriving pathogen.

In our laboratory we performed a survey for isolating fungal root endophytes from natural vegetation under stress in south eastern Spain (Maciá-Vicente et al. 2008a). Twenty-four plant species characteristic of either sandy soils or salt marshes were sampled at 12 sites (7 sandy soils and 5 salt marshes). Plant species included a wide array of botanical families. The sampling sites were divided into coastal or inland locations, regarding their soil characteristics. A collection of ca. 1,900 root endophytic isolates was obtained, comprising a total of 142 fungal species in 57 genera, and sterile mycelia grouped in 177 morphological species. Endophytes included known biocontrol agents of both nematodes and insects, either strict (e.g. *Dactylaria* sp., *Lecanicillium* sp.) or non-strict (e.g. *Fusarium* spp., *Acremonium* spp.).

The endophyte collection has been found to include antagonists economically important fungal root pathogens such as *Fusarium oxysporum* f. sp. *radicis-lycopersici* and *Gaeumannomyces graminis* var. *tritici* (Maciá-Vicente et al. 2008b). We are currently using *F. equiseti* isolates which have been found to produce nematicidal components (Nitao et al. 2001) for experiments of rhizosphere colonisation and biocontrol of nematodes, together with *P. chlamydosporia* (a *bona fide* egg-parasite). Recently developed tools for the study of both these interesting fungal species, such as real time quantitative-PCR techniques for specific quantification, and successful transformation with the green fluorescent protein (GFP) gene for microscopical studies, have evidenced their capabilities as efficient endophyte colonisers of roots (Maciá-Vicente et al. 2009a). Such studies have shown that both *F. equiseti* and *P. chlamydosporia* colonising barley roots elicit plant defence reactions to an extent that do not completely eliminate the endophytes from the internal tissues, but force them to avoid plant reactions. This is again a piece of evidence in support of the balanced antagonism hypothesis for the host plant-endophyte interaction (Schulz et al. 1999).

The former isolates of natural origin would complement studies on the use of mutualistic fungi for suppressiveness, since these have mostly originated from crop plants (Sikora et al. 2008). Recently, a *Fusarium culmorum* isolated from above- and below-ground tissues and seed coats of the dunegrass *Leymus mollis* growing under saline stress, has been proven to be the cause for abiotic stress protection (Rodriguez and Redman 2008). This protection could be transferred to unrelated crop plants when inoculating the endophyte. Our *F. equiseti* isolates obtained from a similar ecosystem are likely to perform symbioses with these characteristics. Therefore a new generation of biocontrol and stress-adapting agents is likely to be developed in the next years.

9.5 Compatibility of FPI with Natural Compounds: Chitosan as a Case Study

FPI have to break their hosts outer barriers for infection. These barriers, as we have discussed (see Sect. 9.2.3) mostly include a chitin-protein structure (e.g. nematode egg-shell, insect cuticle). Furthermore chitin is the second most abundant polymer in nature after cellulose (Cohen-Kupiec and Chet 1998) and occurs in various organisms (e.g. crustaceans, insects, nematodes and most fungi). Chitin waste is an abundant by-product of the crustacean fishing industry worldwide. This waste product has been a target for the development of organic nematicides. The rationale behind this was that a chitin amendment to soil would enhance the chitinolytic microbiota, which would in turn be nematophagous, in view of the biochemical composition of nematode barriers. This resulted in the isolation of nematophagous organisms, including new species of bacteria (Spiegel et al. 1986, 1987, 1988). The experiments of soil organic amendments for biocontrol of soil-borne pathogens have produced various results with the development of composting, and their performance has improved (Trillas et al. 2002, 2006). However, addition of chitin rich waste to soil for nematode control resulted in a build up of ammonia in quantities which may turn to be phytotoxic (Rodriguez-Kabana et al. 1987; Carvajal and Rodriguez-Kabana 1998; Hallmann et al. 1999).

Chitosan is obtained by chitin deacetylation, and consists of polymers of β-1,4-glucosamine subunits, with molecular weight up to 400 kDa (Rabea et al. 2003). It can be produced by chemical or microbial/enzymatic treatments (Tsai et al 2002). Besides, fungi can turn chitin in their cell walls to chitosan. This is achieved by means of chitin deacetylase activity and may block host recognition and degradation of fungal cell walls by plant chitinases in biotrophic fungi (Deising et al. 1995). Commercial chitosan is mainly produced by chemical methods using chitin waste products from seafood industry (Kumar 2000).

Chitosan has several advantages over chitin for combination with biocontrol agents purposes, the main ones are a higher solubility and interesting biological properties. This polymer is no toxic to mammals (Dodane and Vilivalam 1998; Lee et al. 2004) and it elicits plant defence mechanisms (Benhamou et al. 1994; Lafontaine and Benhamou 1996; Ait Barka et al. 2004; Trotel-Aziz et al. 2006), but displays antibiotic activity against microorganisms, both bacteria (Liu et al. 2001b, 2004; Tikhonov et al. 2006) and fungi (Bell et al. 1998; Laflamme et al. 1999; Park et al. 2002; Pascencia-Jatomea et al. 2003; Bautista-Banos et al. 2006; Palma-Guerrero et al. 2008). We have investigated its compatibility with biocontrol agents and are extending these results to the biological control of plant parasitic nematodes.

The study of the fungitoxic effect of chitosan has mostly dealt with colony growth inhibition of plant-pathogenic fungi and associated ultrastructural changes in the hyphae (Laflamme et al. 1999). Much less is known about the effects of chitosan on spore germination. Palma-Guerrero et al. (2008), were the first in comparing the effect of chitosan on biocontrol fungi and plant-pathogenic fungi, considering both hyphal growth and spore germination. Great variations in tolerance to chitosan between the different fungi tested were obtained. Most plant pathogenic fungi tested were highly sensitive to chitosan, whereas nematophagous and entomopathogenic fungi were much less inhibited by chitosan. Mycoparasitic fungi were the only exception among the three types of biocontrol fungi tested, since they were as sensitive to chitosan as the plant pathogenic fungi. The low effect of chitosan on nematophagous and entomopathogenic fungi seemed to be related, at least partly, with their ability to degrade chitosan. These fungi showed the highest chitosan-degrading activity, according to the size of their substrate degradation halos observed in amended solid growth media.

Conidial germination was more sensitive to chitosan than hyphal growth. Conidia of nematophagous and entomopathogenic fungi were again the least sensitive to chitosan. For some isolates, germination in the presence of chitosan was similar to that of untreated controls. Furthermore, germination was not completely inhibited by increasing chitosan concentration. Only one *P. chlamydosporia* isolate (P.c. 4624) out of 9 tested showed the same sensitivity to chitosan as the plant pathogenic fungi tested. In fact this was the nematophagous fungus strain with the lowest chitosanase activity.

Nematophagous fungi showed the highest ability to degrade chitosan. This was confirmed for several *P. chlamydosporia* isolates tested, from worldwide origins. One possible explanation to the little effect of chitosan on the growth of nematophagous and entomopathogenic fungi may be that their chitosan degrading enzymes are more abundant and/or efficient. They could prevent diminish the toxic effect of chitosan to fungal cells. These fungi may even be using chitosan as nutrient in some cases. Although chitinolytic and chitosanolytic activities have been found in plant pathogenic and mycoparasitic fungi (Shimosaka et al. 1993; Nogawa et al. 1998), Palma-Guerrero et al. (2008) could not detect halos of chitosan degradation for the plant pathogenic fungi or oomycetes tested (except for *V. dahliae*). This would suggest low chitosanase activities for these organisms. The "special" chitosan degrading enzymes of nematophagous and entomopathogenic fungi may be related with their multimodal lifestyle as explained above (Sect. 9.2.3). In fact both fungal groups have coevolved to degrade their host cuticles which contain chitin, a similar polymer to chitosan (Palma-Guerrero et al. 2010).

9.6 Concluding Remarks

According to all the results presented, our working hypothesis is that probably most fungal biocontrol agents with a saprophytic lifestyle, are able to colonise endophytically plant tissues. Many fungi can be isolated from surface-sterilised roots in nature, and these may not have a strict mutualistic interaction with the host plant, or just colonise roots for a short time period. This is a way to evade competition with complex microbial communities in the soil. The optimisation of the inoculation processes and plant growth conditions, aided by the addition of natural compounds such as chitosan, can allow us to exploit these abilities to perform perdurable applications of certain organisms, by an artificial "endophytisation" of the plants. Probably this already occurs in nature, and could be related with the existence of some suppressive soils to plant-pathogens.

Acknowledgements We thank Prof. H.-B. Jansson for critical reading of the manuscript. This chapter was supported with funds of the research project AGL2008-00716/AGR from the Spanish Ministry of Science and Innovation.

References

Agrios GH (1997) Plant pathology. Academic, London

Ahman J, Johansson T, Olsson M et al (2002) Improving the pathogenicity of a nematode-trapping fungus by genetic engineering of a subtilisin with nematotoxic activity. Appl Environ Microbiol 68:3408–3415

Ait Barka E, Eullaffroy P, Clement C et al (2004) Chitosan improves development, and protects *Vitis vinifera* L. against *Botrytis cinerea*. Plant Cell Rep 22:608–614

Allen MF (1992) Mycorrhizal functioning: an integrative plant-fungal process. Chapman and Hall, New York

Andersen SO (1979) Biochemistry of insect cuticle. Annu Rev Entomol 24:29–61

Anke H, Sterner O (2002) Insecticidal and nematicidal metabolites from fungi. In: Osiewacz K, Esser K, Bennett JW (eds) The mycota. Springer, Berlin, pp 109–128

Artjariyasripong S, Mitchell JI, Hywel-Jones NL et al (2001) Relationship of the genus *Cordyceps* and related genera, based on parsimony and spectral analysis of partial 18S and 28S ribosomal gene sequences. Mycoscience 42:503–517

Azevedo JL, Maccheroni J, Pereira JO et al (2000) Endophytic microorganisms: a review on insect control and recent advances on tropical plants. Electron J Biotechnol 3:40–65

Bacon CW, Hill NS (1996) Symptomless grass endophytes: products of coevolutionary symbioses and their role in ecological adaptations of grasses. In: Redlin SC, Carris LM (eds) Endophytic fungi in grasses and woody plants. Systematics, ecology, and evolution. APS Press, St Paul, pp 155–178

Bacon CW, Porter JK, Norred WP et al (1996) Production of fusaric acid by *Fusarium* species. Appl Environ Microbiol 62:4039–4043

Barron GL (1992) Ligninolytic and cellulolytic fungi as predators and parasites. In: Carroll GC, Wicklow TD (eds) The fungal community: its organisation and role in the ecosystems. Marcel-Dekker, New York, pp 311–326

Bautista-Banos S, Hernandez-Lauzardo AN, Velazquez-del Valle MG et al (2006) Chitosan as a potential natural compound to control pre and postharvest diseases of horticultural commodities. Crop Prot 25:108–118

Bell AA, Hubbard JC, Liu L et al (1998) Effects of chitin and chitosan on the incidence and severity of *Fusarium* yellows of celery. Plant Dis 82:322–328

Benhamou N, Lafontaine PJ, Nicole M (1994) Induction of systemic resistance to *Fusarium* crown and root rot in tomato plants by seed treatment with chitosan. Phytopathology 84:1432–1444

Bidochka MJ, St Leger RJ, Stuart A et al (1999) Nuclear rDNA phylogeny in the fungal genus *Verticillium* and its relationship to insect and plant virulence, extracellular proteases and carbohydrases. Microbiology 145:955–963

Bills GF, Peláez F, Polishook JD et al (1994) Distribution of zaragozic acids (squalestatins) among filamentous ascomycetes. Mycol Res 98:733–739

Bing LA, Lewis LC (1993) Occurrence of the entomopathogen *Beauveria bassiana* (Balsamo) Vuillemin in different tillage regimes and in *Zea mays* L. and virulence towards *Ostrinia nubilalis* (Hubner). Agric Ecosys Environ 45:147–156

Bird AF (1985) The nature of the adhesion of *Corynebacterium rathayi* to the cuticle of the infective larva of *Anguina agrostis*. Int J Parasitol 15:301–308

Bird AF, Bird J (1991) The structure of nematodes, 2nd edn. Academic, San Diego

Bonants PJM, Fitters PFL, Thijs H et al (1995) A basic serine protease from *Paecilomyces lilacinus* with biological activity against *Meloidogyne hapla* eggs. Microbiology 141:775–784

Bordallo JJ, Lopez-Llorca LV, Jansson HB et al (2002) Colonization of plant roots by egg-parasitic and nematode-trapping fungi. New Phytol 154:491–499

Bourne JM, Kerry BR, De Leij FAAM (1996) The importance of the host plant on the interaction between root-knot nematodes (*Meloidogyne* spp.) and the Nematophagous fungus, *Verticillium chlamydosporium* Goddard. Biocontrol Sci Technol 6:539–548

Boyle C, Gotz M, Dammann-Tugend U et al (2001) Endophyte-host interactions III. Local vs. systemic colonization. Symbiosis 31:259–281

Butt TM (2002) Use of entomogenous fungi for the control of insect pests. In: Esser K, Bennett JW (eds) The mycota. Springer, Berlin/Heidelberg, pp 112–134

Calhoun LA, Findlay JA, Miller JD et al (1992) Metabolites toxic to spruce budworm from balsam fir needle endophytes. Mycol Res 96:281–286

Capellano A, Dequatre B, Valla G et al (1987) Root-nodules formation by *Penicillium* sp. on *Alnus glutinosa* and *Alnus incana*. Plant Soil 104:45–51

Carroll G (1988) Fungal endophytes in stems and leaves - from latent pathogen to mutualistic symbiont. Ecology 69:2–9

Carroll G (1995) Forest endophytes: pattern and process. Can J Bot 73:S1316–S1324

Carvajal JA, Rodriguez-Kabana R (1998) Alginate films for assessment of parasitism of *Meloidogyne incognita* eggs in soils treated with organic amendments. Nematropica 28:41–48

Charnley AK (1997) Entomopathogenic fungi and their role in pest control. In: Esser K, Lemke PA (eds) The mycota. Springer, Berlin/Heidelberg/New York, pp 185–201

Cohen-Kupiec R, Chet I (1998) The molecular biology of chitin digestion. Curr Opin Biotechnol 9:270–279

de Bary A (1866) Morphologie und physiologie der pilze, flechten und myxomyceten. Engelmann, Leipzig

Deising H, Rauscher M, Haug M et al (1995) Differentiation and cell wall degrading enzymes in the obligately biotrophic rust fungus *Uromyces viciae-fabae*. Can J Bot 73:624–631

Dimock M, Turner J, Lampel J (1993) Endophytic microorganisms for delivery of genetically engineered microbial pesticides in plants. In: Kim L (ed) Advanced engineered pesticides. Marcel Dekker, New York/Basel/Hong Kong, pp 85–94

Dodane V, Vilivalam VD (1998) Pharmaceutical applications of chitosan. Pharm Sci Technol Today 1:246–253

Domsch KH, Gams W, Anderson TH (1980) Compendium of soil fungi. Academic, London

Dong LQ, Yang JK, Zhang KQ (2007) Cloning and phylogenetic analysis of the chitinase gene from the facultative pathogen *Paecilomyces lilacinus*. J Appl Microbiol 103:2476–2488

Dreyfuss MM, Chapela IH (1994) Potential of fungi in the discovery of novel, low-molecular weight pharmaceuticals. In: Gullo VP (ed) The discovery of natural products with therapeutic potential. Butterworth-Heinemann, Stoneham, pp 49–80

Ernst M, Mendgen KW, Wirsel SGR (2003) Endophytic fungal mutualists: seed-borne *Stagonospora* spp. enhance reed biomass production in axenic microcosms. Mol Plant Microbe Interact 16:580–587

Fan Y, Fang W, Guo S et al (2007) Increased insect virulence in *Beauveria bassiana* strains overexpressing an engineered chitinase. Appl Environ Microbiol 73:295–302

Fang W, Leng B, Xiao Y et al (2005) Cloning of *Beauveria bassiana* chitinase gene Bbchit1 and its application to improve fungal strain virulence. Appl Environ Microbiol 71:363–370

Gams W, Zare R (2001) A revision of *verticillium* sect. Prostrata. III. Generic classification. Nova Hedwig 72:329–337

Gan Z, Yang J, Tao N et al (2007) Cloning of the gene *Lecanicillium psalliotae* chitinase Lpchi1 and identification of its potential role in the biocontrol of root-knot nematode *Meloidogyne incognita*. Appl Microbiol Biotechnol 76:1309–1317

Gaspard JT, Mankau R (1986) Nematophagous fungi associated with *Tylenchulus semipenetrans* and the citrus rhizosphere. Nematologica 32:359–363

Gloer JB (1997) Application of fungal ecology in the search for new bioactive natural products. In: Esser K, Lemke PA (eds) The mycota. Springer, Berlin/Heidelberg/New York, pp 257–283

Gómez-Vidal S, Lopez-Llorca LV, Jansson HB et al (2006) Endophytic colonization of date palm (*Phoenix dactylifera* L.) leaves by entomopathogenic fungi. Micron 37:624–632

Gómez-Vidal S, Salinas J, Tena M et al (2009) Proteomic analysis of date palm (*Phoenix dactylifera* L.) responses to endophytic colonization by entomopathogenic fungi. Electrophoresis 30:2996–3005

Hajek AE, St Leger RJ (1994) Interactions between fungal pathogens and insect hosts. Annu Rev Entomol 39:293–322

Hallmann J, Rodríguez-Kábana R, Kloepper JW (1999) Chitin-mediated changes in bacterial communities of the soil, rhizosphere and within roots of cotton in relation to nematode control. Soil Biol Biochem 31:551–560

Jansson HB (1993) Adhesion to nematodes of conidia from the nematophagous fungus *Drechmeria coniospora*. J Gen Microbiol 139:1899–1906

Jumpponen A (2001) Dark septate endophytes – are they mycorrhizal? Mycorrhiza 11:207–211

Jumpponen A, Trappe JM (1998) Dark septate endophytes: a review of facultative biotrophic root-colonizing fungi. New Phytol 140:295–310

Kerry BR (2000) Rhizosphere interactions and the exploitation of microbial agents for the biological control of plant-parasitic nematodes. Annu Rev Phytopathol 38:423–441

Kimmons CA, Gwinn KD, Bernard EC (1990) Nematode reproduction on endophyte-infected and endophyte-free tall fescue. Plant Dis 74:757–761

Knudsen IMB, Hockenhull J, Jensen DF et al (1997) Selection of biological control agents for controlling soil and seed-borne diseases in the field. Eur J Plant Pathol 103:775–784

Kuldau GA, Yates IE (2000) Evidence for *Fusarium* endophytes in cultivated and wild plants. In: Bacon W, White JF (eds) Microbial endophytes. Marcel Dekker, New York/Basel, pp 85–120

Kumar MNVR (2000) A review of chitin and chitosan applications. React Funct Polym 46:1–27

Laflamme P, Benhamou N, Bussieres G et al (1999) Differential effect of chitosan on root rot fungal pathogens in forest nurseries. Can J Bot 77:1460–1468

Lafontaine PJ, Benhamou N (1996) Chitosan treatment: an emerging strategy for enhancing resistance of greenhouse tomato plants to infection by *Fusarium oxysporum* f.sp. *radicis-lycopersici*. Biocontrol Sci Technol 6:111–124

Latch GCM (1993) Physiological interactions of endophytic fungi and their hosts. Biotic stress tolerance imparted to grasses by endophytes. Agric Ecosyst Environ 44:143–156

Lee KY, Shibutani M, Takagi H et al (2004) Subchronic toxicity study of dietary N-acetylglucosamine in F344 rats. Food Chem Toxicol 42:687–695

Leinhos GME, Buchenauer H (1992) Hyperparasitism of selected fungi on rust fungi of cereal. Z Pflanzenkrankh Pflanzenschutz 99:482–498

Lewis LC, Bing LA (1991) *Bacillus thuringiensis* Berliner and *Beauveria bassiana* (Balsamo) vuillimen for European corn borer control: program for immediate and season-long suppression. Can Entomol 123:387–393

Lewis LC, Cossentine JE (1986) Season long intraplant epizootics of entomopathogens, *Beauveria bassiana* and *Nosema pyrausta*, in a corn agroecosystem. Entomophaga 31:363–369

Li J, Yang J, Huang X et al (2006) Purification and characterization of an extracellular serine protease from *Clonostachys rosea* and its potential as a pathogenic factor. Process Biochem 41:925–929

Lingham RB, Silverman KC, Bills GF et al (1993) *Chaetomella acutiseta* produces chaetomellic acids A and B which are reversible inhibitors of farnesyl-protein transferase. Appl Microbiol Biotechnol 40:370–374

Liu CH, Zou WX, Lu H et al (2001a) Antifungal activity of *Artemisia annua* endophyte cultures against phytopathogenic fungi. J Biotechnol 88:277–282

Liu XF, Guan YL, Yang DZ et al (2001b) Antibacterial action of chitosan and carboxymethylated chitosan. J Appl Polym Sci 79:1324–1335

Liu H, Du YM, Wang XH et al (2004) Chitosan kills bacteria through cell membrane damage. Int J Food Microbiol 95:147–155

Liu SQ, Meng ZH, Yang JK et al (2007) Characterizing structural features of cuticle-degrading proteases from fungi by molecular modeling. BMC Struct Biol 7:33

Lopez-Llorca LV (1990) Purification and properties of extracellular proteases produced by the nematophagous fungus *Verticillium suchlasporium*. Can J Microbiol 36:530–537

Lopez-Llorca LV, Fry SC (1989) Dityrosine, trityrosine and tetratyrosine, potential cross-links in proteins of plant-parasitic nematodes. Nematologica 35:165–179

Lopez-Llorca LV, Jansson HB (2006) Fungal parasites of invertebrates: multimodal biocontrol agents. In: Robson GD, van West P, Gadd GM (eds) Exploitation of fungi. Cambridge University Press, Cambridge, pp 310–335

Lopez-Llorca LV, Robertson WM (1992) Immunocytochemical localization of a 32-kDa protease from the nematophagous fungus *Verticillium suchlasporium* in infected nematode eggs. Exp Mycol 16:261–267

Lopez-Llorca LV, Bordallo JJ, Salinas J et al (2002) Use of light and scanning electron microscopy to examine colonisation of barley rhizosphere by the nematophagous fungus *Verticillium chlamydosporium*. Micron 33:61–67

Lopez-Llorca LV, Jansson HB, Maciá-Vicente JG et al (2006) Nematophagous fungi as root endophytes. In: Schulz B, Boyle C, Sieber TN (eds) Microbial root endophytes. Springer, Heidelberg, pp 191–206

Lopez-Llorca LV, Maciá-Vicente JG, Jansson HB (2008) Mode of action and interactions of nematophagous fungi. In: Ciancio A, Mukerji KG (eds) Integrated management and biocontrol of vegetable and grain crops nematodes. Springer, Dordrecht, pp 51–76

Maciá-Vicente JG, Jansson HB, Abdullah SK et al (2008a) Fungal root endophytes from natural vegetation in Mediterranean environments with special reference to *Fusarium* spp. FEMS Microbiol Ecol 64:90–105

Maciá-Vicente JG, Jansson HB, Mendgen K et al (2008b) Colonization of barley roots by endophytic fungi and their reduction of take-all caused by *Gaeumannomyces graminis* var. *tritici*. Can J Microbiol 54:600–609

Maciá-Vicente JG, Jansson HB, Talbot NJ et al (2009a) Real time-PCR quantification and live cell imaging of endophytic colonisation of barley (*Hordeum vulgare*) roots by *Fusarium equiseti* and *Pochonia chlamydosporia*. New Phytol 182:213–228

Maciá-Vicente JG, Rosso LC, Ciancio A et al (2009b) Colonisation of barley roots by endophytic *Fusarium equiseti* and *Pochonia chlamydosporia*: effects on plant growth and disease. Ann Appl Biol 155:391–401

Monfort E, Lopez-Llorca LV, Jansson HB et al (2005) Colonisation of seminal roots of wheat and barley by egg-parasitic nematophagous fungi and their effects on *Gaeumannomyces graminis* var. *tritici* and development of root-rot. Soil Biol Biochem 37:1229–1235

Morton CO, Hirsch PR, Peberdy JP et al (2003) Cloning of and genetic variation in protease VCP1 from the nematophagous fungus *Pochonia chlamydosporia*. Mycol Res 107:38–46

Nitao JK, Meyer SLF, Schmidt WF et al (2001) Nematode-antagonistic trichothecenes from *Fusarium equiseti*. J Chem Ecol 27:859–869

Noble HM, Langley D, Sidebottom PJ et al (1991) An echinocandin from an endophytic *Cryptosporiopsis* sp. and *Pezicula* sp. in *Pinus sylvestris* and *Fagus sylvatica*. Mycol Res 95:1439–1440

Nogawa M, Takahashi H, Kashiwagi A et al (1998) Purification and characterization of exo-beta-d-glucosaminidase from a cellulolytic fungus, *Trichoderma reesei* PC-3-7. Appl Environ Microbiol 64:890–895

Obledo EN, Barragan-Barragan LB, Gutierrez-Gonzalez P et al (2003) Increased photosyntethic efficiency generated by fungal symbiosis in Agave victoria-reginae. Plant Cell Tissue Organ Cult 74:237–241

Olivares-Bernabeu CM, Lopez-Llorca LV (2002) Fungal egg-parasites of plant-parasitic nematodes from Spanish soils. Rev Iberoam Micol 19:104–110

Palma-Guerrero J, Jansson HB, Salinas J et al (2008) Effect of chitosan on hyphal growth and spore germination of plant pathogenic and biocontrol fungi. J Appl Microbiol 104:541–553

Palma-Guerrero J, Lopez-Jimenez JA, Pérez-Berná AJ et al (2010) Membrane fluidity determines sensitivity of filamentous fungi to chitosan. Mol Microbiol 75:1021–1032

Park RD, Jo KJ, Jo YY et al (2002) Variation of antifungal activities of chitosans on plant pathogens. J Microbiol Biotechnol 12:84–88

Pascencia-Jatomea M, Viniegra G, Olayo R et al (2003) Effect of chitosan and temperature on spore germination of *Aspergillus niger*. Macromol Biosci 3:582–586

Peláez F, Collado J, Arenal F et al (1998) Endophytic fungi from plants living on gypsum soils as a source of secondary metabolites with antimicrobial activity. Mycol Res 102:755–761

Persmark L, Jansson HB (1997) Nematophagous fungi in the rhizosphere of agricultural crops. FEMS Microbiol Ecol 22:303–312

Persmark L, Nordbring-Hertz B (1997) Conidial trap formation of nematode-trapping fungi in soil and soil extracts. FEMS Microbiol Ecol 22:313–323

Persson C, Jansson HB (1999) Rhizosphere colonization and control of *Meloidogyne* spp. by nematode-trapping fungi. J Nematol 31:164–171

Peterson EA, Katznelson H (1964) Occurrence of nematode-trapping fungi in the rhizosphere. Nature 204:1111–1112

Petrini O (1991) Fungal endophytes of tree leaves. In: Andrews JH, Hirano S (eds) Microbial ecology of leaves. Springer, New York, pp 179–197

Rabea EI, Badawy ME, Stevens CV et al (2003) Chitosan as antimicrobial agent: applications and mode of action. Biomacromolecules 4:1457–1465

Read DJ (1999) Mycorrhiza – the state of the art. In: Varma A, Hock B (eds) Mycorrhiza. Springer, Berlin, pp 3–34

Read JC, Camp BJ (1986) The effect of the fungal endophyte *Acremonium coenophialum* in tall fescue on animal performance, toxicity, and stand maintenance. Agron J 80:811–814

Redman RS, Ranson JC, Rodriguez RJ (1999) Conversion of the pathogenic fungus *Colletotrichum magna* to a nonpathogenic, endophytic mutualist by gene disruption. Mol Plant Microbe Int 12:969–975

Redman RS, Dunigan DD, Rodriguez RJ (2001) Fungal symbiosis from mutualism to parasitism: who controls the outcome, host or invader? New Phytol 151:705–716

Redman RS, Sheehan KB, Stout RG et al (2002) Thermotolerance generated by plant/fungal symbiosis. Science 298:1581

Riekert HF, Tiedt LR (1994) Scanning electron microscopy of *Meloidogyne incognita* juveniles entrapped in maize roots by a nematode-trapping fungus *Arthrobotrys dactyloides*. S Afr J Zool 29:189–190

Rodriguez R, Redman R (2008) More than 400 million years of evolution and some plants still can't make it on their own: plant stress tolerance via fungal symbiosis. J Exp Bot 59:1109–1114

Rodriguez-Kabana R, Morgan-Jones G, Chet I (1987) Biological control of nematodes: soil amendments and microbial antagonists. Plant Soil 100:237–247

Schulz B, Boyle C (2005) The endophytic continuum. Mycol Res 109:661–686

Schulz B, Boyle C (2006) What are endophytes? In: Schulz B, Boyle C, Sieber TN (eds) Microbial root endophytes. Springer, Heidelberg, pp 1–14

Schulz B, Sucker J, Aust HJ et al (1995) Biologically active secondary metabolites of endophytic *Pezicula* species. Mycol Res 99:1007–1015

Schulz B, Guske S, Dammann U et al (1998) Endophyte-host interactions. II. Defining symbiosis of the endophyte-host interaction. Symbiosis 25:213–227

Schulz B, Rommert AK, Dammann U et al (1999) The endophyte-host interaction: a balanced antagonism? Mycol Res 103:1275–1283

Schulz B, Boyle C, Draeger S et al (2002) Endophytic fungi: a source of novel biologically active secondary metabolites. Mycol Res 106:996–1004

Segers R, Butt TM, Kerry BR et al (1994) The Nematophagous fungus *Verticillium chlamydosporium* produces a chymoelastase-like protease which hydrolyzes host nematode proteins *in situ*. Microbiology 140:2715–2723

Shimosaka M, Nogawa M, Ohno Y et al (1993) Chitosanase from the plant-pathogenic fungus, *Fusarium solani* f.sp. *phaseoli* - purification and some properties. Biosci Biotechnol Biochem 57:231–235

Siddiqui ZA, Akhtar MS (2008) Synergistic effects of antagonistic fungi and a plant growth promoting rhizobacterium, an arbuscular mycorrhizal fungus, or composted cow manure on populations of *Meloidogyne incognita* and growth of tomato. Biocontrol Sci Technol 18:279–290

Sieber TN (2002) Fungal root endophytes. In: Waisel Y, Eshel A, Kafkafy U (eds) Plant roots: the hidden half. Marcel-Dekker, Basel/New York, pp 887–917

Sikora RA, Pocasangre L, zum Felde A et al (2008) Mutualistic endophytic fungi and in-planta suppressiveness to plant-parasitic nematodes. Biol Control 46:15–23

Spiegel Y, Cohn E, Chet I (1986) Use of chitin for controlling plant-parasitic nematodes - I. Direct effects on nematode reproduction and plant performance. Plant Soil 95:87–95

Spiegel Y, Chet I, Cohn E (1987) Use of chitin for controlling plant plant-parasitic nematodes - II. Mode of action. Plant Soil 98:337–345

Spiegel Y, Chet I, Cohn E et al (1988) Use of chitin for controlling plant-parasitic nematodes - III. Influence of temperature on nematicidal effect, mineralization and microbial population buildup. Plant Soil 109:251–256

St Leger RJ, Staples RC, Roberts DW (1993) Entomopathogenic isolates of *Metarhizium anisopliae*, *Beauveria bassiana*, and *Aspergillus flavus* produce multiple extracellular chitinase isozymes. J Invert Pathol 61:81–84

St Leger RJ, Joshi L, Bidochka MJ et al (1996a) Characterization and ultrastructural localization of chitinases from *Metarhizium anisopliae*, *M. flavoviride*, and *Beauveria bassiana* during fungal invasion of host (*Manduca sexta*) cuticle. Appl Environ Microbiol 62:907–912

St Leger RJ, Joshi L, Bidochka MJ et al (1996b) Biochemical characterization and ultrastructural localization of two extracellular trypsins produced by *Metarhizium anisopliae* in infected insect cuticles. Appl Environ Microbiol 62:1257–1264

Stirling GR (1991) Biological control of plant-parasitic nematodes. Progress, problems and prospects. CAB International, Wallingford

Stone J, Viret O, Petrini O et al (1994) Histological studies of host penetration and colonization by endophytic fungi. In: Petrini O, Oullette GB (eds) Host wall alterations by parasitic fungi. Phytopathological Society Press, St Paul, pp 115–128

Summerell BA, Leslie JF (2004) Genetic diversity and population structure of plant-pathogenic species in the genus *Fusarium*. In: Gillins M, Holmes A (eds) Plant microbiology. Bios, Oxford, pp 207–223

Tikhonov VE, Lopez-Llorca LV, Salinas J et al (2002) Purification and characterization of chitinases from the nematophagous fungi *Verticillium chlamydosporium* and *V. suchlasporium*. Fungal Genet Biol 35:67–78

Tikhonov VE, Stepnova EA, Babak VG et al (2006) Bactericidal and antifungal activities of a low molecular weight chitosan and its N-/2(3)-(dodec-2-enyl)succinoyl/-derivatives. Carbohydr Polym 64:66–72

Trillas I, Aviles M, Ordovas J et al (2002) Using compost as a methyl bromide alternative. BioCycle 43:64

Trillas MI, Casanova E, Cotxarrera L et al (2006) Composts from agricultural waste and the *Trichoderma asperellum* strain T-34 suppress *Rhizoctonia solani* in cucumber seedlings. Biol Control 39:32–38

Trotel-Aziz P, Couderchet M, Vernet G et al (2006) Chitosan stimulates defense reactions in grapevine leaves and inhibits development of *Botrytis cinerea*. Eur J Plant Pathol 114(4):405–413

Tsai GJ, Su WH, Chen HC et al (2002) Antimicrobial activity of shrimp chitin and chitosan from different treatments and applications of fish preservation. Fish Sci 68:170–177

Tunlid A, Rosen S, Ek B et al (1994) Purification and characterization of an extracellular serine protease from the nematode-trapping fungus *Arthrobotrys oligospora*. Microbiology 140:1687–1695

Varma A, Verma S, Sudha et al (1999) *Piriformospora indica*, a cultivable plant-growth-promoting root endophyte. Appl Environ Microbiol 65:2741–2744

Varma A, Singh A, Sahay NS et al (2000) *Piriformospora indica*: an axenically culturable mycorrhiza-like endosymbiotic fungus. In: Hock B (ed) Mycota. Springer, Heidelberg/New York, pp 101–106

Vega FE (2008) Insect pathology and fungal endophytes. J Invert Pathol 98:277–279

Vega FE, Posada F, Catherine Aime M et al (2008) Entomopathogenic fungal endophytes. Biol Control 46:72–82

Vilich V, Dolfen M, Sikora RA (1998) *Chaetomium* spp. colonization of barley following seed treatment and its effect on plant growth and *Erysiphe graminis* f. sp. *hordei* disease severity. Z Pflanzenkrankh Pflanzenschutz 105:130–139

Wang M, Yang J, Zhang KQ (2006a) Characterization of an extracellular protease and its cDNA from the nematode-trapping fungus *Monacrosporium microscaphoides*. Can J Microbiol 52:130–139

Wang RB, Yang JK, Lin C et al (2006b) Purification and characterization of an extracellular serine protease from the nematode-trapping fungus *Dactylella shizishanna*. Lett Appl Microbiol 42:589–594

West CP, Izekor E, Oosterhuis DM et al (1988) The effect of *Acremonium coenophialum* on the growth and nematode infestation of tall fescue. Plant Soil 112:3–6

Wharton D (1980) Nematode egg-shells. Parasitology 81:447–463

Wilson D (1995) Endophyte - the evolution of a term, and clarification of its use and definition. Oikos 73:274–276

Yang J, Huang X, Tian B et al (2005) Characterization of an extracellular serine protease gene from the nematophagous fungus *Lecanicillium psalliotae*. Biotechnol Lett 27:1329–1334

Yang J, Tian B, Liang L et al (2007) Extracellular enzymes and the pathogenesis of nematophagous fungi. Appl Microbiol Biotechnol 75:21–31

Zhang Y, Liu X, Wang M (2008a) Cloning, expression, and characterization of two novel cuticle-degrading serine proteases from the entomopathogenic fungus *Cordyceps sinensis*. Res Microbiol 159:462–469

Zhang YJ, Feng MG, Fan YH et al (2008b) A cuticle-degrading protease (CDEP-1) of *Beauveria bassiana* enhances virulence. Biocontrol Sci Technol 18:551–563

Zhao ML, Mo MH, Zhang KQ (2004) Characterization of a neutral serine protease and its full-length cDNA from the nematode-trapping fungus *Arthrobotrys oligospora*. Mycologia 96:16–22

Chapter 10
Endophytic Fungi

Johannes Hallmann and Richard A. Sikora

Abstract Endophytic fungi as well as plant-parasitic nematodes probably coevolved with all plant life on earth including cultivated crop plants. While endophytic fungi often form mutualisitc associations to the benefit of the plant, plant-parasitic nematodes can cause detrimental yield losses. Although both groups of organisms interact very closely within the plant tissue, the potential role of endophytic fungi in nematode control was long overseen. Only recently has research on the interrelationships between endophytic fungi and plant-parasitic nematodes gained the interest of science working in plant protection. Numerous non-pathogenic endophytic fungi have been isolated from agronomic crops such as tomato and banana and have shown antagonistic potential towards a diverse spectrum of plant-parasitic nematodes. This chapter reviews the research that has been conducted on the use of endophytic fungi to control plant-parasitic nematodes. The chapter concentrates on several important groups of mutualistic endophytic fungi, in particular the endomycorrhizal fungi, *Piriformospora indica*, grass endophytes and *Fusarium* endophytes. Their biological control potential is presented and possible mode-of-actions are discussed. The review is meant to further stimulate research in this fascinating area and to provide a road map for practical application of endophytic fungi for the integrated management of plant-parasitic nematodes in the near future.

J. Hallmann (✉)
Julius Kühn-Institut, Federal Research Centre for Cultivated Plants, Institute for Epidemiology and Pathogen Diagnostics, Toppheideweg 88, 48161 Münster, Germany
e-mail: johannes.hallmann@jki.bund.de

R.A. Sikora
University of Bonn, Institute for Crop Science and Ressource Conservation, Nematology in Soil Ecosystems, Nußallee 9, 53115 Bonn, Germany
e-mail: rsikora@uni-bonn.de

K. Davies and Y. Spiegel (eds.), *Biological Control of Plant-Parasitic Nematodes: Building Coherence between Microbial Ecology and Molecular Mechanisms*, Progress in Biological Control 11, DOI 10.1007/978-1-4020-9648-8_10,

10.1 Introduction

The enormous fungal biodiversity that has been found colonizing the internal tissues of plants is for the most part made-up of mutualistic endophytes (endon gr. = within, phyton = plant) that do not cause harm to the plant (Schulz and Boyle 2006). Mutualistic endophytic fungi are far removed from the more highly evolved fungi that have developed pathogenic characteristics and can cause severe damage to plant tissue. The importance of the mutualisitc forms are of great interest to science because of their potential beneficial effects on plant growth and plant health (Elmi et al. 1990; Hol and Cook 2005; Sikora et al. 2008).

Recent molecular analysis suggests that green plants were colonized by endophytic fungi long before plants colonized the land (Tehler et al. 2003). The degree of compatibility achieved by endophytic fungi with their host is truly remarkable: they colonize the plant intercellular and intracellular and establish a very close contact with the plant cells (Panstruga 2003). Since plants appear to detect endophytes and pathogens by a similar set of genes, endophytes have the ability to bypass or suppress plant defence mechanisms meant to protect the plant against invaders (Spanu 2006; Pieterse and Dicke 2007). One hypothesis is that endophytic fungi produce effectors to overcome plant defence responses (Zhao and Qi 2008). Unfortunately, only few effectors have been cloned until now, but ongoing sequencing of entire genomes of endophytic as well as pathogenic fungi in combination with bioinformatic approaches will hopefully identify more effectors in the near future (Zhao and Qi 2008). Most likely, effectors of endophytic and pathogenic fungi have similar functions, just like plant defence and plant beneficial interactions use same signalling molecules, such as salicylic acid, jasmonic acid and ethylene (Zhao and Qi 2008). The quantity, composition and timing of these signalling molecules results in a specific set of genes which shapes the outcome of the plant response (Pieterse and Dicke 2007). For example, in beneficial interactions, jasmonic acid seems to play a dominant role in the plant response, whereas obligate fungal pathogens are generally more sensitive to salicylic acid-dependent defences and necrotrophic fungi to jasmonic acid- and ethylene-dependent defences (Pieterse and Dicke 2007).

Endophytic fungi can be divided into two main groups based on their feeding strategy: facultative endophytes which can utilize living and dead organic matter and obligate endophytes which rely on living plant cells. Why and how some endophytic fungi switch to an obligate behaviour is still a phenomenon which cannot be explained. In this regard, Spanu (2006) indicates that evolution is not forward-looking but the result of a balance of fitness at any given time. In other words, the ability to grow and reproduce as a free-living facultative endophyte can have a greater cost than developing obligate host specialisation. A better understanding of the obligate nature of fungal endophytes is expected in the future by comparison of the genomes of obligate endophytes with those of closely related non-obligate endophytes.

With few exceptions (Lambert and Casagrande 2006) endophytic fungi have been isolated from almost all plants and plant parts studied (Amin 1994; Hallmann and Sikora 1994a; Kuldau and Yates 2000; Pocasangre et al. 2000). Although most of the fungi are narrowly distributed, few are widely spread (e.g. non-pathogenic *Fusarium oxysporum*) suggesting a long and close mutually beneficial interaction. Within a certain plant tissue, many different species of endophytic fungi can be found; however, only a small number of all existing endophytic fungi have been identified. Our lack of knowledge is due first, to the low number of studies conducted to date, and second, to difficulties associated with fungal identification such as slow growth, lack of fruiting structures and the existence of non culturable isolates. In general, endophytic fungi can be found in all fungal divisions suggesting that the potential for endophytic colonization has evolved independently on many occasions. Colonization of plants by endophytic fungi usually occurs from the surrounding environment any time after seed germination. However, some endophytic fungi are also transmitted vertically, i.e. via seeds (see Sect. 10.3.2). Vertically transmitted fungal endophytes are usually asexual and use hyphae to colonize the host plant from the roots to seeds and in some cases from the seed over the root back to the seed. Since their life cycle almost completely depends on the host plant, these fungi are often mutualistic. On the contrary, horizontally transmitted fungal endophytes are sexual and use spores for dissemination.

The close association between plants and endophytic fungi provides benefits for both organisms. For endophytic fungi, the plant tissue guarantees a steady supply of nutrients and protection from most abiotic and biotic stressors (Schulz 2006). In exchange, the plant benefits from endophytic colonization by the induction of defence metabolites active against plant pathogens, endophytic secretions of phytohormones, and mobilisation of plant nutrients from the rhizosphere, which may lead among others to induced disease resistance (Benhamou and Garand 2001; Schulz 2006, Vu et al. 2006, Dababat and Sikora 2007a) and improved plant growth (Waller et al. 2005; Schulz 2006; Dai et al. 2008; Saldajeno et al. 2008). The fact that endophytic fungi colonize the same ecological niche within plants as endoparasitic nematodes raised scientific interest as to their importance as candidates for biological control of plant-parasitic nematodes. Although this field of research is still young, some remarkable achievements in nematode control have been already obtained by different fungal endophytes which are reviewed in this chapter.

Aware of the manifold definitions for "endophytes" (reviewed in Schulz and Boyle 2006), the term endophytic fungus is here used for fungi occurring within living plant tissue without causing any symptoms (Carroll 1988; Wilson 1995; Brundrett 2006). This definition encompasses mycorrhiza fungi, grass endophytes, *Fusarium* endophytes and endophytic nematophageous fungi. The objective of this review is to summarize the existing work on the interrelationships between endophytic fungi and plant-parasitic nematodes, with emphasis on the underlying mechanisms of nematode control. A better understanding of the function of endophytic fungi in this unique mutualist-parasite complex is urgently needed to establish a foundation for their development as biocontrol agents to combat plant-parasitic nematodes.

10.2 Mycorrhizal Fungi

The potential that endophytic fungi possess to regulate plant-parasitic nematode populations was first shown for endomycorrhizal fungi by Fox and Spasoff (1972). Mycorrhiza (myc gr. = fungus, rhiza=root) describes the symbiotic association between a fungus and the plant root. Approximately 80% of all plant species harbour one or more mycorrhizal fungi (Wang and Qiu 2006). Of the different types of mycorrhizas the arbuscular mycorrhiza (AM) are the most common and have the greatest impact on plant-parasitic nematodes. They are also considered to be the ancestral form. Molecular analysis indicates that this symbiosis was formed 400–460 million years ago when plants first colonized land (Wang and Qiu 2006). Other forms of mycorrhizas include ericoid mycorrhiza, orchid mycorrhiza and ectomycorrhiza; however, very little is known about their interactions with plant-parasitic nematodes (Villenave and Duponnois 2002).

Arbuscular-mycorrhizal fungi (AMF), formerly known as vesicular-arbuscular mycorrhizas (VAM), can be separated from other endophytic fungi by their obligate symbiotic interaction with the host plant, the presence of a localized interface of specialised hyphae, synchronised plant-fungus development, and fungus-to-plant nutrient transfer (Brundrett 2006). It has been hypothesized that mycorrhizal associations evolved from non-specific endophytic fungi that colonized plant roots and over time culminated in the development of fully functional associations with coordinated development and nutrient transfer (Brundrett 2002).

Arbuscular-mycorrhiza fungi are characterised by the presence of arbuscules, dichotomously-branching invaginations within the plant cell that allow a greater contact of the fungal hypha with the cell cytoplasm for optimum nutrient transfer between fungus and plant cell membrane. The arbuscules are linked to a hyphal network inside and outside the plant root. Within this symbiosis, the fungus provides water and mineral nutrients to the plant through its massive hyphal network that extends into the surrounding soil and the fungus receives in exchange assimilates from the plant for its own growth. As a result, plant growth is enhanced and plants can better tolerate unfavorable conditions such as drought or infestation by soil-borne pathogens and plant-parasitic nematodes. Taxonomically, AMF belong to the phylum Glomeromycota, which currently contains approximately 200 species (Walker and Schüßler 2004).

Since the first reports by Fox and Spasoff (1972) and Baltruschat et al. (1973) describing the beneficial effects of the AM fungi *Endogone gigantea* and *Glomus mossae* to reduce cyst and root-knot nematodes on tobacco, several reviews have addressed the potential of AMF to control endoparasitic as well as ectoparasitic nematodes (Francl 1993; Hussey and Roncadori 1982; Ingham 1988; Smith 1987; Sikora 1995; Pinochet et al. 1996; Hol and Cook 2005; Habte and Schmitt 2005). The information published to date on AMF × plant-parasitic nematode interactions shows a great variation in the final effect on plant health which ranges from low to highly significant levels of nematode control to in some cases even increased susceptibility. Hol and Cook (2005) hypothesized that based on proximity in tissue, the interactions between AMF and endoparasitic nematodes would be stronger and more adverse for the nematode, than those between AMF and ectoparasitic nematodes. However, contrary to this hypothesis the authors found that relative to control plants, AMF-inoculated plants were damaged less by endoparasitic than by ectoparasitic nematodes. Within the group of sedentary endoparasites, they reported that root-knot nematodes were negatively affected more by AMF than where cyst nematodes (Hol and Cook 2005). However, general conclusions are difficult to draw, and therefore the following section reviews in more detail the different types of AMF × plant-parasitic nematode interactions and their possible mode-of-actions.

10.2.1 AMF and Root-Knot Nematodes

Most of the research conducted in the past on AMF × plant-parasitic nematode interactions focused on root-knot nematodes. For example, biological control of root-knot nematodes by AMF has been demonstrated on many crops (Table. 10.1). In mycorrhizal plants egg production of *M. hapla* was reduced up to 75% and disease severity up to 71% (Waceke et al. 2001). Other effects of AMF towards root-knot nematode infestation include reduced juvenile penetration (Sikora 1979), reduced number and size of root-knot galls (Bagyaraj et al. 1979; Kellam and Schenck 1980), reduced nematode reproduction (Sikora 1979) and improved plant growth (Feldmann et al. 2008) (Fig. 10.1).

Fig. 10.1 Development of cucumber plants with (*in front*) and without (*behind*) mycorrhizal inoculation in a greenhouse plot highly infested with *Meloidogne hapla* (Feldmann et al. 2008)

Table 10.1 Examples for root-knot nematode × arbuscular mycorrhizal fungal interactions studied on various crops

Root-knot nematode species	Crop	AM fungus	References
Meloidogyne arenaria	Peanut	*Gigaspora margarita, Glomus etunicatum*	Carling et al. (1996)
Meloidogyne hapla	Carrot	*Glomus mosseae*	Sikora and Schönbeck (1975)
	Clover	*Glomus mosseae, G. fasciculatum, G. tenue, Gigaspora margarita*	Cooper and Grandison (1986)
	Pyrethrum	*Glomus etunicatum, Glomus* sp.	Waceke et al. (2001)
	Onion	*Glomus fasciculatum*	MacGuidwin et al. (1985)
	Tomato	*Glomus etunicatum, G. mosseae, G. fasciculatum, G. tenue, Gigaspora margarita*	Cooper and Grandison (1986) and Feldmann et al. (2008)
Meloidogyne incognita	Cotton	*Glomus fasciculatum*	Saleh and Sikora (1984)
		Glomus intraradices	Smith et al. (1986)
	Grapevine	*Glomus versiforme*	Li et al. (2006)
	Lucerne	*Glomus fasciculatum*	Jain et al. (1998)
	Olive	*Glomus intraradices, G. mosseae, G. viscosum*	Castillo et al. (2006)
	Soybean	*Gigaspora margarita*	Carling et al. (1989)
		Glomus macrocarpum	Kellam and Schenck (1980)
	Tobacco	*Glomus mosseae*	Sikora and Schönbeck (1975)
	Tomato	*Gigaspora margarita*	Cason et al. (1983)
		Glomus intraradices	Bagyaraj et al. (1979) and Reimann et al. (2008)
		Glomus mosseae	Sikora (1978), Cason et al. (1983), and Talavera et al. (2001)
		Glomus fasciculatum	Suresh et al. (1985), Rao et al. (1995), and Nagesh and Reddy (2004)
	Peach	*Gigaspora margarita, Glomus etunicatum*	Strobel et al. (1982)
	Piper nigrum	*Glomus etunicatum*	Sivaprasad et al. (1990)
Meloidogyne javanica	Chickpea	*Gigaspora gigantea*	Diederichs (1987)
	Olive	*Glomus intraradices, G. mosseae, G. viscosum*	Castillo et al. (2006)
	Tomato	*Glomus mosseae*	Al-Raddad (1995)

Although the majority of work supports the potential of AM fungi to control root-knot nematodes, some reports come to contrary results. Cason et al. (1983) studied the interaction of *Gigaspora margarita* and *G. mosseae* with *M. incognita* on tomato. Both AM fungi had no effect on root-knot nematode penetration of roots and reproduction when compared with non-mycorrhizal plants. The rate of mycorrhizal colonization was 37.3% for *G. margarita* and 21.4% for *G. mosseae* at the end of the experiment, suggesting much lower colonization rates at time of nematode inoculation. The rate of mycorrhization at time of nematode infection plays a critical role as root tissue colonized by AM fungi is generally not parasitized by sedentary endoparasitic nematodes (Cooper and Grandison 1986; Diedhiou et al. 2003). These results demonstrate that control may be AMF strain related and that nematode control is not axiomatic with the presence of the symbiont in the root system.

In praxis, AM fungal spores are directly inoculated into the field soil or growth substrate. Under those conditions, nematode control often fails, because AM fungi grow slowly and do not colonize sufficient root tissue. Root colonization by AM fungi takes 2–4 weeks, whereas nematodes penetrate roots in a matter of hours (Talavera et al. 2001). Therefore, early and intensive colonization of roots by AM fungi is essential to achieve good nematode control. Usually, AM fungal colonization rates above 40% are required for nematode control (Saleh and Sikora 1984; Diedhiou et al. 2003; Waceke et al. 2001); however, the level of colonisation will vary between crops and AM fungal species involved. In addition, the exact quantification of mycorrhization is difficult and traditionally depends on estimating the amount of stained fungal mycelium within the root (e.g. Guttenberger 2000). This does not say much about the activity of the fungus which is more determined by the number of arbuscules in the cells. In the future, real-time PCR methods might provide more reliable data on AMF quantification as already demonstrated for *Glomus intraradices* on tomato and the legume *Medicago truncatula* (Alkan et al. 2004).

High colonization rates can be achieved by selection of fast growing and rapid colonizing AM fungal strains. Host plant specificity plays an important role, as AM fungi differ in their preference for certain host plants. In addition, the host plant itself determines whether it will allow colonization by a specific AM fungus or not (Barker et al. 1998). Larkan et al. (2007) identified a gene in tomato that caused reduced mycorrhizal colonisation which they called *rmc*. Tomato plants homozygous for *rmc* were no longer hosts for most AM fungi. Depending on the AM fungal species involved, the interaction with tomato roots results in three distinct root colonization phenotypes: (1) inability to penetrate the epidermis, (2) inability to colonise the cortex, and (3) slower establishment of the symbiosis (Larkan et al. 2007). The *rmc* mutation also caused increased reproduction of *M. javanica* in root organ cultures (Barker et al. 2006). Furthermore, the selection of crops that are known to promote AM fungal populations in the field, such as white clover or leek (Deguchi et al. 2007; Sorensen et al. 2003), might be an option to increase the mycorrhiza inoculum in the soil, thus allowing rapid and high levels of root colonization of the following crop. Another option to enhance mycorrhization is the simultaneous inoculation of mycorrhiza helper bacteria (Frey-Klett et al. 2007). For example, co-inoculation with *Rhizobium etli* promoted the establishment of *Glomus intraradices* on tomato roots and caused a significantly higher reduction of *M. incognita* than inoculation of *G. intraradices* alone (Reimann et al. 2008). Other microorganisms reported to stimulate mycorrhization include non-pathogenic strains of *Fusarium oxysporum* (Diedhiou et al. 2003; García-Romera et al. 1998) and rhizosphere bacteria (von Alten et al. 1993; Perotto and Bonfante 1997; Frey-Klett et al. 2007).

10.2.2 AMF and Cyst Nematodes

Compared with the amount of work done on *Meloidogyne*, few reports describe interactions between AMF and cyst forming nematodes. Tylka et al. (1991) have shown that soybean cv. Wright inoculated with a mixture of *Gigaspora margarita, Glomus etunicatum, G. macrocarpum* and *G. mosseae* responded more tolerant to *Heterodera glycines* infection than non-mycorrhizal plants. However, this effect was transient and only observed in greenhouse experiments but not in the field. In a greenhouse experiment, numbers of *H. glycines* in roots and soil decreased up to 73% at 49 days after planting, but nematode numbers did not vary significantly between mycorrhizal and non-mycorrhizal plants at the end of the experiment. The authors further showed that nematode suppression was not systemic and was independent of soil phosphate fertility (Tylka et al. 1991). The effect of nematode species on the outcome of the AM fungus × plant-parasitic nematode interaction was exemplary demonstrated by Ryan et al. (2000, 2003) on potato. Potato microplants inoculated at transplanting with Vaminoc, a mycorrhizal inoculum consisting of three *Glomus* species, resulted in a significant higher multiplication rate of *G. rostochiensis* than on non-mycorrhized plants, whereas no such difference was observed for *G. pallida* (Ryan et al. 2003). Tuber yield was higher in the presence of the AM fungus as a result of more tubers per plant. But how can the effect be explained that two closely related plant-parasitic nematode species, which often occur concomitantly in the field, react so different to AM fungal inoculation? One explanation could be differences in root exudate composition between different species. Quite interestingly, potato microplants inoculated with Vaminoc increased the hatch of *G. pallida* but not of *G. rostochiensis* indicating that mycorrhizal inoculation stimulated the production of *G. pallida*-selective hatching chemicals, either hatching factors or hatching factor stimulants (Ryan et al. 2000). Another explanation might be found in the behaviour of the two species with *G. rostochiensis* penetration quickly after planting and escaping AMF antagonistic activity while *G. pallida* hatches more slowly over time and is affected by increased AMF colonization over time.

Following up on this work, Ryan and Jones (2004) and Deliopoulos et al. (2007) found that AMF-induced stimulation of *G. pallida* hatch was associated partly with the production of novel hatching factors by AMF and partly with increases in the activity of existed potato-derived hatching factors. In a further study with 27.5 cm diam. pots under conditions resembling the field situation, effect of pre-inoculation of potato with AM fungi on *G. pallida* infestation and tuber yield was studied (Deliopoulos et al. 2008). As a first result, AMF and *G. pallida* turned out to be mutually inhibitory, whereas the inhibition of *G. pallida* by AMF was greater than vice versa. Second, although it would have been expected that AMF-inoculated plants compared to non-mycorrhizal plants might result in higher infestation and multiplication by *G. pallida* due to AMF-stimulated juvenile hatch, the opposite was the case with a lower multiplication rate of *G. pallida* in mycorrhizal plants compared to non-mycorrhizal plants. Due to the authors, those encouraging results open up new avenues for the management of *G. pallida* such as AMF application as part of an integrated potato cyst nematode management system (Deliopoulos et al. 2008).

10.2.3 AMF and Migratory Nematodes

Contrary to sedentary endoparasitic root-knot and cyst nematodes, migratory endoparasitic nematodes, such as *Pratylenchus* spp. and *Radopholus similis*, migrate through the root leaving behind necrotic tissue and thus little space for colonization by AM fungi (Hussey and Roncadori 1978). Such destructive feeding behaviour suggests a negative impact on the obligate AM fungi, and indeed, the majority of studies on the interaction between AM fungi and migratory endoparasitic nematodes show reduced mycorrhization rates in the presence of the nematode (O'Bannon and Nemec 1979; Umesh et al. 1988; Pinochet et al. 1996). For example, on rough lemon the severity of root damage caused by *R. similis* adversely affected growth by *Glomus etunicatus* as indicated by significantly reduced mycorrhizal colonization rates and lower numbers of chlamydospores (O'Bannon and Nemec 1979). Although plant tolerance towards nematode infestation was increased by AMF, the overall beneficial effect was too low to compensate for the damage caused by *R. similis*. Smith and Kaplan (1988) observed increased plant growth on rough lemon inoculated with the AM fungus *Glomus intraradices* and *R. citrophilus* (syn. *R. similis*). They concluded that the observed growth enhancement in the presence of *G. intraradices* appeared to result from improved phosphorus nutrition and not from the antagonism between the fungus and *R. citrophilus*. For cotton it was shown that *Pratylenchus brachyurus* apparently did not adversely affected *Gigaspora magarita* development; nematode reproduction and fungus sporulation were unaffected by the presence of either organism (Hussey and Roncadori 1978). The results suggested that growth of cotton was stimulated more by *G. margarita* than by increased soil fertility and was not retarded by *P. brachyurus* at either fertility level.

The potential of AM fungi to control migratory endoparasitic nematodes has been demonstrated for *G. intraradices* and *R. similis* on rough lemon (Smith and Kaplan 1988), *G. mosseae* and *Pratylenchus vulnus* on apple and peach rootstocks

(Pinochet et al. 1993, 1995), *G. fasciculatum* and *R. similis* on banana (Umesh et al. 1988), *G. mosseae* and *R. similis* and *Pratylenchus coffeae* on banana (Elsen et al. 2003), *G. mosseae* and *P. penetrans* on carrots (Talavera et al. 2001) and *Glomus* spp. on the dune grass *Ammophila arenaria* (de la Peña et al. 2006). Pinochet et al. (1993) pointed out that despite lower density of *P. vulnus* in mycorrhizal apple rootstocks, the final nematode population was higher in these plants due to their larger root systems, capable of supporting a larger nematode population. However, Talavera et al. (2001) reported that soil inoculation with *Glomus* sp. compensated for the damage caused by *P. penetrans* on carrot and caused a 49% decrease in soil nematode numbers. Elsen et al. (2003) indicated that decreased root branching caused by *R. similis* and *P. coffeae* on banana was counterbalanced by the increased branching caused by *Glomus mosseae* leading to an overall better plant growth and significant nematode control. In contrast to those reports, Camprubí et al. (1993) and Calvet et al. (1995) did not observe any effect of *G. mosseae* on the root-lesion nematode *P. vulnus* on plum and cherry rootstocks, respectively. Overall, those results indicated that the interaction between AMF and plant-parasitic nematodes depends on many variables such as plant genotype, AMF specificity, nematode-plant interaction, nematode numbers and AMF inoculum densities.

The results obtained on the interaction between ectoparasitic nematodes and AM fungi are often inconsistent even within certain nematode taxa. This again demonstrates the enormous influence of biotic and abiotic factors on the interaction. For sweet potato, Kassab (1990) reported higher final population densities of *Criconemella* sp. on mycorrhiza versus non-mycorrhiza plants; whereas the opposite was observed for *Tylenchorhynchus* spp. On *Trifolium alexandrinum* high population densities of *Tylenchorhynchus vulgaris* were found to be associated with abundant mycorrhizal counts indicating that both organisms do not affect the development of each other (Hasan and Jain 1987). On the contrary, sorghum roots infected >50% by AM fungi had lower numbers of *T. vulgaris* and *Helicotylenchus dihystera* (Jain and Hasan 1986). Lucerne pre-inoculated with *G. fasciculatum* significantly reduced the adverse effects of *T. vulgaris* on plant biomass (Jain et al. 1998). However, if the plant was inoculated with nematodes prior to the AM fungus, root mycorrhization was reduced.

10.2.4 AMF Mode of Action

From all the work done over the past years on the mode-of-action of the AMF × plant-parasitic nematode interaction it appears that under a wide spectrum of environmental conditions, the antagonistic activity of AMF is related to specific mechanisms of action or a combination of several mechanisms. As nematodes are generally not presented in root segments colonized by AM fungi (Diedhiou et al. 2003), competition for space and feeding sites may occur (Hussey and Roncadori 1978). This hypothesis is supported by the fact that both plant-parasitic nematodes and AM fungi are obligate organisms depending on intact root tissue for development and reproduction. If competition for space would be the sole mechanism one would expect a clear dose response, i.e. increasing rates of mycorrhiza colonization would result in reduced nematode infection. Although de la Peña et al. (2006) have shown that AM fungi can outcompete with *Pratylenchus* spp. when they occur together in the same root zone, a clear dose response was not observed. Diedhiou et al. (2003) reported decreasing nematode numbers with increasing mycorrhization rates only until a certain threshold level was reached, which for tomatoes was at approx 30% mycorrhization. A further increase in mycorrhization rate above that level did not lead to a higher nematode control. Similar effects are also reported by Saleh and Sikora (1984) for mycorrhization rates above 40% by *G. fasciculatum* which gave no additional control of *M. incognita* on cotton.

It has also been suggested that improved host nutrient status and nutrient uptake in the presence of AM fungi enables the host plant to better tolerate nematode infection, outgrow nematode infestation or even become resistant towards nematode attack. AM fungi have been shown to improve the uptake of diffusion-limited nutrients such as phosphorus, copper, zinc and sulphur. Of those nutrients, most attention has been given to phosphorus. But do increased phosphorus levels necessarily improve nematode control? Cooper and Grandison (1986) found that mycorrhizal tomato plants compared to non-mycorrhizal plants were more resistant to *M. hapla* regardless the level of applied phosphorus. However, nematode numbers were higher in the lower P soils whereas at higher P levels nematode numbers were either unaffected or reduced. The authors concluded that control of *M. hapla* was probably due to some alteration in the physiology of the root system but was not entirely a result of better host nutrition and improved phosphorous uptake by mycorrhizal plants (Cooper and Grandisons 1986). In this regard, Graham et al. (1981) reported that root exudation of amino acids and reducing sugars is lower from mycorrhizal sudangrass grown in P-deficient soils than from non-mycorrhizal plants which might cause reduced nematode attraction. Cason et al. (1983) attained higher numbers of *M. incognita* in tomatoes grown at high P levels; but in relation to plant growth, nematode penetration and reproduction was lower compared to low P levels due to significant better plant growth a high P levels. Furthermore, the relative growth benefits attributable to AM fungi were greater when plants were pre-inoculated with AMF than when both AMF and plant-parasitic nematodes were inoculated simultaneously (Cooper and Grandisons 1986). De la Peña et al. (2006) reported that although plant biomass was higher when *Pratylenchus penetrans* was inoculated 5 weeks after infection with AMF than simultaneously, they did not find a higher concentration of C and N in plants that were pre-inoculated with the AM fungus. The authors concluded that the positive effect of AM fungal pre-inoculation might have occurred through nematode suppression and not through increased plant tolerance because the effect of pre-inoculation with AMF resulted in reduced nematode infection. Finally, Elsen et al. (2008) did not observe any growth promotion by AM fungi in their system albeit nematode control levels above 50% were reached. Those results indicate that mechanisms other than enhanced plant growth are more likely to explain the antagonistic effect of AM fungi.

The potential of AM fungi to induce resistance is a well described mechanism known from the control of fungal pathogens (Whipps 2004). But for plant-parasitic nematodes the situation is different. In the literature reports are found that support, as well as deny the involvement of induced resistance in nematode control. For example, in split-root studies on tomatoes where *Gigaspora margarita* was applied to one half of the root system and 2 weeks later, *M. incognita* to the other half of the root system, nematode penetration and reproduction was not altered compared to non-mycorrhizal plants (Cason et al. 1983). In a similar approach using a split-root system of the dune grass *Ammophila arenaria* inoculated with *Glomus* sp. and *P. penetrans*, de la Peña et al. (2006) achieved no nematode control compared to approx. 50% control in single pot experiments. The authors concluded that nematode control by AM fungi did not occur through a systemic response but through local mechanisms. This is supported by the fact, that new or fresh tomato roots which were not colonized by the mycorrhiza were infected by *M. incognita* and developed root galls (Nagesh and Reddy 2004). Those results contradict the effects described by Elsen et al. (2008) for nematode control on banana: Tissue-cultured banana plantlets, grown in a split-root system and inoculated with the AM fungus *Glomus intraradices*, reduced numbers of *Radopholus similis* and *Pratylenchus coffeae* by over 50%, even when the AM fungus and the plant-parasitic nematodes were spatially separated. The latter strongly supports the involvement of induced resistance as control mechanism. However, the underlying mechanisms are far from being understood. Factors affecting nematode penetration such as changes in rhizosphere microflora or root exudates composition (Gerdemann 1968), induced by mycorrhizal roots, might influence plant-parasitic nematodes (Smith 1987). Within mycorrhizal roots, increased lignification and suberization as a result of induced resistance might as well negatively affect nematode migration and feeding (Hayman 1982; Pinochet et al. 1996).

Despite the many reports clearly showing the potential of AM fungi to reduce plant damage caused by plant-parasitic nematodes, the molecular mechanisms of how the symbiosis provides nematode control are still poorly understood. Of the various physiological and molecular changes caused by AM fungi in their host, the expression of chitinases received most attention (Li et al. 2006; Azcón-Aguilar and Barea 1996; Dumas-Gaudot et al. 1992, 1996; Pozo et al. 2002; Spanu et al. 1989; Vierheilig et al. 1994). Plant chitinases catalyze the hydrolysis of chitin, a β-1,4-linked N-acetylglucosamine, which is part of the cell wall of fungi including AMF as well as nematode eggs. Quite interestingly, AM fungi appear to be insensitive to plant chitinase (Arlorio et al. 1992) and transgenic tobacco plants overexpressing chitinase do not affect root mycorrhization (Vierheilig et al. 1995). In general, plant chitinases are considered to be one of the important components of plant defence systems (Li et al. 2006); their activity is induced in response to various abiotic and biotic stressors, including infection by plant-parasitic nematodes (Qiu et al. 1997) but also in the early phase of mycorrhiza symbiosis (Spanu et al. 1989). Li et al. (2006) have demonstrated that both the AM fungus *Glomus versiforme* and, to slightly lower extent, also *M. incognita*, induce expression of the class III chitinase in grapevine. However, mycorrhizal grapevine roots challenged with *M. incognita* expressed the highest chitinase activity. Overall, nematode infection was reduced with increasing colonization rate by *G. versiforme*. The authors concluded that *G. versiforme* induced a defense response against *M. incognita* in the grapevine roots, which appeared to involve transcriptional control of *VCH3* expression throughout the entire root tissue (Li et al. 2006). However, to confirm the hypothesis that plant chitinases systemically trigger nematode control, further experiments in split-root systems are required. Another important factor contributing to increased nematode tolerance in mycorrhizal plants might be enhanced water uptake. Recent molecular studies identified three water channel protein (*MIP* genes) sequence groups in mycorrhizal roots compared to only one in non-mycorrhizal roots (Zeze et al. 2008). Water channel proteins act as water transporting systems in the plants and an increase in their number or even activity might compensate for losses of water transport due to nematode damage of vascular tissue.

Although AM fungi are a major model system to study mutualistic plant fungal interaction, there is little information available on AMF genes involved in the mutualistic interaction. The main handicap is the obligate nature of AM fungi that complicates genetic manipulation (Newman and Reddell 1987). Besides, AMF does not colonize *Arabidopsis thaliana*, an excellent model plant to study plant responses at the genetic level. Those limitations might be circumvented by using the mycorrhiza-like endophytic fungus *Piriformospora indica* as a model organism (see next paragraph).

In conclusion, certain species and strains of AM fungi have biocontrol potential towards sedentary and to a lesser extent also towards migratory endoparasitic nematodes. If used for biocontrol of plant-parasitic nematodes, AM fungi need to be given enough time to establish the symbiosis before the plant is infected by the plant-parasitic nematodes. Therefore, commercial application of AM fungi is most promising in transplanting systems such as vegetables, bananas or fruit tree rootstocks. A better understanding of the molecular mechanisms involved in the symbiotic mutualistic association between plants and AM fungi in controlling plant-parasitic nematodes will hopefully lead to new strategies for better nematode control in the near future.

10.3 Other Endophytic Fungi

10.3.1 Piriformospora indica

Piriformospora indica was described in 1998 by Ajit Verma and collaborators as an axenically cultivable, mycorrhiza-like fungus that forms intercellular and intracellular hyphae within the root epidermal and cortical tissue of their host plants (Verma et al. 1998; Singh and Varma 2000). Following root colonization, *P. indica* improves tremendously plant growth and induces tolerance to salt stress as well as resistance to fungal diseases and to plant-parasitic nematodes (Varma et al. 1999; Waller et al. 2005; Wallner 2007; Daneshkhah 2008). *Piriformospora indica* even stimulates plant colonization by plant growth promoting rhizobacteria such as shown for *Pseudomonas striata* on maize and mungbean (Singh et al. 2009). But most important, *P. indica* colonizes *Arabidopsis thaliana*, the most important model plant for genetic and molecular studies over the last 50 years (Peškan-Berghöfer et al. 2004).

As a member of the mustard (Brassicaceae) family, *A. thaliana* is closely related to important agricultural crops such as canola, cabbage, radish, turnip etc. and results achieved from fundamental research on *A. thaliana* can at least to some degree be transferred to those crops but most likely also to other crops of economic importance. Furthermore, *A. thaliana* is thought to be a promising model plant for studying plant × fungal endophyte interactions; it has a short live cycle of 6 weeks from germination to mature seeds, the plant can be easily cultivated in the lab, the genome is completely sequenced (The Arabidopsis Genome Initiative 2000) and a large number of ready-to-use mutant lines are available for studying gene expression, for example, in studies involving colonization by endophytic fungi and effects on plant-parasitic nematodes. Proteomic approaches have already led to the identification of several *P. indica* responsive *A. thaliana* proteins that are thought to be involved in signal perception or transduction and to be required for the establishment of a compatible association (Shahollari et al. 2005, 2007). As shown for barley, *P. indica* confers systemic resistance towards the foliar pathogen *Blumeria graminis* f.sp. *hordei* (Waller et al. 2005). Results from work with *A. thaliana* signal transduction mutants indicated that fungal resistance followed the principles of induced systemic resistance (Schäfer et al. 2007) instead of systemic acquired resistance. Compared with the numerous studies on fungal pathogens, research on *P. indica* conferred control of plant-parasitic nematodes using *A. thaliana* as model plant has just started. As shown by Grundler and collaborators, colonization of *A. thaliana* by *P. indica* caused a reduction in infection rate by the beet cyst nematode *Heterodera schachtii* (Wallner 2007) and the root-knot nematode *M. incognita* (Daneshkhah 2008). Furthermore, results showed that *P. indica* influences the expression of phloem-specific sucrose transporters in syncytia. In conclusion, the enormous genetic information available on *A. thaliana* in combination with the powerful endophyte *P. indica* therefore seems to be a very promising model system to study the interactions of endophytic fungi × plant-parasitic nematodes at the molecular level.

10.3.2 Grass Endophytes

Cool season grasses such as tall fescue (*Festuca arundiaceae*) and perennial ryegrass (*Lolium perenne*) are commonly colonized by fungal species of the genus *Neotyphodium* (former *Acremonium*). These endophytic fungi are obligate biotrophs which colonize the plant tissue intercellular (Hinton and Bacon 1988; Siegel et al. 1987). They often form close associations with the plant cell walls (Christensen et al. 2002); however, they do not invade the plant cells. *Neotyphodium* endophytes produce several alkaloids forming a specific mutualistic association with the plant. Within this symbiosis, the plant provides protection and nutrition to the fungus in exchange to plant protection against abiotic and biotic stresses, conferred by the fungus (Timper et al. 2005; Sullivan et al. 2007). For several grass × *Neotyphodium* association resistance to some, but not all, plant-parasitic nematodes, has been reported. For example, in tall fescue colonized by the endophyte *N. coenophialum*, nematode numbers of *Meloidogyne marylandi* (Kimmons et al. 1990; Elmi et al. 2000), *M. graminis* (Elmi et al. 1990), *Pratylenchus scribneri* (West et al. 1988; Kimmons et al. 1990), *Helicotylenchus dihystera* (Pedersen et al. 1988), *Paratrichodorus minor* (Pedersen et al. 1988) and *Tylenchorhynchus maximus* (Bernard et al. 1998) are reduced. For perennial ryegrass infected with *Neotyphodium lolii*, suppression of *M. naasi* (Stewart et al. 1993) and *P. scribneri* (Panaccione et al. 2006; Bacetty et al. 2009a) has been described. On the contrary, Cook et al. (1991) did not observe any effect of *N. lolii*-infected perennial ryegrass against *M. naasi* under greenhouse or field conditions when compared with endophyte-free plants. However, root growth of endophyte-infected plants was better than for endophyte-free plants, thus creating better conditions for nematode multiplication which might have accounted for the increased nematode populations.

Within grasses, nematode control is achieved despite the fact that roots are never colonized by the endophytes as they colonize the plant exclusively aboveground. Therefore, the fungus must either change plant physiology or produce toxic or repellent metabolites which are translocated basipedal into the root. So far, the mode of action by which grass endophytes control plant-parasitic nematodes is still not fully understood. At least, direct antagonism can be ruled out due to the spatial separation between endophyte and plant-parasitic nematode. For insect control it has been shown that specific alkaloids, such as peramine, or the ergot alkaloid ergovaline, are responsible for the insecticidal activity observed, although environmental factors and presence of mycorrhizae and nutrients can influence the outcome (Kuldau and Bacon 2008; Vega et al. 2008). Some of those alkaloids have anti-feeding or repellent attributes but they are only rarely detectable in roots, if at all (Panaccione et al. 2006).

Besides alkaloids, *Neotyphodium* endophytes also secrete proteins in the host plant that might play a role in herbivore defence. Li et al. (2004) detected a fungal endochitinase in the apoplastic fluid of *Poa ampla*. This chitinase was encoded by a single gene and turned out to be different in its amino acid sequence to other fungal chitinases. If expressed or translocated into the roots, chitinases could affect plant-parasitic nematodes. However, results achieved by Brants et al. (2000) with transgenic tobacco plants expressing elevated levels of an endochitinase cDNA from *Trichoderma harzianum* showed no effect on *M. hapla*. If the chitinases secreted by *Neotyphodium* sp. behave differently and do affect plant-parasitic nematodes still needs to be confirmed. From comparative studies on resistant and susceptible soybean cultivars, challenged with the root-knot nematode *M. incognita*, it is known that increased chitinase activity is associated with plant resistance (Qiu et al. 1997). However, the mechanisms of how chitinases affect plant-parasitic nematodes within the plant tissue are still unknown. Besides chitinases, application of molecular methods such as DNA microarrays and mRNA differential display reverse transcription (DDRT)-PCR clearly demonstrated that plant parasites cause changes in the transcription of hundreds of genes (Baldwin et al. 2001; Schmidt et al. 2005; Sullivan et al. 2007). In tall fescue colonized by *N. coenophialum*, damage by the fall armyworm, *Spodoptera frugiperda*, caused an up-regulation of *lolC*, a gene required for biosynthesis of the fungal alkaloid loline (Sullivan et al. 2007). The authors suggested that *N. coenophilum* seems to switch its host's defensive strategy from tolerance through compensative growth to insect resistance. Whether or not similar mechanisms also exist for the control of plant-parasitic nematodes is still unknown. There is definitely an enormous need for more fundamental research to better understand the *Neotyphodium* × plant-parasitic nematode interactions. Recent progress in large-scale gene discovery from different *Neotyphodium* species and the following generation of cDNA-based microarrays provide tools for high-throughput transcriptome analysis, including detection of novel endophyte genes and investigation of the host grass × symbiont interaction (Felitti et al. 2006).

For nematode suppression of tall fescue, Timper et al. (2005) emphasized that ergot alkaloids produced by *Neotyphodium coenophialum* are involved in control. In their studies with a mixed population of *Pratylenchus zeae* and *P. scribneri* they found similar nematode numbers in tall fescue containing non-ergot producing strains of *N. coenophialum* and endophyte-free plants, whereas much less nematodes were recorded in plants containing the wild-type *N. coenophialum*, which produces the ergot alkaloid. Furthermore, nematode control seems to depend on the specific plant genotype × fungal strain interaction as the non-ergot strain AR584 induced some resistance towards *Pratylenchus* spp. in the cultivar Georgia 5 but not in the cultivar Jesup. The nematicidal activity of different ergot alkaloids can vary significantly. While ergovaline and alpha-ergocryptine were nematicidal at 5 and 50 μg/ml, respectively, ergonovine and ergocornine were nematistatic at most concentrations (Bacetty et al. 2009a). Ergovaline was also shown to be repellent for *P. scribneri* (Bacetty et al. 2009b). Besides ergot alkaloids, root exudates of *N. coenophialum*-infected tall fescue also showed nematistatic activity. Analysis of those root exudates identified several polyphenols such as chlorogenic acid, 3,5-dicaffeoylquinic acids and caffeic acid (Bacetty et al. 2009a). For perennial ryegrass, Panaccione et al. (2006) rejected the involvement of ergot alkaloids in nematode suppression. They compared the effect of the wild-type *N. lolii* and a genetically modified form completely lacking ergot alkaloids to control *P. scribneri*. Both the wild type and non-ergot producing strains significantly reduced nematode numbers compared with endophyte-free plants which were highly infested with *P. scribneri*. Similarly, Ball et al. (1997) observed significantly fewer *M. marylandi* in perennial ryegrass infected with a strain of *N. lolii* not producing ergovaline than in endophyte-free plants. These results suggest that the role of endophyte alkaloids in nematode control might differ in different grass × *Neotyphodium* associations. Quite interestingly, *N. lolii* endophytes can influence the biology of entomopathogenic nematode, i.e. *Steinernema carpocapsae*, parasitizing herbivorous insects. When black cutworm, *Agrotis ipsilon* (Kunkel et al. 2004), or fall armyworm, *Spodoptera frugiperda* (Richmond et al. 2004), fed on endophyte-colonized perennial ryegrass they were less susceptible for the entomopathogenic nematode, *S. carpocapsae*. The authors concluded that herbivores capable of developing on endophytic grasses may acquire some level of resistance against infection by *S. carpocapsae*. Most likely, it is due to uptake of the endophyte alkaloid ergocristine which seems to be toxic for the symbiotic *Xenorhabdus nematophila*, the bacterium responsible for the lethal effect of *S. carpocapsae* against insect pests.

Other effects might contribute to nematode suppression. West et al. (1988) observed substantially lower soil populations of *Pratylenchus scribneri* and *Tylenchorhynchus acutus* in plots containing endophyte-infected tall fescue than in plots containing non-infected tall fescue. In addition, drought tolerance was higher in endophyte-infected tall fescue which the authors felt may be at least partially mediated through enhanced resistance to soil-borne nematodes. This is in line with Elmi et al. (2000) who concluded that endophyte-enhanced persistence of tall fescue in *M. marylandi*-infested soils proned to drought may be explained at least partly by endophyte protection of roots from nematode damage. As shown by Richardson et al. (1990), colonization of grasses by *Neotyphodium* endophytes increases the rate and length of root growth which might support drought tolerance but also nematode tolerance. Also, changes of the root surface, resulting in less attractiveness for plant-parasitic nematodes, could contribute to nematode control. Following endophyte infection, the Fe^{3+} reducing activity on the root surface and total phenolic concentration in roots of tall fescue increased substantially, especially under phosphorus limiting conditions (Malinowski et al. 2004). Elucidating the mechanisms by which the fungal endophytes imparts resistance to plant-parasitic nematodes may help define traits for selecting beneficial endophyte strains that will improve grass performance.

A potential use of endophyte-grass associations for nematode control is seen especially for sports greens. Recently, the number of reports on damage caused by plant-parasitic nematodes such as *Meloidogyne* spp. and *Pratylenchus* spp. on golf and football greens is increasing (Hallmann, personal communication 2010). Effective and economical fungal endophyte application could simply be done by seed application and once established, the fungus would continue to be transmitted via seeds.

10.3.3 Fusarium Endophytes

Of all plant species investigated for the occurrence of endophytic fungi, most were colonized by members of the genus *Fusarium* (Kuldau and Yates 2000; Bacon and Yates 2006; Maciá-Vicente et al. 2008). Among them, *F. oxysporum* was the most cosmopolitan endophytic species isolated so far. Although some strains are known for their pathogenicity, the majority of strains in nature are non-pathogenic saprophytes. Many of these non-pathogenic *F. oxysporum* strains have been described as antagonists of fungal pathogens, plant-parasitic nematodes and insects

(Alabouvette and Couteaudier 1992; Hallmann and Sikora 1994b; Griesbach 1999). Biocontrol potential of non-pathogenic stains of *F. oxysporum* towards plant-parasitic nematodes has been reported among others for *Meloidogyne incognita* on tomato (Hallmann and Sikora 1994b), *Pratylenchus goodeyi* on banana (Speijer 1993) and *Radopholus similis* on banana (Amin 1994; Niere 2001; Paparu et al. 2007; Pocasangre et al. 2000). Hallmann and Sikora (1994b), working with tomato inoculated with the non-pathogenic *F. oxysporum* strain 162, achieved 50% fewer galls by *M. incognita* than in endophyte-free tomatoes. Similar reductions were obtained for number of galls and egg masses per g root. More over, *F. oxysporum* strain 162 demonstrated a high ability to colonize roots of tomato cultivars resistant to the Fusarium wilt pathogen (Dababat et al. 2008). The latter finding is of practical importance as many tomato varieties on the market are resistant to Fusarium wilt disease. The level of nematode control in Fusarium wilt resistant cultivars was found similar or higher compared to susceptible cultivars, indicating that the genetic plant resistance to wilt strains did not interfere with the biocontrol antagonistic strain, *F. oxysporum* strain 162. The fungus was re-isolated up to 100% from surface-disinfested root sections of resistant, as well as susceptible, tomato plants, demonstrating its high level of root colonization. Fungal densities reached 2.3×10^6 cfu g^{-1} root fresh weight compared to densities below the detection level of 1.0×10^2 cfu g^{-1} in noninoculated control plants. Application of the endophyte alone had no effect on shoot and root fresh weight or on root length and thus can be considered non-pathogenic. Application time and inoculum density of the fungal antagonist are crucial for its control efficacy. For *F. oxysporum* strain 162, application at sowing always resulted in a higher reduction of *M. incognita* penetration of tomato roots than application at transplanting (Dababat and Sikora 2007b). However, combining application at sowing and transplanting did not result in an additional reduction of juvenile penetration compared with sole application at sowing. Significant nematode control, up to 75%, was achieved at 10^4 cfu g^{-1} soil (Dababat and Sikora 2007b). Increasing the inoculum density of *F. oxysporum* strain 162 from 10^4 to 10^5 cfu g^{-1} soil did not improve nematode control.

Pocasangre et al. (2000) recovered a total of 132 isolates of endophytic fungi from different banana cultivars sampled in Central America. Species of *Fusarium* including *F. oxysporum* were the predominant fungi in all countries surveyed and in all localities within a country (Pocasangre 2000). Five out of 28 *Fusarium* spp. tested significantly reduced the number of *R. similis* per g^{-1} root by 80% and higher on banana tissue culture plantlets under greenhouse conditions. Furthermore, high nematode control was shown to occur on four banana cultivars: Gran Enano (AAA), Williams (AAA), Gros Michel (AAA) and FHIA-23 (AAAA). Similar studies were conducted in Uganda on East African Highland banana and on the commercial Cavendish banana cultivar 'Grand Nain' (Niere 2001). Population densities of *R. similis* on tissue culture plantlets inoculated with different strains of non-pathogenic *F. oxysporum* were reduced between 49% and 79% (Niere 2001). The observed variation in nematode control is probably caused by genetic variability of the fungal strains and banana cultivars used. For optimum control, it is recommended to identify suitable cultivar × endophyte strain combinations (Paparu et al. 2006). Banana tissue culture plantlets can then be easily inoculated with the endophyte before planting in the field. Also fungal colonization is high in the beginning, fungal densities decrease over time, whereas persistence in the roots is better than in rhizomes (Paparu et al. 2008).

Several attempts have been made to improve the control potential of non-pathogenic *F. oxysporum*. Within this respect, Paparu et al. (2009) reported for banana that application with two non-pathogenic *F. oxysporum* strains significantly improved plant growth and reduced nematode infestation compared with single fungal application. A further attempt is the combined application of non-pathogenic *F. oxysporum* with other antagonists, having complementary mode-of-actions, such as AM fungi or antagonistic bacteria. Diedhiou et al. (2003) studied possible interactions between the AM fungus *Glomus coronatum* and non-pathogenic *F. oxysporum* strain Fo162 for the control of *M. incognita* on tomato. Both fungal antagonists stimulated plant growth and reduced *M. incognita* infestation. However, combined application of the AM fungus and Fo162 did not increase overall nematode control or plant growth although mycorrhization of tomato roots was enhanced. The authors conclude that synergisms between fungal antagonists with different mode-of-actions seem to be of minor importance compared to their effects on different growth stages of the pathogen. Mendoza and Sikora (2009) combined the mutualistic endophyte *F. oxysporum* strain Fo162 with the egg pathogen *Paecilomyces lilacinus* strain 251 and the bacterium *Bacillus firmus* in hope of increasing the biocontrol of the burrowing nematode *Radopholus similis* on banana. Although the antagonists proved to be compatible and gave improved levels of control, no synergistic effects were detected.

A further aspect underlining the importance of endophytic *F. oxysporum* in disease control are suppressive soils. Non-pathogenic, endophytic strains of *F. oxysporum* isolated from suppressive soils are used as biocontrol agents to control fungal pathogens (Alabouvette et al. 1979; Lemanceau and Alabouvette 1991; Postma and Rattink 1991; Schneider 1984). In each case, these fungi were endophytes of the hosts they protected (Kuldau and Yates 2000; Benhamou and Garand 2001). More recently, it was shown that *F. oxysporum* from the endorhiza of banana growing in *R. similis* suppressive soil gave high levels of *R. similis* control in greenhouse trials (zum Felde et al. 2005; Sikora et al. 2008). This work demonstrated that the suppressiveness observed in the fields was at least partly plant based, and not limited to a suppressive soil-borne microflora.

One of the first steps in describing the mechanism of nematode control by endophytic *F. oxysporum* isolates was to look for secondary metabolites. In vitro suppression of plant-parasitic nematodes by secondary metabolites of *F. oxysporum* was demonstrated for the plant-parasitic nematode species *R. similis, M. incognita* and *G. pallida* (Amin 1994; Hallmann and Sikora 1996; Athman 2006). However, the degree of nematode suppression varied between feeding type and trophic group. Complete nematode inactivation following treatment with secondary metabolites of the non-pathogenic *F. oxysporum* strain 162 was achieved for juveniles of the sedentary endoparasites *Heterodera schachtii, M. arenaria, M. incognita* and *M. javanica*, whereas inactivation only reached 60% for mixed stages of the migratory endoparasites *R. similis* and *P. zeae* and no inactivation was observed for the mycophagous species *Aphelenchoides composticola* and the microphagous species *Panagrellus redivivus*. The reduced effect against migratory endoparasites might partly be due to the concomitant occurrence of adult nematodes in the test suspension which are more resistant to culture filtrates. The observed selectivity against plant-parasitic nematodes but not free-living nematodes could be due to differences in cuticular composition or nematode physiology between those trophic groups. Sedentary nematodes have not been naturally exposed to high levels of toxins and have therefore failed to develop resistance to such compounds (Sikora et al. 2007). However, it needs to be considered that in those studies fungal metabolites were produced in nutrient-rich artificial media with a completely different nutrient composition than the root apoplast. Unfortunately, it is still unknown if fungal metabolites toxic to plant-parasitic nematodes are released within the plant tissue. Even though nematicidal metabolites might not be the responsible mode-of-action under natural condition, it does create a potential use of endophytic fungi for screening new nematicidal key structures.

At the future, if it will be realized that endophytic fungi might express secondary metabolites at levels toxic to plant-parasitic nematodes within the plant tissue, the fungi might not be acceptable for biocontrol purpose due to food safety aspects. However, if those metabolites are non-toxic and released in plant roots, they might interfere with nematode attraction and penetration or even nematode development. The effect of *F. oxysporum*-colonized banana plants on nematode attraction and penetration was studied by Vu (2005) in a linked twin-pot chamber. Two pots, one planted with a *F. oxysporum*-colonized banana plant and the other with a non-colonized plant, were connected with a pipe filled with growth substrate. The centre of the pipe was than inoculated with *R. similis*. Nematode attraction and penetration into *F. oxysporum*-colonized banana roots was reduced over 50% compared with non-colonized roots 2 weeks later. The level of reduction was similar for three different *F. oxysporum* strains. Work by Dababat and Sikora (2007c), using the same linked twin-pot chamber and one of the *F. oxysporum* strains described above, confirmed those results for *M. incognita* on *F. oxysporum*-colonized tomatoes. Nematode reduction ranged from 36% to 56% in repeated experiments.

Quite interestingly, physical presence of the fungus does not even seem to be required for nematode control. In an advanced choice test where one side of a soil chamber was treated with tomato root exudates collected from *F. oxysporum*-colonized plants and the other side with root exudates of an untreated plant, 80% of the nematodes moved to the side containing root exudates of the untreated plant (Dababat and Sikora 2007c). When both sides of the chamber received the same treatment, nematode numbers were similar. This indicates that root exudates from *F. oxysporum*-colonized plants were either less attractive to the nematode or they contained substances repellent to *M. incognita*. Chemical analysis using high performance liquid chromatography (HPLC) for root exudates collected from tomato plants treated with *F. oxysporum* strain 162 versus non-treated plants illustrated that the chemical properties of root exudates were changed by the fungus (Schouten and Selim, personal communication 2009). Furthermore, *F. oxysporum* strain 162 stimulated the accumulation of metabolites different to control plants especially 2–3 weeks after inoculation. The levels of metabolite accumulation were higher within the plant tissue than in the root exudates and higher in the shoots than in the roots. The fact that endophytic *F. oxysporum* are capable of altering root exudation has also been reported by Steinkellner et al. (2008) studying fungal pathogens. Tomato plants inoculated with non-pathogenic strains of *F. oxysporum* caused reduced microconidial germination of pathogenic isolates of *F. oxysporum* in the rhizosphere.

Several studies emphasized nematicidal properties of fungal metabolites to be involved in nematode control. Vu et al. (2004) found that if *F. oxysporum* was applied to the soil, numbers of *R. similis* were significantly reduced by 16–30% in the absence of a plant. As direct parasitism could be ruled out by the authors, this reduction in nematode numbers can only be explained by fungal metabolites produced during decomposition of organic matter in the soil that interfere with nematode activity. Under field conditions, the behaviour of endophytic *F. oxysporum* in the bulk soil is of minor importance since the preferred choice of fungal application is not to the soil but target-delivery to seeds, seedlings or tissue culture plantlets to foster rapid root colonization and allow nematode control in the early stages of plant growth.

Another mechanism often discussed to explain the antagonistic effect of non-pathogenic *F. oxysporum* against plant-parasitic nematodes is competition for nutrients and colonisation sites. In that case the fungal antagonist needs to colonize the root tissue before the nematode invades. The potential of non-pathogenic *F. oxysporum* to rapidly colonize the root interior is well documented (Bacon and Yates 2006; Benhamou and Garand 2001; Hallmann and Sikora 1994b). Growth of endophytic *F. oxysporum* in roots is usually extensive and can be inter- and/or intracellular (Benhamou and Garand 2001). On pea roots, Benhamou and Garand (2001) showed that growth of non-pathogenic *F. oxysporum* was restricted to the epidermis and outer cortex while pathogenic *F. oxysporum* rapidly colonized the epidermis, cortex, endodermis and paratracheal parenchyma cells. The inability of non-pathogenic *F. oxysporum* to colonize the vascular tissue might explain its missing pathogenicity.

The involvement of induced systemic resistance (ISR) caused by *F. oxysporum* on plant-parasitic nematodes was investigated in split-root systems on banana and tomato (Fig. 10.2). Vu et al. (2006) showed that pre-inoculation of banana plantlets with several endophytic *F. oxysporum* isolates on one half of the root system significantly reduced root penetration of the migratory endoparasitic nematode *R. similis* on the non-treated half of the root by 30–40%. Comparable reductions in nematode numbers on the responder site of a split-root system were obtained by Dababat and Sikora (2007a) for the root-knot nematode *M. incognita* on tomato, following inoculation of the inducer site of a split-root system with the endophyte *F. oxysproum* stain 162. Nematode penetration was reduced by 26–45%, number of root galls by 21–36% and number of egg masses by 22–26%. Unfortunately, the mechanism behind the induced systemic resistance is still unknown. However, recent results obtained by Vu et al. (2006) indicated that challenge-inoculation of banana plantlets with non-pathogenic *F. oxysporum* changes root attractiveness in the responder side of the split root-system. This means that ISR initiates systemic

Fig. 10.2 Split-root chamber designed to study *Fusarium oxysporum* strain 162-mediated induced resistance against *Meloidogyne incognita* on tomato (Dababat and Sikora 2007a)

changes in the plant, with the plant reacting with altered exudate secretion into the rhizosphere (Sikora et al. 2007). In fact, HPLC data confirmed increased expression of certain root compounds (Schouten, unpublished). On the contrary, studies by Athman (2006) and Athman et al. (2007) imply that *F. oxysporum* colonization of banana does not interfere with host recognition by *R. similis* but reduces nematode reproduction after host plant infestation. Paparu et al. (2007) studied the expression of defence-related genes in a tolerant and susceptible banana cultivar inoculated with non-pathogenic *F. oxysporum* strain V5w2 and challenged with *R. similis*. In the tolerant cultivar Kayinja, gene expression of phenylalanine ammonia lyase (PAL), pectin acetylesterase (PAE), β-1,3-glucanases and catalase were up-regulated 33 days after endophyte inoculation and 3 days after nematode inoculation, respectively (Paparu et al. 2007), indicating that these genes might be involved in plant protection against nematode infestation. PAL and PEA contribute directly or indirectly to cell wall strengthening (Yalpani and Raskin 1993; Savary et al. 2003). Catalase activity might be associated with signal transduction during plant-defence (Chen et al. 1993) and β-1,3-glucanase is reported to release elicitors for phytoalexin synthesis, which than can be toxic to nematodes (Zinoveva et al. 2001).

An increase in phenolic compounds has been reported to occur in banana roots inoculated with non-pathogenic *F. oxysporum* and than challenged with *R. similis* (Athman 2006).The up-regulation of those defence-related genes was not observed in the susceptible cultivar Nabusa, although the non-pathogenic *F. oxysporum* strain V5w2 has been reported to suppress *R. similis* (Paparu et al. 2007). The authors speculate that either other genes are induced by V5w2 or up-regulation of the plant-defence genes occurs at a later stage not covered in that study. In the absence of *R. similis*, certain genes are down-regulated such as PAE, PAL, peroxidase (PIR7A) and lectin (Paparu et al. 2007). Although the reasons for the down-regulation of those genes are not clear, the authors presume that the down-regulated genes might not be required for the mutualistic relationship between the endophyte and banana or their down-regulation is necessary for the establishment and development of the endophyte-banana symbiosis.

Those results are encouraging and promise a better understanding of the genetic background of *F. oxysporum*-mediated resistance towards plant-parasitic nematodes in the near future. Further research is needed especially on the nature of the inducing factors, the signal pathways involved and the true mechanisms affecting nematode behaviour where induced resistance is a factor.

10.3.4 Nematophagous Fungi

The recent observation that nematophagous fungi colonize plant roots endophytically (Lopez-Llorca et al. 2002) adds a broader aspect to the concept of biological control of plant-parasitic nematodes. Nematophagous fungi are common soil inhabitants infecting nematodes by trapping, endoparasitism, egg parasitism or toxin production and further details of this group of fungi can be found in Chaps. 6 and 9.

Acknowledgement Thanks to Florian Grundler from the University of Natural Resources and Applied Life Sciences, Vienna, for sharing unpublished data on *Piriformospora indica* × plant-parasitic nematode interactions in *Arabidopsis* and to Abd El-Fattah Dababat, Andreas Kurtz and Alexander Schouten for critical revision of the manuscript.

References

Alabouvette C, Couteaudier Y (1992) Biological control of *Fusarium* wilts with non-pathogenic *Fusarium*. In: Tjamos EC, Cook RJ, Papavizas GC (eds) Biological control of plant diseases. Plenum, New York, pp 415–426

Alabouvette C, Roucel F, Louvet J (1979) Characteristics of *Fusarium* wilt-suppressive soils and prospects for their utilization in biological control. In: Schippers B, Gams W (eds) Soil-borne plant pathogens. Academic, London, pp 165–182

Alkan N, Gadkar V, Goburn J et al (2004) Quantification of arbuscular mycorrhizal fungus *Glomus intraradices* in host tissue using real-time polymerase chain reaction. New Phytol 161:877–885

Al-Raddad AM (1995) Interaction of *Glomus mosseae* and *Paecilomyces lilacinus* on *Meloidogyne javanica* of tomato. Mycorrhiza 5:233–236

Amin N (1994) Untersuchungen über die Bedeutung endophytischer Pilze für die biologishe Bekämpfung des wandernden Endoparasiten *Radopholus similis* (Cobb) Thorne an Bananen (Investigations on the importance of endophytic fungi for the biological control of the migratory endoparasite *Radopholus similis* (Cobb) Thorne on banana). Ph.D. thesis (German), University of Bonn

Arlorio M, Ludwig A, Boller T et al (1992) Effects of chitinase and β-1, 3 glucanase from pea on the growth of saprophytic, pathogenic and mycorrhizal fungi. G Bot Ital 126:956–958

Athman SY (2006) Host-endophyte-pest interactions of endophytic *Fusarium* antagonistic to *Radopholus similis* in banana (*Musa* spp.). Ph.D. dissertation, Pretoria, University of Pretoria

Athman SY, Dubois CD et al (2007) Effect of endophytic *Fusarium oxysporum* on root penetration and reproduction of *Radopholus similis* in tissue culture-derived banana (*Musa* spp.) plants. Nematology 9:599–607

Azcón-Aguilar C, Barea JM (1996) Arbuscular macorrhizas and biocontrol of soil-borne plant pathogens – an overview of the mechanisms involved. Mycorrhiza 6:457–464

Bacetty AA, Snook ME, Glenn AE et al (2009a) Toxicity of endophyte-infected tall fescue alkaloids and grass metabolites on *Pratylenchus scribneri.* Phytopathology 99:1336–1345

Bacetty AA, Snook ME, Glenn AE et al (2009b) Chemotaxis disruption in *Pratylenchus scibneri* by tall fescue root extracts and alkaloids. J Chem Ecol 35:844–850

Bacon CW, Yates IE (2006) Endophytic root colonization by *Fusarium* species: histology, plant interactions, and toxicity. In: Schulz B, Boyle C, Sieber TN (eds) Soil biology vol 9: microbial root endophytes. Springer, Berlin, pp 133–152

Bagyaraj DJ, Manjunath A, Reddy DDR (1979) Interaction of vesicular arbuscular mycorrhia with root knot nematodes in tomato. Plant Soil 51:397–403

Baldwin IT, Halitschke R, Kessler A et al (2001) Merging molecular and ecological approaches in plant-insect interactions. Curr Opin Plant Biol 4:351–358

Ball OJP, Bernard EC, Gwinn KD (1997) Effect of selected *Neotyphodium lolii* isolates on root-knot nematode (*Meloidogyne marylandi*) numbers in perennial ryegrass. In: Proceedings of the 50th New Zealand plant protection conference, Lincoln University, Canterbury, 8–21 Aug 1997, pp 65–68

Baltruschat H, Sikora RA, Schönbeck F (1973) Effect of VA-mycorrhiza (*Endogone mossae*) on the establishment of *Thielaviopsis basicola* and *Meloidogyne incognita* in tobacco. In: Second International congress of plant Pathology, Minneapolis, 661 (Abstract)

Barker SJ, Stummer B, Gao L, Dispain I et al (1998) A mutant in *Lycopersicon esculentum* Mill. with highly reduced VA mycorrhizal colonization: isolation and preliminary characterisation. Plant J 15:791–797

Barker SJ, Edmonds-Tibbett TL, Forsyth L et al (2006) Root infection of the reduced mycorrhizal colonization (*rmc*) mutant of tomato reveals genetic interaction between symbiosis and parasitism. Physiol Mol Plant Path 67:277–283

Benhamou N, Garand C (2001) Cytological analysis of defense-related mechanisms induced in pea root tissues in response to colonization by the nonpathogenic *Fusarium oxysporum* Fo47. Phytopathology 91:730–740

Bernard EC, Gwinn KD, Griffin GD (1998) Forage grasses. In: Barker KR, Pederson GA, Windham GL (eds) Plant and nematode interactions. American Society of Agronomy, Madison, pp 427–454

Brants A, Brown CR, Earle ED (2000) *Trichoderma harzianum* endochitinase does not provide resistance to *Meloidogyne hapla* in transgenic tobacco. J Nematol 32:289–296

Brundrett MC (2002) Coevolution of roots and mycorrhizas of land plants. New Phytol 154:275–304

Brundrett MC (2006) Understanding the roles of multifunctional mycorrhizal and endophytic fungi. In: Schulz B, Boyle C, Sieber T (eds) Soil biology, vol 9: Microbial root endophytes. Springer, Berlin, pp 281–298

Calvet E, Pinochet J, Camprubí A et al (1995) Increased tolerance to the root-lesion nematode *Pratylenchus vulnus* in mycorrhizal micropropagated BA-29 quince rootstock. Mycorrhiza 5:253–258

Camprubí A, Pinochet J, Calvet E et al (1993) Effects of the root-lesion nematode *Pratylenchus vulnus* and the vesicular arbuscular mycorrhizal fungus *Glomus mosseae* on the growth of three plum rootstocks. Plant Soil 153:223–229

Carling DE, Roncadori RW, Hussey RS (1989) Interactions of vesicular-arbuscular mycorrhizal fungi, root-knot nematode, and phosphorus fertilization on soybean. Plant Dis 73:730–733

Carling DE, Roncadori RW, Hussey RS (1996) Interactions of arbuscular mycorrhizae, *Meloidogne arenaria*, and phosphorus fertilization on peanut. Mycorrhiza 6:9–13

Carroll GC (1988) Fungal endophytes in stems and leaves: from latent pathogen to mutualistic symbiont. Ecology 69:2–9

Cason KM, Hussey RS, Roncadori RW (1983) Interaction of vesicular-arbuscular mycorrhizal fungi and phosphorus with *Meloidogyne incognita* on tomato. J Nematol 15:410–417

Castillo P, Nico AI, Azcón-Aguilar C et al (2006) Protection of olive planting stocks against parasitism of root-knot nematodes by arbuscular mycorrhizal fungi. Plant Pathol 55:705–713

Chen Z, Ricigliano JW, Klessing DF (1993) Purification and characterization of a soluble salicyclic acid-binding protein from tobacco. Proc Natl Acad Sci USA 90:9733–9537

Christensen MJ, Bennett RJ, Schmid J (2002) Growth of *Epichlo /Neotyphodium* and p-endophytes in leaves of *Lolium* and *Festuca* grasses. Mycol Res 106:93–106

Cook R, Lewis GC, Mizen KA (1991) Effects on plant-parasitic nematodes of infection of perennial ryegrass, *Lolium perenne*, by the endophytic fungus, *Acremonium lolii*. Crop Prot 10:403–407

Cooper KM, Grandison GS (1986) Interaction of vesicular-arbuscular mycorrhizal fungi and root knot nematode on cultivars of tomato and white clover susceptible to *Meloidogyne hapla*. Ann Appl Biol 108:555–565

Dababat AA, Sikora RA (2007a) Induced resistance by the mutualisitc endophyte, *Fusarium oxysporum* strain 162, toward *Meloidogyne incognita* on tomato. Biocontrol Sci Technol 17:969–975

Dababat AA, Sikora RA (2007b) Importance of application time and inoculum density of *Fusarium oxysporum* 162 for biological control of *Meloidogyne incognita* on tomato. Nematropica 37:267–275

Dababat AA, Sikora RA (2007c) Influence of the mutualistic endophyte *Fusarium oxysporum* 162 on *Meloidogyne incognita* attraction and invasion. Nematology 9:771–776

Dababat AA, Selim ME, Saleh AA et al (2008) Influence of *Fusarium* wilt resistant tomato cultivars on root colonization of the mutualisitc endophyte *Fusarium oxysporum* strain 162 and its biological control efficacy toward the root-knot nematode *Meloidogyne incognita*. J Plant Dis Prot 115:273–278

Dai C-C, Yu B-Y, Li X (2008) Screening of endophytic fungi that promote the growth of *Euphorbia pekinensis*. Afr J Biotech 7:3505–3510

Daneshkhah R (2008) Untersuchung der Auswirkung von *Piriformospora indica* auf Arabidopsis *thaliana* und Tomate (*Solanum lycopersicum* L) in der Interaktion mit *Meloidogyne incognita* (Study of effects of *Piriformospora indic*a on *Arabidopsis thaliana* and tomato (*Solanum lycopersicum* L) in the interaction with *Meloidogyne incognita*). MSc thesis, University of Natural Resources and Applied Life Sciences, Vienna

De La Peña E, Rodriguez Echeverria S, Van der Putten WH et al (2006) Mechanism of control of root-feeding nematodes by mycorrhizal fungi in the dune grass *Ammophila arenaria*. New Phytol 169:829–840

Deguchi S, Shimazaki Y, Uozumi S et al (2007) White clover living mulch increases the yield of silage corn via arbuscular mycorrhizal fungus colonization. Plant Soil 291:291–299

Deliopoulos T, Devine KJ, Haydock PPJ et al (2007) Studies on the effect of mycorrhization of potato roots on the hatching activity of potato root leachate towards the potato cyst nemato*des,* Globodera pallida and G. rostochiensis. Nematology 9:719–729

Deliopoulos T, Haydock PJ, Jones PW (2008) Interaction between arbuscular mycorrhizal fungi and the nematicide aldicarb on hatch and development of the potato cyst nematode, *Globodera pallida*, and yield of potatoes. Nematology 10:783–799

Diederichs C (1987) Interaction between five endomycorrhizal fungi and the root-knot nematode *Meloidogyne javanica* on chickpea under tropical conditions. Trop Agric 64:353–355

Diedhiou PM, Hallmann J, Oerke E-C et al (2003) Effects of arbuscular mycorrhizal fungi and a non-pathogenic *Fusarium oxysporum* on *Meloidogyne incognita* infestation of tomato. Mycorrhiza 13:199–204

Dumas-Gaudot E, Furlan V, Grenier J et al (1992) New acidic chitinase isoforms induced in tobacco roots by vesicular arbuscular mycorrhizal fungi. Mycorrhiza 1:133–136

Dumas-Gaudot E, Slezack S, Dassi B et al (1996) Plant hydrolytic enzymes (chitinase and β-1, 3-glucanase) in root reactions to pathogenic and symbiotic microorganisms. Plant Soil 185:211–221

Elmi AA, West CP, Kirkpatrick TL et al (1990) *Acremonium* endophyte inhibits root-knot nematode reproduction in tall fescue. Ark Farm Res 39:3

Elmi AA, West PC, Robbins RT et al (2000) Endophyte effects on reproduction of a root-knot nematode (*Meloidogyne marylandi*) and osmotic adjustment in tall fescue. Grass Forage Sci 55:166–172

Elsen A, Baimey H, Swennen R et al (2003) Relative mycorrhizal dependency and mycorrhiza-nematode interaction in banana cultivars (*Musa* spp.) differing in nematode susceptibility. Plant Soil 256:303–313

Elsen A, Gervacio D, Swennen R et al (2008) AMF-induced biocontrol against plant parasitic nematodes in *Musa* sp.: a systemic effect. Mycorrhiza 18:251–256

Feldmann F, Hallmann J, Wagner S (2008) Mycorrhizal fungi as biological components of integrated cucumber production (BIOMYC) – promising results for mycorrhizal technology transfer to horticultural practice. In: Feldmann F, Kapulnik Y, Baar JKG (eds) Mycorrhiza works. DPG Selbstverlag, Braunschweig, pp 25–38

Felitti S, Shields K, Ramsperger M et al (2006) Transcriptome analysis of *Neotyphodium* and *Epichloe* grass endophytes. Fungal Genet Biol 43:465–475

Fox JA, Spasoff L (1972) Interaction of *Heterodera solanacearum* and *Endogone gigantea* on tobacco. J Nematol 4:224–225 (abstr)

Francl LJ (1993) Interactions of nematodes with mycorrhizae and mycorrhizal fungi. In: Khan MW (ed) Nematode interactions. Chapman and Hall, London, pp 203–216

Frey-Klett P, Garbaye J, Tarkka M (2007) The mycorrhiza helper bacteria revisited. New Phytol 176:22–36

García-Romera I, García-Garrido JM, Martin J et al (1998) Interactions between saprophytic *Fusarium* strains and arbuscular mycorrhizas of soybean plants. Symbiosis 24:235–246

Gerdemann JW (1968) Vesicular arbuscular mycorrhiza and plant growth. Annu Rev Phytopathol 6:396–418

Graham JH, Leonard RT, Menge JA (1981) Membrane-mediated decrease in root exudation responsible for phosphorus inhibition of vesicular-arbuscular mycorrhiza formation. Plant Physiol 68:548–552

Griesbach M (1999) Occurrence of mutualistic fungal endophytes in bananas (*Musa* spp.) and their potential as biocontrol agents of the banana weevil *Cosmopolites sordidus* (Germar) (Coleoptera: Curculionidae) in Uganda. Ph.D. thesis, University of Bonn

Guttenberger M (2000) A rapid staining procedure for arbuscules of living arbuscular mycorrhizas using neutral red as acidotropic dye. Plant Soil 226:211–218

Habte M, Schmitt D (2005) Antagonistic interactions between plant parasitic nematodes and arbuscular mycorrhizal fungi. In: Mehrotra VS (ed) Mycorrhiza: role and application. Allied Publishers, New Delhi, pp 183–203

Hallmann J, Sikora RA (1994a) Occurrence of plant-parasitic nematodes and non-pathogenic species of *Fusarium* in tomato plants in Kenya and their role as mutualisitc synergists for biological control of root-knot nematodes. Int J Pest Manag 40:321–325

Hallmann J, Sikora RA (1994b) Influence of *Fusarium oxysporum*, a mutualistic fungal endophyte, on *Meloidogyne incognita* infection of tomato. J Plant Dis Prot 101:475–481

Hallmann J, Sikora RA (1996) Toxicity of fungal endophyte secondary metabolites to plant-parasitic nematodes and soil-borne plant pathogenic fungi. Eur J Plant Pathol 102:155–162

Hasan N, Jain RK (1987) Parasitic nematodes and vesicular-arbuscular mycorrhizal VAM fungi associated with berseem *Trifolium alexandrinum* L in Bundelkhand region India. Ind J Nematol 17:184–188

Hayman D (1982) The physiology of vesicular-arbuscular endomycorrhizal symbiosis. Can J Bot 6:944–963

Hinton DM, Bacon CW (1988) The distribution of the endophyte of toxic tall fescue. Can J Bot 63:36–42

Hol WHG, Cook R (2005) An overview of arbuscular mycorrhizal fungi-nematode interactions. Basic Appl Ecol 6:489–503

Hussey RS, Roncadori RW (1978) Interaction of *Pratylenchus brachyurus* and *Gigaspora margarita* on cotton. J Nematol 10:16–20

Hussey RS, Roncadori RW (1982) Vesicular-arbuscular mycorrhizae may limit nematode activity and improve plant growth. Plant Dis 66:9–14

Ingham RE (1988) Interactions between nematodes and vesicular-arbuscular mycorrhizae. Agric Ecosys Environ 24:169–182

Jain RK, Hasan N (1986) Association of vesicular-arbuscular mycorrhizal (VAM) fungi and plant parasitic nematodes with forage sorghum (*Sorghum bicolor* L.). Sorghum Newsl 29:84

Jain RK, Hasan N, Singh RK et al (1998) Interaction between plant-parasitic nematodes, *Meloidogyne incognita*, *Tylenchorhynchus vulgaris* and VAM fungi, *Glomus fasciculatum* on lucerne. Ann Plant Prot Sci 6:37–40

Kassab AS (1990) Aspects of the host-parasite relationship of nematodes and sweet potato II: relationship between initial population densities of *Criconemella* ssp., *Rotylenchulus reniformis* or *Tylenchorhynchus* spp. and yield of mycorrhizal and non-mycorrhizal plants. Ann Agric Sci 35:1107–1116

Kellam MK, Schenck NC (1980) Interactions between a vesicular-arbuscular mycorrhizal fungus and root-knot nematode on soybean. Phytopathology 70:293–296

Kimmons CA, Gwinn KD, Bernard EC (1990) Nematode reproduction on endophyte-infected and endophyte-free tall fescue. Plant Dis 74:757–761

Kuldau G, Bacon C (2008) Clavicipitaceous endophytes: their ability to enhance resistance of grasses to multiple stresses. Biol Control 46:57–71

Kuldau GA, Yates IE (2000) Evidence for *Fusarium* endophytes in cultivated and wild plants. In: Bacon CW, White JF (eds) Microbial endophytes. Marcel Dekker, New York, pp 85–120

Kunkel BA, Grewal PS, Quigley MF (2004) A mechanism of acquired resistance against an entomopathogenic nematode by *Agrotis ipsilon* feeding on perennial ryegrass harboring a fungal endophyte. Biol Control 29:100–108

Lambert AM, Casagrande RA (2006) No evidence of fungal endophytes in native and exotic *Phragmites australis*. Northeast Nat 13:561–568

Larkan NJ, Smith SE, Barker SJ (2007) Position of the reduced mycorrhizal colonisation (*rmc*) locus on the tomato genome map. Mycorrhiza 17:311–318

Lemanceau P, Alabouvette C (1991) Biological control of *Fusarium* diseases by fluorescent *Pseudomonas* and non-pathogenic *Fusarium*. Crop Prot 10:279–286

Li HM, Sullivan R, Moy M et al (2004) Expression of a novel chitinase by the fungal endophyte in *Poa ample*. Mycologia 96:526–536

Li H-Y, Yang G-D, Shu H-R et al (2006) Colonization by the arbuscular mycorrhizal fungus *Glomus versiforme* induces a defence response against the root-knot nematode *Meloidogyne incognita* in the grapevine (*Vitis amurensis* Rupr.), which includes transcriptional activation of the class III chitinase gene *VCH3*. Plant Cell Physiol 47:154–163

Lopez-Llorca LV, Bordallo JJ, Salinas J et al (2002) Use of light and scanning electron microscopy to examine colonisation of barley rhizosphere by the nematophagous fungus *Verticillium chlamydosporium*. Mircon 33:61–67

MacGuidwin AE, Bird GW, Safir GR (1985) Influence of *Glomus fasciculatum* on *Meloidogyne hapla* infecting *Allium cepa*. J Nematol 17:389–395

Maciá-Vicente JG, Jansson H-B, Abdullah SK et al (2008) Fungal root endophytes from natural vegetation in Mediterranean environments with special references to *Fusarium* spp. FEMS Microbiol Ecol 64:90–105

Malinowski DP, Alloush GA, Belesky DP (2004) Evidence for chemical changes on the root surface of tall fescue in response to infection with the fungal endophyte *Neotyphodium coenophialum*. Plant Soil 205:1–12

Mendoza A, Sikora RA (2009) Biological control of *Radopholus similis* in banana by co-application of the mutualistic endophyte *Fusarium oxysporum* strain 162, the egg pathogen *Paecilomyces lilacinus* strain 251 and the antagonistic bacteria *Bacillus firmus*. Biocontrol 54:263–272

Nagesh M, Reddy P (2004) Biochemical changes in *Glomus fasciculatum* colonized roots of *Lycopersicon esculentum* in presence of *Meloidogyne incognita*. Ind J Exp Biol 42:721–727

Newman EI, Reddell P (1987) The distribution of mycorrhizas among families of vascular plants. New Phytol 106:745–751

Niere BI (2001) Significance of non-pathogenic isolates of *Fusarium oxysporum* Schlecht: fries for the biological control of the burrowing nematode *Radopholus similis* (Cobb) Thorne on tissue cultured banana. Ph.D. thesis, University of Bonn

O'Bannon JH, Nemec S (1979) The response of citrus limon seedlings to a symbiont, *Glomus etunicatus*, and a pathogen, *Radopholus similis*. J Nematol 11:270–275

Panaccione DG, Kotcon JB, Schardl CL et al (2006) Ergot alkaloids are not essential for endophytic fungus-associated population suppression of the lesion nematode, *Pratylenchus scribneri*, on perennial ryegrass. Nematology 8:583–590

Panstruga R (2003) Establishing compatibility between plants and obligate biotrophic pathogens. Curr Opin Plant Biol 6:320–326

Paparu P, Dubois T, Gold CS et al (2006) Colonization pattern of non-pathogenic *Fusarium oxysporum*, a potential biological control agent, in roots and rhizomes of tissue-cultured *Musa* plantlets. Ann Appl Biol 149:1–8

Paparu P, Dubois T, Coyne D et al (2007) Defense-related gene expression in susceptible and tolerant bananas (*Musa* spp.) following inoculation with non-pathogenic *Fusarium oxysporum* endophytes and challenge with *Radopholus similis*. Physiol Mol Plant Pathol 71:149–157

Paparu P, Dubois T, Gold CS et al (2008) Screenhouse and field persistence of non-pathogenic endophytic *Fusarium oxysporum* in *Musa* tissue culture plants. Microb Ecol 55:561–568

Paparu P, Dubois T, Coyne D et al (2009) Dual inoculation of *Fusarium oxysporum* endophytes in banana: effect on plant colonization, growth and control of the root burrowing nematode and the banana weevil. Biocontrol Sci Technol 19:639–655

Pedersen JF, Rodríguez-Kábana R, Shelby RA (1988) Ryegrass cultivars and endophyte in tall fescue affect nematodes in grass and succeeding soybean. Agron J 80:811–814

Perotto S, Bonfante P (1997) Bacterial associations with mycorrhizal fungi: close and distant friends in the rhizosphere. Trends Microbiol 5:496–501

Peškan-Berghöfer T, Shahollaria B, Giong PH et al (2004) Association of *Piriformospora* indica with *Arabidopsis thaliana* roots represent a novel system to study beneficial plant-microbe interactions and involves early plant protein modifications in the endoplasmic reticulum and at the plasma membrane. Physiol Plant 122:465–477

Pieterse CMJ, Dicke M (2007) Plant interactions with microbes and insects: from molecular mechanisms to ecology. Trends Plant Sci 12:564–569

Pinochet J, Camprubí A, Calvet C (1993) Effects of the root-lesion nematode *Pratylenchus vulnus* and the mycorrhizal fungus *Glomus mosseae* on the growth of EMLA-26 apple rootstock. Mycorrhiza 4:79–83

Pinochet J, Calvet C, Camprubí A et al (1995) Growth and nutritional response of Nemared peach rootstock infected with *Pratylenchus vulnus* and the mycorrhizal fungus *Glomus mosseae*. Fund Appl Nematol 17:205–210

Pinochet J, Calvet C, Camprubí A et al (1996) Interactions between migratory endoparasitic nematodes and arbuscular mycorrhizal fungi in perennial crops: a review. Plant Soil 185:183–190

Pocasangre LE (2000) Biological enhancement of banana tissue culture plantlets with endophytic fungi for the control of the burrowing nematode *Radopholus similis* and the Panama disease (*Fusarium oxysporum* f. sp. *cubense*). Ph.D. thesis, University of Bonn

Pocasangre LE, Sikora RA, Vilich V et al (2000) Survey of banana endophytic fungi from Central America and screening for biological control of the burrowing nematode (*Radopholus similis*). Acta Hortic 531:283–289

Postma J, Rattink H (1991) Biological control of *Fusarium* wilt of carnation with a non-pathogenic isolate of *Fusarium oxysporum*. Can J Bot 70:1199–1205

Pozo MJ, Cordier C, Dumas-Gaudot E et al (2002) Localized versus systemic effect of arbuscular mycorrhizal fungi on defence responses to *Phytophthora* infection in tomato plants. J Exp Bot 368:525–534

Qiu J, Hallmann J, Kokalis-Burelle N et al (1997) Activity and differential induction of chitinase isozymes in soybean cultivars resistant or susceptible to root-knot nematodes. J Nematol 29:523–530

Rao MS, Reddy PP, Mohandas S (1995) Effect of integration of endomycorrhiza (*Glomus mossae*) and neem cake on the control of root-knot nematode on tomato. J Plant Dis Prot 102:526–529

Reimann S, Hauschild R, Hildebrand U et al (2008) Interrelationships between *Rhizobium etli* G12 and *Glomus interradices* and multitropic effects in the biological control of the root-knot nematode *Meloidogyne incognita* on tomato. J Plant Dis Prot 115:108–113

Richardson MD, Hill NS, Hoveland CS (1990) Rooting patterns of endophyte infected tall fescue grown under drought stress. Agron Abstr 81:129

Richmond DS, Kunkel BA, Somasekhar N et al (2004) Top-down and bottom-up regulation of herbivores: *Spodoptera frugiperda* turns tables on endophyte-mediated plant defence and virulence of an entomopathogenic nematode. Ecol Entomol 29:353–360

Ryan NA, Jones P (2004) The effect of mycorrhization of potato roots on the hatching chemicals active towards the potato cyst nematodes, Globodera pallida and G. rostochiensis. Nematology 6:335–342

Ryan NA, Duffy EM, Cassells AC et al (2000) The effect of mycorrhizal fungi on the hatch of potato cyst nematodes. Appl Soil Ecol 15:233–240

Ryan NA, Deliopoulos T, Jones P et al (2003) Effects of a mixed-isolate mycorrhizal inoculum on the potato – potato cyst nematode interaction. Ann Appl Biol 143:111–119

Saldajeno MGB, Chandanie WA, Kubota M et al (2008) Effects of interactions of arbuscular mycorrhizal fungi and beneficial saprophytic mycoflora on plant growth and disease protection. In: Siddiqui ZA; Akhtar MS, Futai K (eds) Mycorrhizae: sustainable agriculture and forestry. Springer, Dordrecht, pp 211–226

Saleh HM, Sikora RA (1984) Relationship between *Glomus fasciculatum* root colonisation of cotton and its effect on *Meloidogyne incognita*. Nematologica 30:230–237

Savary BJ, Nunez A, Liu LS (2003) Pectin acetylesterase-analysis and application for beet pectin utilization. In: Proceedings of the first joint international beet research-American society of sugar beet technologists Congress. Beet Sugar Development Foundation Co, Denver, pp 917–920

Schäfer P, Khatabi B, Kogel K-H (2007) Root cell death and systemic effects of *Piriformospora indica*: a study on mutualism. FEMS Microbiol Lett 275:1–7

Schmidt DD, Voelckel C, Hartl M et al (2005) Specificity in ecological interactions. Attack from the same lepidopteran herbivore results in species-specific transcriptional responses in two solanaceous host plants. Plant Physiol 138:1763–1773

Schneider RW (1984) Effects of non-pathogenic strains of *Fusarium oxysporum* on celery root infection by *F. oxysporum* f. sp. *apii* and a novel use of the Lineweaver-Burk double reciprocal plot technique. Phytopathology 74:646–653

Schulz B (2006) Mutualistic interactions with fungal root endophytes. In: Schulz BJE, Boyle JC, Sieber TE (eds) Microbial root endophytes, soil biology, vol 9. Springer, Berlin, pp 247–260

Schulz B, Boyle C (2006) What are endophytes. In: Schulz BJE, Boyle JC, Sieber TE (eds) Microbial root endophytes, soil biology, vol 9. Springer, Berlin, pp 1–13

Shahollari B, Varma A, Oelmüller R (2005) Expression of a receptor kinase in *Arabidopsis* roots is stimulated by the basidiomycete *Piriformospora indica* and the protein accumulates in Triton X-100 insoluble plasma membrane microdomains. J Plant Physiol 162:945–958

Shahollari B, Vadassery J, Varma A et al (2007) A leucin-rich repeat protein is required for growth promotion and enhanced seed production mediated by the endophytic fungus *Piriformospora indica* in *Arabidopsis thaliana*. Plant J 50:1–13

Siegel MR, Jarlfors U, Latch GCM et al (1987) Ultrastructure of *Acremonium coenophialum*, *Acremonium lolii*, and *Epichloë typhina* endophytes in host and nonhost *Festuca* and *Lolium* species of grasses. Can J Bot 65:357–2367

Sikora RA (1978) Effect of the endotrophic mycorrhizal fungus, *Glomus mosseae*, on the host-parasite relationship of *Meloidogyne incognita* in tomato. J Plant Dis Prot 185:197–202

Sikora RA (1979) Predisposition to *Meloidogyne* infection by the endotrophic mycorrhizal fungus *Glomus mosseae*. In: Lamberti F, Taylor CE (eds) Root-knot nematode (*Meloidogyne* species) systematics, biology and control. Academic, New York, pp 399–404

Sikora RA (1995) Vesicular-arbuscular mycorrhizae: their significance for biological control of plant-parasitic nematodes. Biocontrol 1:29–33

Sikora RA, Schönbeck F (1975) Effect of vesicular-arbuscular mycorrhiza (*Endogone mosseae*) on the population dynamics of the root-knot nematodes (*Meloidogyne incognita* and *Meloidogyne hapla*). In: 8th International plant protection congress, Moscow (Abstract)

Sikora RA, Schäfer K, Dababat AA (2007) Mode of action associated with microbially induced *in planta* suppression of plant-parasitic nematodes. Aust Plant Pathol 36:124–134

Sikora RA, Pocasangre L, Zum Felde A (2008) Mutualistic endophytic fungi and in-planta suppressiveness to plant-parasitic nematodes. Biol Control 46:15–23

Singh A, Varma A (2000) Mycorrhizal biology. In: Mukerji KG, Chamola BP, Singh J (eds) Orchidaceous mycorrhizal fungi. Kluwer, New York, pp 265–288

Singh G, Singh N, Marwaha TS (2009) Crop genotype and a novel symbiotic fungus influences the root endophytic colonization potential of plant growth promoting rhizobacteria. Physiol Mol Biol Plants 15:87–92

Sivaprasad P, Jacob A, Nair SKet al (1990) Influence of VA mycorrhizal colonization on root-knot nematode infestation in *Piper nigrum* L. In: Jalali BL, Chand H (eds) Trends in mycorrhizal research. Proceedings of the National Conference on Mycorrhiza, Haryana Agricultural University, pp 100–101

Smith GS (1987) Interactions of nematodes with mycorrhizal fungi. In: Veech JA, Dickson DW (eds) Vistas on nematology. Society of Nematologists, Hyattsville, pp 292–300

Smith GS, Kaplan DT (1988) Influence of mycorrhizal fungus, phosphorus and burrowing nematode interactions on growth of rough lemon citrus seedlings. J Nematol 20:539–544

Smith GS, Roncadori RW, Hussey RS (1986) Interaction of endomycorrhizal fungi, superphosphate, and *Meloidogyne incognita* on cotton in microplot and field studies. J Nematol 18:208–216

Sorensen JN, Larsen J, Jakobsen I (2003) Management strategies for capturing the benefits oy mycorrhizas in the production of field-grown vegetables. Acta Hortic 627:65–71

Spanu PD (2006) Why do some fungi give up their freedom and become obligate dependants on their host? New Phytol 171:447–450

Spanu PD, Boller T, Ludwig A et al (1989) Chitinase in roots of mycorrhizal *Allium porrum*: regulation and localization. Planta 177:447–455

Speijer PR (1993) Interrelationship between *Pratylenchus goodeyi* Sher & Allen and strains of non-pathogenic *Fusarium oxysporum* Schl. Emd. Snyd. & Hans. in roots of two banana cultivars. Ph.D. thesis, University of Bonn

Steinkellner S, Mammerler R, Vierheilig H (2008) Germination of *Fusarium oxysporum* in root exudates from tomato plants challenged with different *Fusarium oxysporum* strains. Eur J Plant Pathol 122:395–401

Stewart TM, Mercer CF, Grant JL (1993) Development of *Meloidogyne naasi* on endophyte-infected and endophyte-free perennial ryegrass. Aust Plant Pathol 22:40–41 (abstr)

Strobel NE, Hussey RS, Roncadori RW (1982) Interactions of vesicular-arbuscular mycorrhizal fungi, *Meloidogyne incognita*, and soil fertility on peach. Phytopathology 72:690–694

Sullivan TJ, Rodstrom J, Vandop J et al (2007) Symbiont-mediated changes in *Lolium arundiaceum* inducible defenses: evidence from changes in gene expression and leaf composition. New Phytol 176:673–679

Suresh DK, Bagyaraj DJ, Reddy DDR (1985) Effect of vesicular-arbuscular mycorrhiza on survival, penetration and development of root-knot nematode in tomato. Plant Soil 87:305–308

Talavera M, Kenji Itou K, Mizukubo T (2001) Reduction of nematode damage by root colonization with arbuscular mycorrhiza (*Glomus* spp.) in tomato-*Meloidogyne incognita* (Tylenchida: Meloidogynidae) and carrot-*Pratylenchus penetrans* (Tylenchida: Pratylenchidae) pathosystems. Appl Entomol Zool 36:387–392

The Arabidopsis Genome Initiative (2000) Analysis of the genome sequence of the flowering plant *Arabidopsis thaliana*. Nature 408:796–815

Tehler A, Little DP, Farris JS (2003) The full-length phylogenetic tree from 1551 ribosomal sequences of chitinous fungi. Mycol Res 107:901–916

Timper P, Gates RN, Bouton JH (2005) Response of *Pratylenchus* spp. in tall fescue infected with different strains of the fungal endophyte *Neotyphodium coenophialum*. Nematology 7:105–110

Tylka GL, Hussey RS, Roncadori RW (1991) Interactions of vesicular-arbuscular mycorrhizal fungi, phosphorus, and *Heterodera glycines* on soybean. J Nematol 23:122–133

Umesh KC, Krishnappa K, Bagyaraj DJ (1988) Interaction of burrowing nematode *Radopholus similis* (Cobb 1893) Thorne 1949 and VA mycorrhiza, *Glomus fasciculatum* (Thaxt) Gerd and Trappe in banana (*Musa acuminate* Colla.). Ind J Nematol 18:6–11

Varma A, Savita V, Sudha Sahay N et al (1999) *Piriformospora indica*, a cultivable plant-growth-promoting root endophyte. Appl Environ Microbiol 65:2741–2744

Vega FE, Posada F, Aime MC et al (2008) Entomopathogenic fungal endophytes. Biol Control 46:72–82

Verma S, Varma A, Rexer K-H et al (1998) *Piriformospora indica*, gen. et sp. nov., a new root-colonizing fungus. Mycologia 90:896–903

Vierheilig H, Alt M, Mohr U et al (1994) Ethylene biosynthesis and activities of chitinase and β-1, 3-glucanase in the roots of host and non-host plants of vesicular-arbuscular mycorrhizal fungi after inoculation with Glomus mosseae. J Plant Physiol 143:337–343

Vierheilig H, Alt M, Lange J et al (1995) Colonization of tobacco constitutively expressing pathogenesis-related proteins by the vesicular-arbuscular mycorrhizal fungus *Glomus mosseae*. Appl Environ Microbiol 61:3031–3034

Villenave C, Duponnois R (2002) Interactions between ectomycorrhizal fungi, plant-parasitic and free-living nematodes and their effects on seedlings of the hardwood *Afzelia africana* Sm. Pedobiologia 46:176–187

von Alten H, Lindemann A, Schönbeck F (1993) Stimulation of vesicular-arbuscular mycorrhiza by fungicides or rhizosphere bacteria. Mycorrhiza 2:167–173

Vu TT (2005) Modes of action on non-pathogenic *Fusarium oxysporum* endophytes for bioenhancement of banana toward *Radopholus similis*. Ph.D. thesis, University of Bonn

Vu TT, Sikora RA, Hauschild R (2004) Effects of endophytic *Fusarium oxysporum* towards *Radopholus similis* activity in absence of banana. Commun Agric Appl Biol Sci 69:381–385

Vu TT, Hauschild R, Sikora RA (2006) *Fusarium oxysporum* endophytes induced systemic resistance against *Radopholus similis* on banana. Nematology 8:847–852

Waceke JW, Waudo SW, Sikora RA (2001) Suppression of *Meloidogyne hapla* by arbuscular mycorrhiza fungi (AMF) on pyrethrum in Kenya. Int J Pest Manag 47:135–140

Walker C, Schüßler A (2004) Nomenclatural clarifications and new taxa in the *Glomeromycota*. Mycol Res 108:981–982

Waller F, Achatz B, Baltruschat H et al (2005) The endophytic fungus *Piriformospora indica* reprograms barley to salt-stress tolerance, disease resistance, and higher yield. Proc Natl Acad Sci USA 102:13386–13391

Wallner V (2007) Auswirkungen von *Piriformospora indica* auf die Entwicklung von *Arabidopsis thaliana* und *Heterodera schachtii* (Effects of *Piriformospora indica* on the development of *Arabidopsis thaliana* and *Heterodera schachtii*). M.Sc thesis, University of Natural Resources and Applied Life Sciences, Vienna

Wang B, Qiu YL (2006) Phylogenetic distribution and evolution of mycorrhizas in land plants. Mycorrhiza 16:299–363

West CP, Izekor E, Oosterhuis DM et al (1988) The effect of *Acremonium coenophialum* on the growth and nematode infestation of tall fescue. Plant Soil 112:3–6

Whipps JM (2004) Prospects and limitations for mycorrhizals in biocontrol of root pathogens. Can J Bot 82:1198–1227

Wilson D (1995) Endophyte – the evolution of a term, and clarification of its use and definition. Oikos 73:274–276

Yalpani N, Raskin I (1993) Salicyclic acid: a systemic signal in induced plant resistance. Trends Microbiol 3:88–92

Zeze A, Brou YC, Meddich A et al (2008) Molecular identification of MIP genes expressed in the roots of an arbuscular mycorrhizal *Trifolium alexandrinum* L. under water stress. Afr J Agric Res 3:78–83

Zhao SZ, Qi XQ (2008) Signaling in plant disease resistance and symbiosis. J Integr Plant Biol 50:799–807

Zinoveva SV, Udalova ZV, Vasileva IS et al (2001) Role of isoprenoids in plant adaptation to biogenic stress induced by parasitic nematodes. Appl Biochem Microbiol 37:456–463

Zum Felde A, Pocasangre L, Sikora RA (2005) The potential use of microbial communities inside suppressive banana plants for banana root protection. In: Turner DW, Rosales FE (eds) Banana root system: towards a better understanding for its productive management. Proceedings of an international symposium, San Jose, 3–5 Nov 2003, pp 169–177

Chapter 11
Utilization of Biological Control for Managing Plant-Parasitic Nematodes

Patricia Timper

Abstract Biological control of plant-parasitic nematodes can be accomplished either by application of antagonistic organisms, conservation and enhancement of indigenous antagonists, or a combination of both strategies. The application of biological control has been inconsistent in suppressing nematode populations because the efficacy of antagonists is influenced by other soil organisms and the host-plant. Integration of biological control with nematicides, solarization, organic amendments, and crop rotation has also had varied success. Progress in biological control of nematodes has been hampered by the opaque nature of soil, the microscopic size of nematodes and their antagonists, and the complex interactions among soil organisms. Molecular biology offers new tools that will aid in determining which organisms are involved in naturally-suppressive soils, the fate of introduced antagonists, and how populations of indigenous and introduced antagonists change seasonally and with different crop production practices. Moreover, organisms have been engineered to over-express traits that enhance their activity against plant-parasitic nematodes.

11.1 Current Status of Biological Control

Management of plant-parasitic nematodes in crop production systems currently relies primarily on nematicides, host-plant resistance, and crop rotation. Although many advances have been made in biological control of plant-parasitic nematodes in the last 20 years, it is still scarcely used in nematode management. When we consider the use of biological control for managing nematodes, we typically

P. Timper (✉)
Crop Protection and Management Research Unit, USDA-ARS,
P.O. Box 748, Tifton GA 31793, USA
e-mail: Patricia.Timper@ars.usda.gov

K. Davies and Y. Spiegel (eds.), *Biological Control of Plant-Parasitic Nematodes: Building Coherence between Microbial Ecology and Molecular Mechanisms*, Progress in Biological Control 11, DOI 10.1007/978-1-4020-9648-8_11,

envision applying some formulated product to the seed, planting furrow, or transplant medium. Historically, there have been few commercial products registered for biological control of plant-parasitic nematodes. If one excludes products containing toxins derived from microorganisms and counts only those products containing viable organisms, then the list is even shorter (Dong and Zhang 2006). Of the eight commercial products containing viable organisms, at least two have been discontinued and three others are formulations of the same fungus (*Paecilomyces lilacinus*, strain 251). Stirling (1991) provides an in depth analysis of the commercial and organizational barriers to the development of biological control products.

In addition to the lack of commercial biological control organisms, the unreliability and relatively low efficacy of nematode antagonists are major obstacles to the use of biological control for managing plant-parasitic nematodes (Stirling 1991). From a practical standpoint, most growers seek to maximize their profits by selecting nematode management options that provide the greatest increase in yield while keeping input costs low. While it is understood that all management options have a risk of failure, host-plant resistance, rotations with non-host plants, and nematicides typically provide more reliable and effective nematode suppression than biological control. Moreover, nematicides and crop rotation can reduce populations of other plant pests (Timper et al. 2001). Reliability is essential for all nematode management options for which there are input costs because failure to reduce nematode populations can lead to greater monetary losses than if no action was taken to control the nematode. The greater the input cost, the greater the expectation for successful nematode control and yield increase. For any management option, including use of nematode antagonists, low or partial nematode control is less problematic than unreliable control. In the case of partial control, an antagonist could be combined either sequentially (i.e., in different seasons) or simultaneously with other management options to achieve acceptable nematode control (Roberts 1993). Research aimed at understanding the environmental factors affecting reliable and effective biological control of nematodes, as well as research to improve the effectiveness of specific antagonists will be presented later in this chapter.

Though there are major barriers to the utilization of commercially-produced antagonists, evidence suggests that some level of biological control is occurring naturally in many agricultural fields. There are a few well-documented cases of field sites where plant-parasitic nematodes are maintained at very low population densities by one or more indigenous microorganisms (Stirling 1991; Westphal 2005). Suppressive field sites are initially identified because nematode populations are inexplicably low despite conducive soil characteristics and cropping history. However, this phenomenon is not restricted to a few unique field sites. Many agricultural soils may contain organisms which keep nematode populations at a level below that which would occur if those organisms were removed, but because the level of nematode suppression is not dramatic, they may not be readily identified as suppressive. Stirling (1991) states "The possibility that every nematode population is affected to some extent by natural enemies, and that all nematode problems would be much worse in the absence of these antagonists has rarely been seriously considered." Several studies have identified low to moderate levels of nematode

Table 11.1 Agricultural fields tested for suppression of plant-parasitic nematodes

Location	No. suppr. fields (total sampled)	Nematode (stage suppressed)	% Suppr.	References
Texas, Mississippi, Louisiana, USA	10 (22)	*Rotylenchulus reniformis* (vermiform/g soil)	37–93%	Robinson et al. (2008)
California, USA	5 (20)	*Meloidogyne incognita* (J2 soil + hatched J2 roots)	35–97%	Gaspard et al. (1990)
California, USA	4 (12)	*M. incognita* (eggs/g soil)	28–63%	Pyrowolakis et al. (2002)
Florida, USA	5 (5)	*Heterodera glycines* (eggs/g soil)	56–92%	Chen et al. (1996a)
Georgia, USA	2 (5)	*M. incognita, M. arenaria* (eggs/g root)	54–76%	Timper, unpublished
Florida, USA	1 (2)	*M. incognita* (hatched J2/g root)	83%	McSorley et al. (2006)
Minas Gerais, Brazil	1 (1)	*M. incognita* (egg masses/root system)	51%	Santos et al. (1992)

suppression in agricultural soils (Table 11.1). The biological nature of the suppression was determine by comparing nematode multiplication in untreated soil with multiplication in fumigated (Santos et al. 1992; Pyrowolakis et al. 2002) or pasteurized soil (Chen et al. 1996a; McSorley et al. 2006). In one study, a small quantity of test soil was mixed with sterilised soil and then nematode multiplication was compared with that in sterilised soil (Robinson et al. 2008). In many of these field sites, suppression of nematodes was not expected, nor was the suppression clearly attributable to any organism or group of organisms. Evidently, far more soils contain organisms capable of suppressing plant-parasitic nematodes than previously recognized. Are we relying on biological control without being aware of it?

If many agricultural soils contain indigenous organisms capable of reducing populations of plant-parasitic nematodes, then it may be possible to conserve or enhance these organisms by modifying or adopting certain farming practices. Such a strategy is commonly employed in the biological control of insects (Barbosa 1998; Pickett and Bugg 1998) and has also been used in biological control of soil-borne plant pathogens (Mazzola 2007). Therefore, biological control of plant-parasitic nematodes can be accomplished either by introduction of antagonistic organisms to the nematode's habitat, manipulation of the habitat to conserve and enhance the activity of indigenous antagonists, or a combination of both strategies.

Progress in biological control of nematodes, whether it be via introduction or conservation and enhancement of antagonists, is hampered by the opaque nature of soil, the microscopic size of nematodes and their antagonists, and the complex interactions among soil organisms (Stirling 1991). There have been few tools that would allow nematologists to determine which organisms are involved in naturally-suppressive soils, the fate of introduced antagonists, and how populations of native and introduced antagonists change seasonally and with different crop production practices. In recent years, molecular tools have been developed and are beginning to be used to answer critical questions related to biological control of nematodes. Moreover, organisms can be engineered to over-express certain compounds that enhance their activity against plant-parasitic nematodes.

11.2 Suppressive Soils

11.2.1 Identifying the Organisms Involved

Before tackling a difficult and complex task such as the biological control of nematodes, it is helpful to study systems where antagonistic organisms are regulating populations of plant-parasitic nematodes. The case histories of several classic nematode-suppressive soils are described in detail by Stirling (1991); they include suppression of *Heterodera avenae* in cereals by *Pochonia chlamydosporia* and *Nematophthora gynophila*, *Meloidogyne* spp. on peach by *Dactylella oviparasitica*, and *M. javanica* on grape by *Pasteuria penetrans*.

Westphal (2005) has recently reviewed techniques for determining whether a soil contains organisms suppressive to nematodes. However, once a soil has been deemed suppressive to nematodes, identifying the causal organisms can be difficult, with the possible exception of *P. penetrans*. Second-stage juveniles (J2) of *Meloidogyne* spp. with attached endospores of *P. penetrans* are readily extracted from soil and there is a good correlation between endospores per J2 and suppression of egg production by the bacterium (Minton and Sayre 1989; Chen et al. 1997; Meyer 2003). Because endospores of *P. penetrans* are very resistant to environmental extremes, drying and heating of soil can be used to selectively eliminate invertebrate predators and fungal parasites of nematodes, respectively, while autoclaving soil eliminates all organisms including spore-forming bacteria. Weibelzahl-Fulton et al. (1996) used such a technique to demonstrate that *P. penetrans* was responsible for suppression of *Meloidogyne* spp. in tobacco.

In most cases, the organisms responsible for nematode suppression are not obvious. Kluepfel et al. (1993) identified two sites in a peach orchard, one suppressive and the other conducive to reproduction of *Mesocriconema xenoplax*. Compared to steam-heated soil, population densities of the nematode were reduced by 64% and 98% in soil from the conducive and suppressive sites, respectively. Of the 290 pseudomonads isolated from the rhizosphere of peach trees in the suppressive site, seven suppressed populations of *M. xenoplax* in glasshouse assays. However, no single strain reduced nematode populations to the level found in the suppressive site. The low populations of *M. xenoplax* in the suppressive site may be due to the concerted action of several antagonistic pseudomonads or to some entirely different organism. This study illustrates the difficulty in assigning causal agents to suppressive soils. Bacteria antagonistic to plant-parasitic nematodes can be isolated from the rhizospheres of many plant species (Kloepper et al. 1992). The presence of antagonistic bacteria, or any other organism for that matter, does not necessarily indicate that they are suppressing nematode populations under field conditions because density of the antagonist and other organisms in rhizosphere can influence the level of biological control (Siddiqui and Ehteshamul-Haque 2001; Siddiqui and Shaukat 2003a, 2005; Weller et al. 2007). The role of the isolated pseudomonads in suppression of *M. xenoplax* would be strengthened by demonstrating (1) that a subset of these bacteria can suppress the nematode to a similar level as observed in the suppressive site, and (2) that these bacteria are either not present or are present at significantly lower densities in the peach rhizosphere in the conducive site, which was actually moderately suppressive to the nematode.

Recently, a three-phase approach was used to identify the organisms involved in suppression of *Heterodera schachtii* in a research field (9E) at the University of California, Riverside (Borneman and Becker 2007). The suppressive nature of this field site had been extensively documented (Westphal and Becker 1999, 2000, 2001b). Although several nematode-parasitic fungi including *Fusarium oxysporum*, *Fusarium* sp., *Dactylella oviparasitica*, and *P. lilacinus* were isolated from cysts in field 9E, it was not clear if one or more of these fungi were responsible for suppressing *H. schachtii* populations (Westphal and Becker 2001a). Ultimately, a population-based approach was used to identify the organism involved. This approach relied on creating soils with varying levels of suppressiveness and then correlating the abundance of microbial taxa with nematode suppression (Borneman and Becker 2007). In order to reduce the scope of fungal taxa to identify, Yin et al. (2003) focused on the cysts which had been previously shown to harbor the suppressive organism (Westphal and Becker 2001a). In the first phase of the study, oligonucleotide fingerprinting of rRNA genes showed that *D. oviparasitica* was the dominant fungus in cysts from the two most suppressive soils (Yin et al. 2003). In the second phase of the study, the association between this fungus and nematode suppression was confirmed by developing sequence-selective PCR primers for the three dominant fungal species. Again, *D. oviparasitica* was the most abundant fungus in the most suppressive soils, but was at low to non-detectable levels in the least suppressive soils. In the final phase of the study, *D. oviparasitica* isolated from field 9E (strain 50) was introduced into fumigated soil where it suppressed the number of eggs per gram soil of *H. schachtii* to the same level as the suppressive soil (82%) after 11 weeks in glasshouse pots (Olatinwo et al. 2006c). In fumigated field microplots, *D. oviparasitica* reduced egg densities of *H. schachtii* to 91% after 19 weeks compared to microplots without the fungus (Olatinwo et al. 2006b). After an additional 16 weeks, the soil inoculated with *D. parasiticia* was still as suppressive as the nonfumigated 9E soil (98%). The fungus also reduced populations of *H. schachtii* by 94–97% in two of four nonfumigated field soils (Olatinwo et al. 2006a). The field soils in which *D. oviparasitica* did not reduce nematode populations were already highly suppressive to *H. schachtii* relative to their fumigated counterparts; these soils were collected from fields with a cropping history that included host-plants of the nematode.

11.2.2 Factors Involved in Development of Suppressive Soils

The one characteristic that all of the well-documented nematode-suppressive soils have in common is that they developed in situations where a host-plant for the nematode was present over an extended time such as continuous cultivation of annual crops or in perennial crops (Kluepfel et al. 1993; Weibelzahl-Fulton et al. 1996; Westphal and Becker 1999; Timper et al. 2001; see Stirling 1991 for additional citations). Presumably, the continuous presence of a particular plant-parasitic nematode, initially at high population densities, leads to the build-up of specialized antagonists of that nematode (Kerry and Crump 1998). It is, therefore, not surprising that the organisms typically involved in nematode-suppressive soils are either host-specific *Pasteuria* spp. or fungal biotypes specialized for parasitizing eggs and sedentary females of cyst and root-knot nematodes (Mauchline et al. 2004; Morton et al. 2003; Siddiqui et al. 2009). Yet suppressive soils do no develop in all perennial systems or in all situations where continuous cropping is practiced (Olatinwo et al. 2006a; Robinson et al. 2008). Is it that these nematode-conducive soils lack key antagonists or is there something in the environment (physical, chemical, or biological) that is limiting the antagonistic organisms?

Very little is known about the organisms involved in or the conditions contributing to moderately suppressive soils. Moderately suppressive soils sometimes have no history of the nematodes they suppress and are not necessarily associated with long-term presence of a host plant (Santos et al. 1992; Chen et al. 1996a; Pyrowolakis et al. 2002). Cook and Baker (1983) differentiate between specific and general soil suppressiveness for plant pathogens. General suppression is caused by the total biological activity of a soil and is a characteristic of most soils, whereas specific suppression is due to an individual or select group of organisms antagonistic to a specific pathogen. With regard to nematodes, there is little evidence for or against a suppressive soil community. Because plant-parasitic nematodes do not compete for organic matter with other microorganisms, they may be less affected by saprophytic organisms than many facultative plant pathogens. Moreover, although suppressive soils are not rare, they are not found in the majority of tested field sites (Table 11.1). Other than the magnitude of nematode suppression, there may be little difference between highly suppressive and moderately suppressive soils; in both cases, suppression may be caused by an individual or a select group of antagonists. The population-based approach used to identify *D. oviparasitica* as the organism responsible for suppression of *H. schachtii* in field 9E could be used to identify the organisms involved in moderately suppressive soils. Following a survey of six agricultural fields, Bent et al. (2008) identified one soil that suppressed *M. incognita* populations by 80–89% compared to fumigated soil. Using several different methods for creating a range of nematode-suppressive environments, reductions in *M. incognita* populations had the strongest negative correlation with *P. chlamydosporia* based on oligonucleotide fingerprinting of rRNA genes. Sequence-selective PCR primers confirmed the association between *P. chlamydosporia* rRNA and suppression of *M. incognita* densities. Further studies are needed to show that this fungus is capable of reducing *M. incognita* populations to the same level as the suppressive soil.

11.3 Application of Antagonists

There are a large number of studies conducted in glasshouse pots demonstrating high levels of nematode suppression with antagonistic organisms. Most of these studies utilized heat-treated or fumigated soil to eliminate resident plant-parasitic nematodes and plant pathogens. While studies using heated-treated or fumigated soil are regarded as a necessary first step toward identifying potential biological control organisms, they can provide unrealistic expectations for nematode suppression. Many fungi and bacteria grow and survive better in soil that has been partially or completely sterilised because of reduced competition, predation, and antibiotic production, and because of increased organic substrates from dead organisms. Furthermore, most planting pots restrict the biological control arena and provide a greater opportunity for the antagonist and nematode to interact than under field conditions. Therefore, in this section, only studies conducted in natural soil will be presented, with emphasis on microplot and field applications published after 1990. Stirling (1991) reviewed earlier attempts to release antagonistic organisms for biological control of nematodes.

11.3.1 Bacteria

Pasteuria penetrans has been the most commonly applied bacterium for the biological control of plant-parasitic nematodes (Chen and Dickson 1998). Application of endospores or dried plant material containing spore-filled females have been used to infest field and microplot soil because of the difficulty of in vitro culture of this fastidious organism. However, recent advances in fermentation culture of *Pasteuria* spp. may lead to large-scale applications of endospores (Smith et al. 2004). In microplots infested with *M. arenaria*, application of 100,000 and 10,000 endospores/g soil reduced root galling of peanut by 81% and 61%, respectively, 2 years after initial application (Chen et al. 1996b). After 3 years, root galling was reduced even in plots initially infested with only 1,000 and 3,000 endospores/g soil (Chen and Dickson 1998). Kariuki and Dickson (2007) used dried roots from an infested field site to transfer *P. penetrans* to another field site. Three years after infestation of the new field site, root galling on peanut was reduced to the same level as in plots fumigated with 1,3-dichloropropene. In a large multi-national project, eight microplot and field studies were conducted to test the hypothesis that intensive cropping of *Meloidogyne*-susceptible crops would lead to an increase in abundance of *P. penetrans* endospores and suppression of the nematode population (Trudgill et al. 2000). However, nematode suppression was only documented in three trials where an exotic isolate of *P. penetrans* had been introduced to supplement an indigenous isolate present at low background levels. Because the indigenous *P. penetrans* in the trials failed to increase following repeated cropping of a host, the authors speculated that the nematode populations had undergone selection for reduced attachment of endospores (Tzortzakakis et al. 1996) leading to low equilibrium levels of parasitism. In two other trials, application of an exotic isolate did not suppress *Meloidogyne* populations suggesting that the environment may not have been conducive for the bacterium.

In addition to *P. penetrans*, a diverse group of bacteria have been applied for control of plant-parasitic nematodes. Some of these bacteria are referred to as rhizobacteria because of their close association with plant roots. In a glasshouse experiment, two strains of *Burkoldera cepacia* suppressed the numbers of *M. incognita* eggs on bell pepper by 60–69% (Meyer et al. 2001). However, in two separate field experiments, a commercial preparation of *B. cepacia* failed to reduce populations of *H. glycines* on soybean (Noel 1990). Although *B. cepacia* is considered a rhizosphere colonizer, a foliar application of a commercial formulation reduced the number of *Aphelenchoides fragariae* on hosta foliage under glasshouse conditions (Jagdale and Grewal 2002). In a microplot study, the rhizobacteria *Pseudomonas fluorescens* strain CHA0 and *P. aeruginosa* strain IE-6S$^+$, and the root-nodulating bacterium *Bradyrhizobium japonicum* suppressed the number of galls on tomato caused by *M. javanica* by 28–43% (Siddiqui and Shaukat 2002). Similarly, in a field trial, seed treatments with two isolates of *P. aeruginosa* reduced *Heterodera cajani* in sesame by up to 58% and increased yield (Kumar et al. 2009). Populations of *Pratylenchus penetrans* in glasshouse pots were suppressed by *P. chloroaphis* strain Sm3 on strawberry in six different field soils; however, suppression only averaged 28% compared to soils without the bacterium (Hackenberg et al. 2000). Chen et al. (2000) demonstrated that both *Streptomyces costaricanus* and a nematode-antagonistic strain of *Bacillus thuringiensis* were able to reduce galling and egg production of *M. hapla* and increase lettuce head weight in microplots. In a field study, tomato and pepper were grown in a potting mix containing strains of rhizobacteria formulated with chitin before transplanting in a field infested with *M. incognita*. None of five bacterial formulations were able to suppress the nematode on tomato; however, one formulation containing *Bacillus subtilis* strain GBO3 and *B. cereus* strain C4 suppressed root galling on pepper (Kokalis-Burelle et al. 2002). Similarly, a commercial formulation containing *B. subtilis* strain GBO3, *B. amyloliquefaciens* strain GB99, and chitin reduced galling by *Meloidogyne* sp. on tomato in field plots (Kokalis-Burelle and Dickson 2003). In both studies, only slight reductions in galling were observed on the pepper and tomato. In a commercial glasshouse naturally infested with *M. incognita*, Giannakou et al. (2004) showed in three separate experiments that a commercial formulation of *B. firmus* suppressed galling and numbers of juveniles in soil by 52–64%. A broadcast application of the formulation was more effective than a banding application. In another study, a wettable powder formulation of *B. firmus* reduced galling by 54–65% in a tomato nursery when used at the recommended rates (Terefe et al. 2009). The formulations of *B. firmus* used in these studies contained 97% plant and animal extracts; therefore, it is unclear whether nematode suppression was due to the bacterium, stimulation of other antagonistic organisms, or toxic products from the degradation of organic matter.

11.3.2 Fungi

Most field and microplot studies testing fungi for biological control of nematodes after 1990 have been conducted with parasites of sedentary stages such as the eggs, developing juveniles and females of cyst and root-knot nematodes. *Paecilomyces lilacinus*, *P. chlamydosporia*, and *Trichoderma* spp. are all common soil inhabitants and some strains are aggressive parasites of sedentary stages of nematodes (Siddiqui and Mahmood 1996; Sharon et al. 2001, 2007). *Trichoderma* spp. may also produce toxic metabolites (Khan and Saxena 1997; Sharon et al. 2001).

Paecilomyces lilacinus strain 251 is registered for biological control of nematodes in several countries (Atkins et al. 2005). An overview of biological control attempts from 1991 to 1995 using strains of this fungus has been published (Siddiqui and Mahmood 1996). Lara Martez et al. (1996) showed that *P. lilacinus* reduced numbers of *M. incognita* J2 in field-grown tomato by 70% and 41% when applied at transplant and 2 weeks after transplanting, respectively. However, the fungus was not able to suppress populations of *R. reniformis* or *Helicotylenchus dihystera*. In another field study, *P. lilacinus* suppressed galling of tomato by *M. incognita* by 39% when applied at transplant (Goswami et al. 2008). On golf-coarse greens, a commercial formulation of *P. lilacinus* failed to reduce densities of *M. marylandi* in two experiments (Starr et al. 2007). Similarly, in greenhouse soil heavily infested with *M. incognita*, strain 251 did not reduce galling in tomato (Kaşkavalci et al. 2009). However, in a commercial plastic house, the fungus was as effective as oxamyl in reducing J2 densities at mid season and harvest of cucumber compared to the untreated control (Anastasiadis et al. 2008).

Pochonia chlamydosporia is associated with nematode-suppressive soils and has been effective in the biological control of root-knot and cyst nematodes in glasshouse pots (Kerry 1995, 2001). Siddiqui and Mahmood (1996) have reviewed studies utilizing this fungus for control of nematodes from 1991 to 1996. When applied at planting as a kaolin formulation, *P. chlamydosporia* was unable to suppress root galling from *Meloidogyne* spp. or numbers of J2 in four field experiments with tomato (Stirling and Smith 1998). Sorribas et al. (2003) applied two isolates, one native and the other exotic, of *P. chlamydosporia* for control of *M. javanica* in plastic houses infested with the nematode. A single application of the fungus 10 weeks after planting tomato had no effect on root galling or egg production by the nematode. When the fungus was applied weekly for 6 weeks, both isolates were equally effective in reducing galling on tomato, but the native isolate parasitized more eggs than the exotic isolate (30% vs 5%) and reduce densities of healthy eggs by 50%, whereas the exotic strain had no effect on egg densities. Nevertheless, root-gall ratings were quite high despite significant suppression by the fungus. Colonization of the rhizosphere of tomato by the native isolate was 15X greater than the exotic isolate suggesting that the former was better adapted to the local habitat. In a field site infested with *Globodera pallida*, *P. chlamydosporia* strain B1357 reduced final nematode numbers by 48–51% but did not increase potato yield relative to the untreated control (Tobin et al. 2008b). Wei et al. (2009)

used a screening strategy based on protease and chitinase production to identify fungi with the greatest potential for nematode suppression. Three of the isolates selected using this strategy, two *P. lilacinus* and one *P. chlamydosporia*, suppressed root galling from *Meloidogyne* sp. in field-grown tomato by 48–61% and increased yields by a similar percentage.

Various species of *Trichoderma* are antagonistic to plant-parasitic nematodes (Sharon et al. 2007). In microplots, *T. harzianum* did not reduce galling of *M. incognita* on eggplant, but 30% of the females in the roots were infected by the fungus (Rao et al. 1998). Parasitism of females was increased to 51% when the fungus was formulated with castor cake extract. In a field site infested with *M. incognita*, *T. harzianum* reduced galling on tomato roots by 47% compared to untreated plots (Goswami et al. 2008). Application of *T. pseudokoningii* did not reduce galling on soybean from *M. incognita* or increase grain yield in a field study (Oyekanmi et al. 2007). However, in a pot experiment, the same fungus reduced the number of egg masses even though galling was not reduced. Two endophytic strains of *T. atroviride* suppressed populations of *Radopholus similis* in banana (Pocasangre et al. 2007). Maehara (2008) demonstrated that *Trichoderma* sp. 3, when inoculated into pine logs, decreased the number of *Bursaphelenchus xylophilus* carried by *Monochamus* beetles. *Trichoderma* spp. appear to have an indirect effect on *B. xylophilus* in pine logs by competing with the blue-stain fungus which is an ideal food source for the nematode, but may also have a direct antagonistic effect on the nematode.

The mobile vermiform stages of nematodes have also been targets for biological control. Conidia of *Hirsutella rhossiliensis* adhere to the cuticle of passing nematodes and penetrate the cuticle via a germ tube. Tedford et al. (1993) introduced *H. rhossiliensis* into microplots in the form of infected nematodes. Although the fungus became established in the microplots, it failed to reduce the number of *H. schachtii* in sugarbeet or *M. javanica* in tomato. The authors speculated that exposure of J2 to the adhesive conidia was limited because of the short distance the juveniles exiting from egg masses needed to travel to re-infect the root. In another microplot study, *H. rhossiliensis,* formulated as hyphae in alginate pellets, was tested for its ability to reduce infection of cabbage seedlings by *H. schachtii* (Jaffee et al. 1996). In this test, the infective juveniles had to move through soil because they were not hatching from egg masses on the root. However, the fungus failed to reduce root invasion. In observation chambers containing field soil, the pelletized hyphae sometimes appeared to have been eaten and the fungal colonies growing from the pellets were smaller than in chambers containing heat-treated soil, suggesting biotic inhibition of the fungus. In a third microplot study, pelletized hyphae of *H. rhossiliensis* was compared to pelletized hyphae of two trapping fungi, *Monacrosporium gephyropagum* and *M. ellipsosporum* (Jaffee and Muldoon 1997). These two fungi trap nematodes by adhesive hyphae and adhesive knobs, respectively. Of the three fungi, only *M. gephyropagum* suppressed invasion of tomato seedlings by *M. javanica* and improved seedling emergence and root growth. Alginate pellets containing the fungi did not persist over the 20 day observation period. The effectiveness of *M. gephyropagum* in this study may be due to its rapid growth and capture of nematodes before the pellets were consumed by grazing microfauna.

11.3.3 Nematodes

The potential of predatory nematodes for biological control of plant-parasitic nematodes has been reviewed by Khan and Kim (2007). Predatory nematodes have not received much attention for biological control of plant-parasitic nematodes because of the difficulty in mass culturing due to low fecundity, long life cycle, complex culture conditions, and cannibalism. Diplogasterid nematodes have advantages over other predatory nematodes in that they have high reproductive rates, short life cycles, and can be cultured on bacteria (Bilgrami et al. 2008). Recently, the diplogasterid nematode *Mononchoides gaugleri* was evaluated in field microplots for suppression of plant-parasitic nematodes in turf grass. The predator reduced total populations of plant-parasitic nematodes 30 days after application, but individual genera were differentially affected. Populations of *Ditylenchus* sp., *Aphelenchoides* sp., *Tylenchorhynchus* sp., and *Tylenchus* sp. were reduced by 45%, 40%, 35%, and 20%, respectively; whereas *Hoplolaimus* sp. and *Helicotylenchus* sp. were not affected by the predator.

Entomopathogenic nematodes in the genera *Steinernema* and *Heterorhabditis* can suppress populations of plant-parasitic nematodes. Suppression may involve one or more of the following mechanisms: interference competition at the root surface (Bird and Bird 1986), stimulation of nematode antagonists (Ishibashi and Kondo 1986), allelochemicals from the symbiotic bacteria associated with these nematodes, or induction of systemic resistance in the plant (Jagdale et al. 2002, 2009). In turf grass plots, a mixture of *S. carpocapsae* and *H. bacteriophora* reduced *Tylenchorhynchus* spp. by 50–59% under irrigated but not non-irrigated conditions 5 weeks after application (Smitley et al. 1992). Application of *S. riobrave* to turf grass at two golf courses suppressed populations of *Meloidogyne* sp., *Belonolaimus longicaudatus*, and *Criconemella* sp. by 84–100% at 4 and 8 weeks after treatment (Grewal et al. 1997). In two different studies, *S. riobrave* and *S. carpocapsae* reduced several genera of plant-parasitic nematodes on boxwood for a least 30 days following treatment (Jagdale et al. 2002; Perez and Lewis 2006). However, *S. riobrave* failed to reduce the populations of *M. xenoplax* on peach grown in glasshouse pots and pecan in microplots (Nyczepir et al. 2004).

11.3.4 Biotic and Abiotic Factors Modifying Efficacy

The environment to which a biological control organism is introduced can play a large role in the success or failure of that organism to reduce populations of plant-parasitic nematodes. Antagonism from other organisms is often cited as the cause of poor nematode control in field soils compared to partially or completely sterilised soil. The organisms involved in antagonism are mostly unknown, but are assumed to be competitors for organic matter, a supplemental food source for many biological control organisms (Mankau 1962; Cook and Baker 1983). However, competition is not the only hostile encounter biological control organisms face when applied to soil. In microplot experiments, collembolans and enchytraeid worms were observed in the vicinity of partially consumed pellets containing nematophagous fungi (Jaffee and Muldoon 1997; Jaffee et al. 1996). Using soil cages of different mesh sizes, Jaffee et al. (1997) and Jaffee (1999) demonstrated that exclusion of enchytraeids and microarthropods increased the persistence of nematophagous fungi growing from alginate pellets. However, smaller organisms (e.g., fungi, bacteria, nematodes, protozoa, etc.), which were not excluded, still reduced persistence of the fungi compared to heat-treated soil.

Microorganisms also release compounds which can inhibit biological control. Diffusible compounds from two soil communities reduced growth of *P. chlamydosporium* and *P. lilaciunus* (Monfort et al. 2006). *Bacillus* sp. strain H6, isolated from a fungistatic soil, produces iturin A-like compounds which caused swelling in the conidia and germ tubes of nematophagous fungi (Li et al. 2007). The egg masses of *Meloidogyne* spp. may also harbor microflora inhibitory to biological control. Kok et al. (2001) isolated 122 bacteria and 19 fungi from egg masses and found that 23% and 74%, respectively, were antagonistic to *P. chlamydosporia*. The production of DAPG (2,4-diacetylphloroglucinol) by *P. fluorescens* is involved in suppression of cyst and root-knot nematodes (Cronin et al. 1997; Siddiqui and Shaukat 2003c). Metabolites from several common soil fungi have been shown to inhibit expression of DAPG (Notz et al. 2002; Siddiqui and Shaukat 2003a, 2005; Siddiqui et al. 2004). Moreover, the presence of some of these fungi (*Fusarium solani*, *Rhizoctonia solani*, and *Aspergillus quadrilineatus*) in soil reduced the ability of the bacterium to suppress populations of *Meloidogyne* spp. on tomato (Siddiqui and Shaukat 2003a, 2005; Siddiqui et al. 2004).

Isolates of nematophagous fungi differ in their sensitivity to biological inhibition. Saprotrophic growth of five *P. chlamydosporia* isolates was compared in two soils (Monfort et al. 2006). In both soils, isolate 5 was the least affected by soil microorganisms, with a reduction in growth of 57–72% compared to sterilised soil. Growth of isolate 4624 was suppressed less in the Lancelin than in the Biar soil; growth of all other isolates was suppressed by 83–98% compared to sterilised soil. In another study, microbial inhibition of *P. chlamydosporia* isolates was very low, ranging from 0% to 37% when tested in two different soils (Siddiqui et al. 2009). There was also a negative correlation between saprotrophic growth and parasitism of eggs suggesting that there may be a trade-off between these two traits. However, in another study, there was no correlation between saprotrophic and parasitic abilities (data presented in Siddiqui et al. 2009).

Soil microorganisms can sometimes enhance biological control of plant-parasitic nematodes. In attachment assays, the presence of some bacterial isolates originating from both soil and gall tissue increased attachment of *P. penetrans* endospores to J2 of *Meloidogyne* spp. (Duponnois and Ba 1998; Duponnois et al. 1999). In a glasshouse experiment, one of the bacterial isolates, *Enterobacter cloacae*, when combined with *P. penetrans* for control of *M. incognita* on tomato, reduced the number of egg masses on the roots by 36% and increased the number of endospores produced in roots compared to treatments with only *P. penetrans*

(Duponnois et al. 1999). Although the mechanism is unclear, enzymes produced by *E. cloacae* and other bacteria may modify either the nematode cuticle or the endospore sporangial wall or exosporium to increase attachment. While a number of fungi can inhibit expression of DAPG by *P. fluorescens* strain CHA0, *Pythium ultimum*, *Aspergillus niger*, and *T. harzianum* can enhance expression of the antibiotic (Notz et al. 2001; Siddiqui and Shaukat 2004; Siddiqui et al. 2004). However, neither *A. niger* or *T. harzianum* were able to significantly increase suppression of *M. javanica* on tomato compared to the bacterium alone (Siddiqui and Shaukat 2004; Siddiqui et al. 2004).

Biological control of *Meloidogyne* spp. by parasites of sedentary stages has been suggested to be more effective on plants that are poor hosts for the nematode because small galls leave egg masses exposed on the root surface and fewer eggs are produced than on good hosts (Stirling et al. 1979; De Leij and Kerry 1991). Bourne et al. (1996) demonstrated this principle with *P. chlamydosporia*, which provided greater suppression of *M. incognita* on the poorer host potato than on the better host tomato despite greater fungal colonization of the tomato roots. Although colonization of the rhizosphere by *P. chlamydosporia* differs among host-plants, there was no relationship between abundance of the fungus on roots and the rate of parasitism (Bourne and Kerry 1999). Colonization of roots by *P. fluorescens* strain CHA0 differed among host-plants and among cultivars of soybean, but degree of colonization was not related to suppression of root galling by *M. incognita* (Siddiqui and Shaukat 2003b). Strain CHA0 suppressed galling on all crops except chili, which was a relatively poor host compared to the other crops.

Abiotic factors that can influence the level of biological control include temperature, soil type, moisture, and rainfall/irrigation. In glasshouse pots using field soil, a commercial formulation of *Paecilomyces lilacinus* strain 251 suppressed galling and egg masses of *M. hapla* on tomato by 66–90% when daytime temperatures were 23–25°C, but was much less effective when the daytime temperature was 21°C (Kiewnick and Sikora 2006). Establishment in the rhizosphere of *P. chlamydosporia* and nematode suppression by the fungus was greater in peaty sand than in loamy sand or sand (De Leij et al. 1993); however, in another study, there was no difference in colonization of a compost, sandy loam, and loamy sand soil by the fungus (Siddiqui et al. 2009). Soil type can also influence retention of *Pasteuria penetrans* endospores in the root zone and acquisition by *Meloidogyne* J2. The bacterium occurs more frequently in sandy soils than in finer-textured soils (Spaull 1984); the mobility of J2 in sandy soils likely allows for greater acquisition of endospores. In a pot experiment, the percentage of *M. incognita* females infected with *P. penetrans* was greater in a sandy soil than in a sandy clay soil (Carneiro et al. 2007). However, leaching of endospores is also greater in sandy soils than in finer-textured soils. Under a drip system, 76% of endospores leached 10 cm after 24 h in sand, and with increasing clay content, there was a decrease in the percentage of endospores leached (Dabire and Mateille 2004). Soils with clay content between 10% and 30% were considered optimal for biological control with *P. penetrans*.

11.3.5 Integration of Biological Control with Other Management Tactics

Integrated pest management utilizes multiple management tactics within a growing season or in different seasons to reduce pest populations (Roberts 1993). Because nematode control in integrated management systems does not rely solely on one management tactic, partially effective tactics such as biological control can be combined to lower nematode populations below the damage threshold. However, attempts to integrate biological control with nematicides, host-plant resistance, crop rotation, solarization of soil, other antagonists, and soil amendments have generated mixed results.

Nematophagous fungi and *P. penetrans* are generally compatible with non-fumigant nematicides and some fumigants such as 1,3-dichloropropene. Nematicides do not usually have an adverse effect on these organisms (Mankau and Prasad 1972; Jacobs et al. 2003) and may even enhance parasitism. Brown and Nordmeyer (1985) suggested that aldicarb and carbofuran increased movement of *M. javanica* J2 and acquisition of endospores leading to a synergistic reduction in galling when the nematicides were combined with *P. penetrans*. However, the frequency of endospores attached to J2 in a field study was not influenced by the application of aldicarb (Timper et al. 2001). Applications of oxamyl and *P. penetrans* to tomato had an additive effect on reducing egg production by *Meloidogyne* spp., but acted synergistically in the subsequent cucumber crop (Tzortzakakis and Gowen 1994). Fungal parasites of sedentary stages cannot protect plant seedlings from nematode invasion and early-season damage, but will often proliferate in the rhizosphere during the growing season (Stirling and Smith 1998; Sorribas et al. 2003; Tobin et al. 2008b). Nematicides could be used in conjunction with these fungi to reduce initial nematode populations while the antagonist reduces egg production and viability leading to lower nematode populations for the succeeding crop. In three separate studies evaluating the combined application of oxamyl and *P. chlamydosporia* (Tzortzakakis 2000; Tzortzakakis and Petsas 2003; Verdejo-Lucas et al. 2003), only one study demonstrated that the fungus provided additional suppression of *M. javanica* galling and egg production over the nematicide alone (Verdejo-Lucas et al. 2003). In a field study, both fosthiazate and *P. chlamydosporia* suppressed final population densities of potato cyst nematodes, but there was no additive effect of the two control tactics (Tobin et al. 2008b).

Very little research has been done to evaluate the effectiveness of combining biological control with host-plant resistance and crop rotation. A nematode antagonist could be applied to a moderately resistant cultivar or to a susceptible cultivar following rotation with a resistant cultivar or non-host crop. Samac and Kinkel (2001) tested a strain of *Streptomyces* sp. for biological control of *Pratylenchus penetrans* on resistant and susceptible alfalfa. Nematode suppression by the resistant cultivar and *Streptomyces* sp. was additive and together they provided >90% control. In plastic houses where susceptible tomato followed resistant tomato in a single growing season, *P. chlamydosporia* failed to suppress *M. javanica* on the

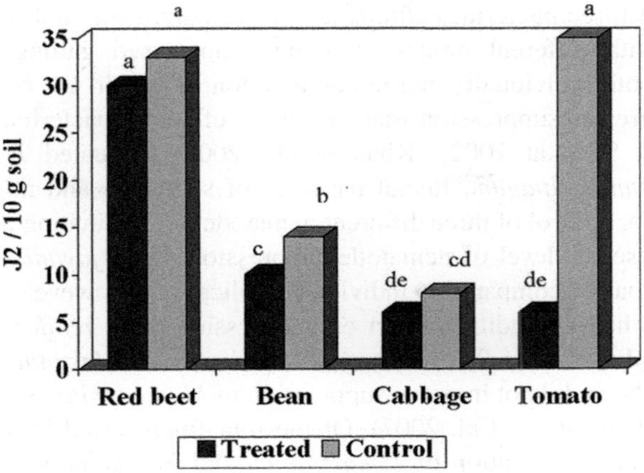

Fig. 11.1 Numbers of second-stage juveniles of *Meloidogyne incognita* in untreated soil and treated with *Pochonia chlamydosporia* var. *catenulate* applied before the bean crop in the rotation (Atkins et al. 2003c)

susceptible tomato (Tzortzakakis and Petsas 2003). The fungus, however, was effective in reducing *M. incognita* on tomato when it was used in a rotation system involving two poor hosts of the nematode, bean and Chinese cabbage (Atkins et al. 2003c). In that study, *P. chlamydosporia*, var. *catenulate* was applied before the bean crop, where it suppressed final densities of J2 in the soil, persisted through the cabbage crop, and prevented population increase on the tomato crop (Fig. 11.1). Egg parasitism on tomato was >70%.

Soil solarization can not only reduce pest populations, but also alters the microbial community, both qualitatively and quantitatively, which may lead to less competition and antagonism of an introduced biological control organism (Katan and DeVay 1991). Kluepfel et al. (2002) used such a strategy to enhance survival and efficacy of *Pseudomonas synxantha* strain BG33R, a bacterial antagonist of *M. xenoplax*. Populations of the nematode were lower in plots that received both solarization and BG33R than in solarization alone. *Pasteuria penetrans* and solarization were also additive in suppression of root-galling and egg production by *Meloidogyne* spp. on cucumber. However, *Bacillus firmus* did not provide any additional control of *M. incognita* on cucumber when combined with solarization even though the bacterium suppressed root galling without solarization (Giannakou et al. 2007). *Paecilomyces lilacinus* reduced final populations of *Meloidogyne* spp. on cucumber, but was not effective in suppressing nematode populations when combined with solarization (Anastasiadis et al. 2008).

Combining different antagonists of nematodes may improve the level and consistency of biological control. Selected combinations may vary in their mode of action, the stage of nematode affected, activity under different soil conditions, and ability to control different pests. A review of studies combining biological control organisms prior to 2002 has been published (Meyer and Roberts 2002). In a field study, application of three bacteria, *P. fluorescens*, *P. aeruginosa*, and *Bradyrhizobium japonicum*, with different modes of action, suppressed galling of tomato by *M. javanica* both individually and in combination; however, the combinations did not provide greater suppression than the most effective bacterium in the group (Siddiqui and Shaukat 2002). Khan et al. (2006) evaluated *P. lilacinus* and *Monacrosporium lysipagum*, fungal parasites of sedentary and migratory stages, respectively, for control of three different nematodes. Combination of the two fungi did not increase the level of nematode suppression of *M. javanica* on tomato or *H. avenae* on barley compared to individual applications; however, it appeared that the two fungi had an additive effect on suppression of *R. reniformis* on banana. Combined applications of *B. japonicum*, *Trichoderma pseudokoningii*, and *Glomus mossae* to soybean did not improve suppression of *M. incognita* over single species applications (Oyekanmi et al. 2007). Of the four fungi tested by Goswami et al. (2008), only the combination of *T. harzianum* and *Acromonium strictum* had an additive effect on suppression of root galling by *M. incognita*.

Organic amendments have been used to suppress plant-parasitic nematodes (Akhtar and Malik 2000). Though the mechanism of suppression is not always clear, it can involve release of toxic compounds and stimulation of antagonistic organisms. A number of studies have examined dual applications of organic amendments and biological control organisms for integrated management of plant-parasitic nematodes. Amendments specifically used to enhance the survival and proliferation of biological control organisms will be covered in the next section. Application of neem cake and *T. harzianum* had an additive effect on suppression of *Tylenchulus semipenetrans* on citrus in pots (Parvatha Reddy et al. 1996). In a field study, however, the combined application of neem and the fungus was not different from application of neem alone in suppression of *M. incognita* galling on eggplant (Rao et al. 1998). Suppression of *M. hapla* in soil amended with chitin was not increased by the application of single or multiple species of antagonistic fungi and bacteria (Chen et al. 1999). Likewise, the efficacy of *B. thuringiensis*, *Paecilomyces marquandii*, and *Streptomyces costaricanus* was not increased by any of the organic amendments, including chitin, though each organism alone reduced galling and reproduction of *M. hapla* on lettuce (Chen et al. 2000).

11.4 Conservation and Enhancement of Indigenous and Introduced Antagonists

Where integrated management seeks to supplement biological control with other nematode control tactics, the goal of conservation and enhancement is to avoid practices that are harmful to antagonists (conservation) and promote practices that increase the survival, abundance, and activity of antagonists (enhancement). A large proportion of the research effort in biological control of nematodes has been directed toward application of antagonists for nematode control during a single cropping cycle and little effort has been given to determining which agricultural practices have positive or negative effects on indigenous and introduced antagonists.

11.4.1 Pesticides

In biological control of insect pests, conservation has emphasized reduced application of broad spectrum insecticides which negatively impact parasitoids and predators (Ruberson et al. 1998). However, there is little information, particularly from field studies, on the impact of pesticides on antagonists of nematodes (Stirling 1991). Fungicide applications could potentially reduce the activity of nematophagous fungi. Recently, Tobin et al. (2008a) demonstrated that the fungicide azoxystrobin had a negative impact on densities of *P. chlamydosporia* in the soil and rhizosphere, but the fungus showed some recovery 49 days after application. The impact of the fungicide on nematode suppression by *P. chlamydosporia* was not tested. Application of captafol resulted in greater nematode reproduction compared to untreated soil, presumably because the fungicide reduced fungal antagonists of the nematode (Muller 1985). Egg parasitism was not affected by captafol; however, in the untreated soil, a significant proportion of juveniles were parasitized by *H. rhossiliensis*. When egg and juvenile parasitism (primarily by *H. rhossiliensis*) was evaluated 8 months after fumigation with 1,3-dichloropropene (1,3-D), there was no difference in parasitism of either stage between fumigated and unfumigated treatments.

Pasteuria penetrans is tolerant of many pesticides including 1,3-D (Chen and Dickson 1998; Mankau and Prasad 1972; Stirling 1984). The bacterium, however, is sensitive to chloropicrin. Kariuki and Dickson (2007) found that the percentage of females infected by *P. pentrans* in plots treated with chloropicrin was less than half the percentage in untreated plots. Moreover, root galling from *M. arenaria* on peanut was greater in the chloropicrin than in the untreated plots (1.1 vs 4.2 on a 0–10 scale).

Nematicides can reduce abundance of omnivorous and predatory nematodes. Population densities of these nematodes were severely suppressed following application of 1,3-D (Fig. 11.2). Populations of omnivorous nematodes partially recovered by mid season, but populations of predatory nematodes remained low and even showed residual effects from application of the fumigant in the previous spring (Timper, Jagdale, Davis, unpublished). It is not known whether the omnivores and predators were regulating populations of plant-parasitic nematodes and if they were, whether suppression was disrupted by 1,3-D.

11.4.2 Organic Amendments

The application of organic amendments is the most commonly used tactic for enhancing the abundance and activity of antagonists of nematodes. This topic has been reviewed by Akhtar and Malik (2000). The organic matter can be used as a substrate for growth of antagonists or it can stimulate populations of microbivorous nematodes which can serve as hosts or prey for antagonists (van den Boogert et al. 1994; Jaffee 2006). Nevertheless, application of a manure/sawdust mixture did not enhance the activity of *Arthrobotrys dactyloides* or *P. chlamydosporia* (Stirling

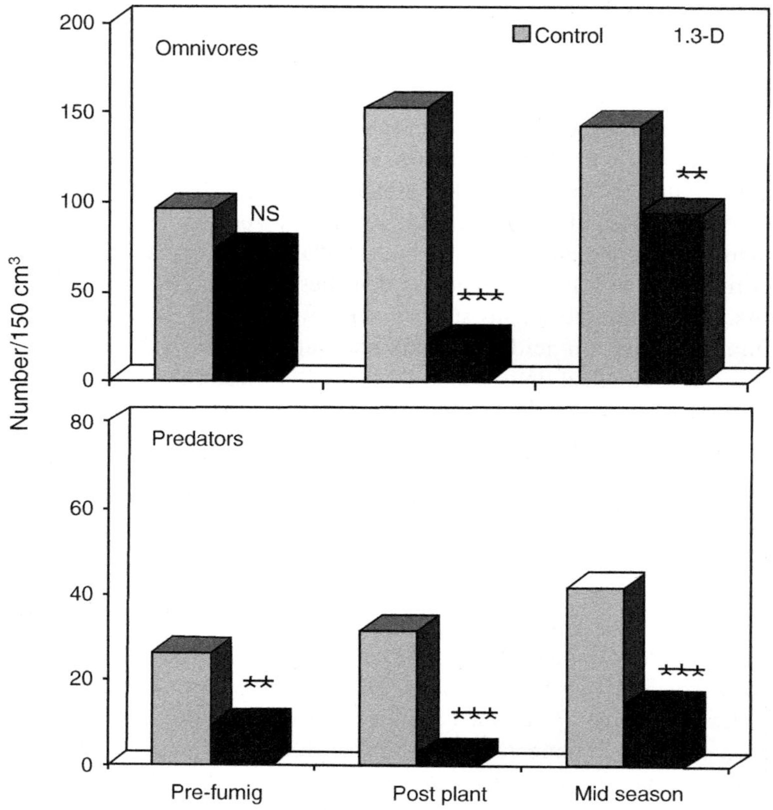

Fig. 11.2 Population densities of omnivorous and predatory nematodes in cotton plots without nematicide and treated with 1,3-dichloropropene (1,3-D). The fumigant was applied 2 weeks before planting in the spring. Nematodes were sampled immediately before fumigation, after planting, and midway through the season. The fumigant had also been applied the previous spring. Differences between the control and 1,3-D are indicated by ** ($P<0.01$) and *** ($P<0.001$) (Timper, Jagdale, Davis, unpublished)

and Smith 1998). Likewise, various plant and manure amendments did not increase parasitism of *M. xenoplax* by *H. rhossiliensis* (Jaffee et al. 1994). Incorporation of *Crotolaria juncea* into soil increased the abundance of nematode-trapping fungi, particularly in soil with high organic matter (Wang et al. 2002, 2003, 2004). Suppression of *R. reniformis* by *C. juncea* amendments in six soils was correlated with nematode-trapping fungi and egg parasitism by fungi (Wang et al. 2003). In another study, incorporation of mustard, oil radish, and rape increased parasitism of *H. schachtii* eggs in one field site, but decreased it in another field site (Pyrowolakis et al. 1999). The enhancing effect of the three crucifers in the one field site may be due to the greater fungal diversity in that site compared to the other site. In vineyard soil, addition of dried grape or alfalfa leaves to soil increased microbivorous nematodes, but did not have a consistent effect on trapping activity of nematophagous fungi (Jaffee 2002, 2004). Although abundance of *Arthrobotrys oligospora* increased with addition of leaves, trapping activity did not increase. The response of *Dactylellina candidum* (=*D. haptotyla*) to the organic matter was more erratic. Abundance of the fungus was correlated with trapping activity; however, the leaf material did not always stimulate abundance or activity. Amending soil with sugarcane trash reduced population densities of *Pratylenchus zeae* and *Tylenchorhynchus annulatus* and increased densities of omnivorous and predatory nematodes three to eightfold (Stirling et al. 2005). An unidentified trapping fungus was also found only in soil amended with sugarcane trash suggesting a possible involvement in suppression of the plant parasites.

11.4.3 Crop Rotation

Population densities of biological control organisms can be influenced by the species of crop planted. The most straight-forward example of this is the continuous cultivation of a crop leading to an increase in specific antagonists of a plant-parasitic nematode on that crop. Rotating non-host crops for *Meloidogyne* spp. resulted in lower densities of *P. penetrans* relative to continuous cropping of a host crop (Madulu et al. 1994; Timper et al. 2001). Other plant-antagonist interactions are more unexpected. In a nematode suppressive soil, later shown to be caused by *D. oviparasitica*, suppression of *H. schachtii* was decreased following both wheat and fallow, but not following nematode resistant sugar beet or radish indicating these plants could support the fungus in the absence of nematodes (Westphal and Becker 2001b). Rumbos and Kiewnick (2006) determined the effect of different plant species on the persistence of *P. lilacinus* and found that only bean significantly reduced persistence compared to fallow soil. Perhaps the bean rhizosphere contained organisms antagonistic to the fungus.

11.4.4 Tillage

Tillage changes the physical, chemical, and biological components of soil (Kladivko 2001). However, few studies have examined the effect of tillage on antagonists of nematodes. Bernard et al. (1996) sampled six different tillage treatments for fungi associated with *Heterodera glycines* and for rates of egg and female parasitism. In the monthly samples, no tillage treatment consistently supported more egg parasitism. When the monthly samples were combined, disc-tilled plots had greater egg parasitism than no-till plots. *Paecilomyces lilacinus*, the most prevalent fungus, parasitized more eggs in disc than in no till plots, whereas *P. chlamydosporia* parasitized more eggs in plots that where moldboard plowed. Tillage may have a negative impact on *P. penetrans*, particularly early in the season when root growth is shallow. At planting, tillage reduced the density of endospores in the upper 10 cm of soil, but tended to increase endospore densities below 10 cm (Talavera et al. 2002). At harvest, the effects of tillage on endospore densities in the upper 10 cm disappeared.

11.4.5 Organic Production Systems

A few studies have evaluated the impact of substantial changes in production practices, such as organic farming, on the level of biological control or on densities of antagonists. Organic farming replaces synthetic fertilizers and pesticides with organic fertilizers (plant material and animal manure), crop rotation, and resistant cultivars. Persmark (1997) sampled 11 pairs of organically and conventionally managed farms and found no difference between the two management systems in either the densities of nematode-trapping fungi, numbers of nematodes in the rhizosphere of pea, or organic matter. In a field plot experiment, organically managed plots had more species of nematophagous fungi and two species, *Arthrobotrys dactyloides* and *Nematoctonus leiosporus*, were more abundant than in conventionally managed plots (Jaffee et al. 1998). However, soils from organic and conventionally managed plots did not differ in level of suppression of *M. javanica*. In another similar study, the number of species of nematophagous fungi was not different in organically and conventionally managed plots (Timm et al. 2001). The only two fungi, *N. leiosporus* and *Meristacrum* sp. that were found more frequently in the organic plots were present at very low densities.

11.5 Using Molecular Techniques to Improve Biological Control

11.5.1 Detection and Quantification of Antagonists and Their Biological Control Activity

From the preceding sections, it is apparent that the abundance and biological control activity of antagonists can be influenced by other soil organisms, plant species, and agricultural practices such as pesticide application, organic amendments, tillage, and crop rotation. Rapid, sensitive and reliable methods for quantifying population densities of antagonists are needed to advance our knowledge of the environmental factors affecting biological control of nematodes. Ultimately, such knowledge will improve the efficacy and consistency of nematode suppression. Non-molecular techniques for detecting and quantifying antagonists of nematodes include extraction of spores, selective media, most probable number procedures, bioassays, and enzyme-linked immunosorbent assays (Stirling 1991; Fould et al. 2001;

Schmidt et al. 2003). All of these techniques have one or more limitations; they can be time and labor intensive, or lack suitable specificity or sensitivity. Competitive and real-time PCR techniques have the potential for rapid, sensitive, culture independent and highly specific quantification of antagonists (Okubara et al. 2005). Sufficiently pure DNA can be extracted from soil and plant tissue using relatively simple and rapid methods utilizing commercial extraction kits. Recently, sequence-specific primers have been developed for either the ITS region or for specific genes from *P. chlamydosporia*, *P. lilacinus*, *Plectosphaerella cucumerina*, *D. oviparasitica*, *H. rhossiliensis*, nematode-trapping fungi (Orbiliales), and *Pasteuria penetrans* (Hirsch et al. 2001; Atkins et al. 2003c, 2005; Yin et al. 2003; Schmidt et al. 2004; Zhang et al. 2006; Smith and Jaffee 2009). These primers showed a high degree of specificity for the organisms for which they were developed. When quantitative PCR techniques were compared with direct plating onto selective media for quantification of *P. chlamydosporia*, *P. lilacinus*, *P. cucumerina*, and nematode-trapping fungi (Orbiliales) the PCR techniques were found to be more sensitive (Mauchline et al. 2002; Atkins et al. 2003a, 2005; Smith and Jaffee 2009). However, all four of these studies emphasized the importance of using quantitative PCR techniques in conjunction with plating onto selective media or bioassays.

Perhaps the greatest advantage of DNA-based detection methods is the potential for differentiating biotypes and strains of a biological control organism. With *P. penetrans*, Schmidt et al. (2004) found greater sequence heterogeneity in the sporulation gene *sigE* than in 16S rDNA and suggested that species- and biotype-specific probes could be developed from this gene. Specific primer sets have been developed to distinguish between two morphologically similar varieties of *P. chlamydosporia*, var. *catenulate* and var. *chlamydosporia* (Atkins et al. 2003b; Hirsch et al. 2000). Siddiqui et al. (2009) developed PCR primers based on the *vcp1* gene to differentiate biotypes of *P. chlamydosporia* from cyst and root-knot nematodes. These primers were able to identify the original nematode host from which the fungus was isolated. PCR fingerprinting with arbitrary primers has also been used to determine variation within populations of *P. chlamydosporia* var. *chlamydo-sporia* (Manzanilla-López et al. 2009a). Unexpectedly, little genetic variation was detected in populations of the fungus at two different locations where it was parasitizing eggs of *M. incognita*. Biotype-specific probes and PCR fingerprinting could be used to determine which biotypes prevail against different host nematodes and under different environmental conditions (Atkins et al. 2009; Manzanilla-López et al. 2009b). Recently, SCAR-PCR primers were developed to detect specific strains of *P. lilacinus* and *P. chlamydosporia* (Zhu et al. 2006). Further research is needed to determine whether these markers can discriminate these strains from background populations of the same species in the field.

Quantitative PCR (qPCR) techniques are not without limitations. DNA from moribund or recently dead propagules can be amplified leading to an overestimation of viable propagules. Extraction of RNA from soil could be combined with DNA extraction to provide a more accurate assessment of viable cells, but further research in this area is necessary (Atkins et al. 2003a). Interpreting the results of qPCR for filamentous fungi is also complicated by the presence of multiple stages such as conidia, hyphae, and chlamydospores. When plating or direct counting techniques are used, chlamydospores and mycelial fragments are counted as single propagules; however, with qPCR each of the cells making up the structure contribute DNA. In other words, qPCR quantifies fungal biomass whereas dilution plating quantifies propagules (or cfu). Germination of chlamydospores and subsequent sporulation may not increase the amount of DNA detected, but would increase the number of propagules (Mauchline et al. 2002). Such shifts in fungal life stages could only be deduced with a combination of plating and qPCR. The level of biological control activity cannot be determined with either qPCR or plating (except of infected cadavers). Based on qPCR and plating on selective medium, populations of *P. lilacinus* were found to be greater in the Spalding location than in the Ely location; however, parasitism of *G. pallida* eggs in a bioassay was similar in both locations (Atkins et al. 2005). In contrast, there was a strong correlation between results of qPCR and assay nematodes parasitized by *H. rhossiliensis* (Zhang et al. 2006). In this comparison, the soil for both the parasitism assay and the qPCR was inoculated with the fungus in the laboratory and left undisturbed until the assay nematodes were extracted. More realistically, soil would be collected from a field site, a process which inactivates the conidia of *H. rhossiliensis* (McInnis and Jaffee 1989). Following such a disturbance, fungal reserves must be used to produce fewer new conidia which would then be quantified in the parasitism assay. However, qPCR would be able to detect both the hyphae and detached conidia from freshly collected soil.

Abundance of an organism is not always an indication of biological control activity. This is predominantly an issue with organisms that are competitive soil saprophytes because they may not depend on nematodes for nutrition. For some of these organisms, such as parasites of sedentary stages and certain trapping fungi, biological control activity can be monitored by recovering infected stages of nematodes (Atkins et al. 2009). However, quantifying biological control activity of bacteria that produce toxins and some trapping fungi is difficult (Jaffee 2004), particularly in field studies. Reporter genes could be used to monitor gene expression involved in antibiotic production, trap formation, and parasitism. The reporter gene *lacZ,* encoding for B-galactosidase, has been used to study expression of the antibiotic DAPG by *Pseudomonas fluorescens* in the rhizosphere of plants (Notz et al. 2001, 2002). Using a strain of the bacterium carrying a translational *phlA* '-' *lacZ* fusion, expression of DAPG was found to be greater in monocots than dicots, influenced by plant cultivar and age, stimulated in the presence of *Pythium ultimum*, and depressed in the presence of fusaric acid-producing strains of *Fusarium oxysporum*. Recently, the gene encoding for green fluorescent protein was used along with flow cytometry to visualize and quantify expression of DAPG in situ on plant roots (de Werra et al. 2008). With improved knowledge of the genes involved in trap formation and nematode infection, reporter genes may be used to monitor biological control activity of nematophagous fungi. For example, a reporter gene could be used with the *PII* gene in *A. oligospora*, which encodes for an extracellular serine protease and is involved in nematode trapping (Ahman et al. 2002), to determine the conditions under which the fungus becomes parasitic. Ahren et al.(2005)

used microarray analysis to determine which genes were up-regulated in the adhesive knobs of *Monacrosporium haptotylum* (syn. *Dactylellina candidum*). A reporter gene could be fused with one of the genes specifically expressed in the adhesive knobs to quantify trap formation under different production practices (e.g., organic vs conventional production).

11.5.2 Trait Enhancement

Improvements in biological control have been achieved by genetically engineering organisms for overexpression of traits involved in pathogenicity or nematicidal activity. Transgenic lines of *Trichoderma atroviride* carrying multiple copies of the *prb1* gene, which encodes for a 31-kDa proteinase (Prb 1), were tested for suppression of *M. javanica* on tomato (Sharon et al. 2001). Of the four transformed strains, only P-2 was more effective than the wild-type strain in reducing root galling. The P-2 strain was similar to the wild-type strain in nematicidal activity, but showed improved ability to penetrate egg masses and colonize eggs in vitro. *Arthrobotrys oligospora* produces an extracellular serine protease designated PII which is thought to be involved in penetration of the nematode cuticle or tissue digestion within the host (Ahman et al. 2002). Transformed strains of the fungus containing additional copies of the *pII* gene produced more traps and had an increase rate of capture compared to the wild-type strain. Siddiqui and Shaukat (2003c) demonstrated that a DAPG over-producing strain of *P. fluorescens* was more effective in reducing root galling from *M. javanica* in tomato than the wild-type strain CHA0. In addition to improving the effectiveness of biological control, these enhanced strains of antagonists may be able to suppress nematode populations when applied at much lower rates, and cost, than wild-type strains.

References

Ahman J, Johansson T, Olsson M et al (2002) Improving the pathogenicity of a nematode-trapping fungus by genetic engineering of a subtilisin with nematotoxic activity. Appl Environ Microbiol 68:3408–3415

Ahren D, Tholander M, Fekete C et al (2005) Comparison of gene expression in trap cells and vegetative hyphae of the nematophagous fungus *Monacrosporium haptotylum*. Microbiology 151:789–803

Akhtar M, Malik A (2000) Roles of organic soil amendments and soil organisms in the biological control of plant-parasitic nematodes: a review. Bioresour Technol 74:35–47

Anastasiadis IA, Giannakou IO, Prophetou-Athanasiadou DA et al (2008) The combined effect of the application of a biocontrol agent *Paecilomyces lilacinus*, with various practices for the control of root-knot nematodes. Crop Prot 27:352–361

Atkins SD, Clark IM, Sosnowska D et al (2003a) Detection and quantification of *Plectosphaerella cucumerina*, a potential biological control agent of potato cyst nematodes, by using conventional PCR, real-time PCR, selective media, and baiting. Appl Environ Microbiol 69:4788–4793

Atkins SD, Hidalgo-Diaz L, Clark IM et al (2003b) Approaches for monitoring the release of *Pochonia chlamydosporia* var. *catenulata*, a biocontrol agent of root-knot nematodes. Mycol Res 107:206–212

Atkins SD, Hidalgo-Diaz L, Kalisz H et al (2003c) Development of a new management strategy for the control of root-knot nematodes (*Meloidogyne* spp) in organic vegetable production. Pest Manag Sci 59:183–189

Atkins SD, Clark IM, Pande S et al (2005) The use of real-time PCR and species-specific primers for the identification and monitoring of *Paecilomyces lilacinus*. FEMS Microbiol Ecol 51:257–264

Atkins SD, Peteira B, Clark IM (2009) Use of real-time quantitative PCR to investigate root and gall colonisation by co-inoculated isolates of the nematophagous fungus *Pochonia chlamydosporia*. Ann Appl Biol 155:143–152

Barbosa P (1998) Conservation biological control. Academic, San Diego

Bent E, Loffredo A, McKenry MV et al (2008) Detection and investigation of soil biological activity against *Meloidogyne incognita*. J Nematol 40:109–118

Bernard EC, Self LH, Tyler DD (1996) Fungal parasitism of soybean cyst nematode, *Heterodera glycines* (Nemata: Heteroderidae), in differing cropping-tillage regimes. Appl Soil Ecol 5:57–70

Bilgrami AL, Brey C, Gaugler R (2008) First field release of a predatory nematode, *Mononchoides gaugleri* (Nematoda: Diplogastrida), to control plant-parasitic nematodes. Nematology 10:143–146

Bird AF, Bird J (1986) Observations on the use of insect parasitic nematodes as a means of biological control of root-knot nematodes. Int J Parasitol 16:511–516

Borneman J, Becker JO (2007) Identifying microorganisms involved in specific pathogen suppression in soil. Annu Rev Phytopathol 45:153–172

Bourne JM, Kerry BR (1999) Effect of the host plant on the efficacy of *Verticillium chlamydosporium* as a biological control agent of root-knot nematodes at different nematode densities and fungal application rates. Soil Biol Biochem 31:75–84

Bourne JM, Kerry BR, De Leij F (1996) The importance of the host plant on the interaction between root-knot nematodes (*Meloidogyne* spp.) and the nematophagous fungus, *Verticillium chlamydosporium* Goddard. Biocontrol Sci Technol 6:539–548

Brown SM, Nordmeyer D (1985) Synergistic reduction in root galling by *Meloidogyne javanica* with *Pasteuria penetrans* and nematicides. Rev Nematol 8:285–286

Carneiro R, de Mesquita LFG, Cirotto PAS et al (2007) The effect of sandy soil, bacterium dose and time on the efficacy of *Pasteuria penetrans* to control *Meloidogyne incognita* race 1 on coffee. Nematology 9:845–851

Chen ZX, Dickson DM (1998) Review of *Pasteuria penetrans*: biology, ecology, and biological control potential. J Nematol 30:313–340

Chen SY, Dickson DW, Mitchell DJ (1996a) Population development of *Heterodera glycines* in response to mycoflora in soil from Florida. Biol Control 6:226–231

Chen ZX, Dickson DW, McSorley R et al (1996b) Suppression of *Meloidogyne arenaria* race 1 by soil application of endospores of *Pasteuria penetrans*. J Nematol 28:159–168

Chen ZX, Dickson DW, Mitchell DJ et al (1997) Suppression mechanisms of *Meloidogyne arenaria* race 1 by *Pasteuria penetrans*. J Nematol 29:1–8

Chen J, Abawi GS, Zuckerman BM (1999) Suppression of *Meloidogyne hapla* and its damage to lettuce grown in a mineral soil amended with chitin and biocontrol organisms. J Nematol 31:719–725

Chen J, Abawi GS, Zuckerman BM (2000) Efficacy of *Bacillus thuringiensis*, *Paecilomyces marquandii*, and *Streptomyces costaricanus* with and without organic amendments against *Meloidogyne hapla* infecting lettuce. J Nematol 32:70–77

Cook RJ, Baker KF (1983) The nature and practice of biological control of plant pathogens. The American Phytopathological Society, St. Paul

Cronin D, MoenneLoccoz Y, Fenton A et al (1997) Role of 2, 4-diacetylphloroglucinol in the interactions of the biocontrol pseudomonad strain F113 with the potato cyst nematode *Globodera rostochiensis*. Appl Environ Microbiol 63:1357–1361

Dabire KR, Mateille T (2004) Soil texture and irrigation influence the transport and the development of *Pasteuria penetrans*, a bacterial parasite of root-knot nematodes. Soil Biol Biochem 36:539–543

De Leij FAAM, Kerry BR (1991) The nematophagous fungus *Verticillium chlamydosporium* as a potential biological control agent for *Meloidogyne arenaria*. Rev Nematol 14:157–164

De Leij FAAM, Kerry BR, Dennehy JA (1993) *Verticillium chlamydosporium* as a biological control agent for *Meloidogyne incognita* and *Meloidogyne hapla* in pot and micro-plot tests. Nematologica 39:115–126

de Werra P, Baehler E, Huser A et al (2008) Detection of plant-modulated alterations in antifungal gene expression in *Pseudomonas fluorescens* CHA0 on roots by flow cytometry. Appl Environ Microbiol 74:1339–1349

Dong LQ, Zhang KQ (2006) Microbial control of plant-parasitic nematodes: a five-party interaction. Plant Soil 288:31–45

Duponnois R, Ba AM (1998) Influence of the microbial community of a Sahel soil on the interactions between *Meloidogyne javanica* and *Pasteuria penetrans*. Nematologica 44:331–343

Duponnois R, Amadou MB, Mateille T (1999) Beneficial effects of *Enterobacter cloacae* and *Pseudomonas mendocina* for biological control of *Meloidogyne incognita* with the endospore-forming bacterium *Pasteuria penetrans*. Nematology 1:95–101

Fould S, Dieng AL, Davies KG et al (2001) Immunological quantification of the nematode parasitic bacterium *Pasteuria penetrans* in soil. FEMS Microbiol Ecol 37:187–195

Gaspard JT, Jaffee BA, Ferris H (1990) *Meloidogyne incognita* survival in soil infested with *Paecilomyces lilacinus* and *Verticillium chlamydosporium*. J Nematol 22:176–181

Giannakou IO, Dimitrios G, Karpouzas DG et al (2004) A novel non-chemical nematicide for the control of root-knot nematodes. Appl Soil Ecol 26:69–79

Giannakou IO, Anastasiadis IA, Gowen SR et al (2007) Effects of a non-chemical nematicide combined with soil solarization for the control of root-knot nematodes. Crop Prot 26:1644–1654

Goswami J, Pandey RK, Tewari JP et al (2008) Management of root knot nematode on tomato through application of fungal antagonists, *Acremonium strictum* and *Trichoderma harzianum*. J Environ Sci Heal B 43:237–240

Grewal PS, Martin WR, Miller RW et al (1997) Suppression of plant-parasitic nematode populations in turfgrass by application of entomopathogenic nematodes. Biocontrol Sci Technol 7:393–399

Hackenberg C, Muehlchen A, Forge T et al (2000) *Pseudomonas chlororaphis* strain Sm3, bacterial antagonist of *Pratylenchus penetrans*. J Nematol 32:183–189

Hirsch PR, Mauchline TH, Mendum TA et al (2000) Detection of the nematophagous fungus *Verticillium chlamydosporium* in nematode-infested plant roots using PCR. Mycol Res 104:435–439

Hirsch PR, Atkins SD, Mauchline TH et al (2001) Methods for studying the nematophagous fungus *Verticillium chlamydosporium* in the root environment. Plant Soil 232:21–30

Ishibashi N, Kondo E (1986) *Steinernema feltiae* (DD-136) and *S. glaseri*: persistence in soil and bark compost and their influence on native nematodes. J Nematol 18:310–316

Jacobs H, Gray SN, Crump DH (2003) Interactions between nematophagous fungi and consequences for their potential as biological agents for the control of potato cyst nematodes. Mycol Res 107:47–56

Jaffee BA (1999) Enchytraeids and nematophagous fungi in tomato fields and vineyards. Phytopathology 89:398–406

Jaffee BA (2002) Soil cages for studying how organic amendments affect nematode-trapping fungi. Appl Soil Ecol 21:1–9

Jaffee BA (2004) Do organic amendments enhance the nematode-trapping fungi *Dactylellina haptotyla* and *Arthrobotrys oligospora*? J Nematol 36:267–275

Jaffee BA (2006) Interactions among a soil organic amendment, nematodes, and the nematode-trapping fungus *Dactylellina candidum*. Phytopathology 96:1388–1396

Jaffee BA, Muldoon AE (1997) Suppression of the root-knot nematode *Meloidogyne javanica* by alginate pellets containing the nematophagous fungi *Hirsutella rhossiliensis, Monacrosporium cionopagum* and *M. ellipsosporum*. Biocontrol Sci Technol 7:203–217

Jaffee BA, Ferris H, Stapleton JJ et al (1994) Parasitism of nematodes by the fungus *Hirsutella rhossiliensis* as affected by certain organic amendments. J Nematol 26:152–161

Jaffee BA, Muldoon AE, Westerdahl BB (1996) Failure of a mycelial formulation of the nematophagous fungus *Hirsutella rhossiliensis* to suppress the nematode *Heterodera schachtii*. Biol Control 6:340–346

Jaffee BA, Muldoon AE, Didden WAM (1997) Enchytraeids and nematophagous fungi in soil microcosms. Biol Fertil Soils 25:382–388

Jaffee BA, Ferris F, Scow KM (1998) Nematode-trapping fungi in organic and conventional cropping systems. Phytopathology 88:344–350

Jagdale GB, Grewal PS (2002) Identification of alternatives for the management of foliar nematodes in floriculture. Pest Manag Sci 58:451–458

Jagdale GB, Somasekhar N, Grewal PS et al (2002) Suppression of plant-parasitic nematodes by application of live and dead infective juveniles of an entomopathogenic nematode, *Steinernema carpocapsae*, on boxwood (*Buxus* spp.). Biol Control 24:42–49

Jagdale GB, Kamoun S, Grewal PS (2009) Entomopathogenic nematodes induce components of systemic resistance in plants: biochemical and molecular evidence. Biol Control 51:102–109

Kariuki GM, Dickson DW (2007) Transfer and development of *Pasteuria penetrans*. J Nematol 39:55–61

Kaşkavalci G, Tüzel Y, Dura O et al (2009) Effects of alternative control methods against *Meloidogyne incognita* in organic tomato production. Ekoloji 18:23–31

Katan J, DeVay JE (1991) Soil solarization. CRC Press, Boca Raton

Kerry BR (1995) Ecological considerations for the use of the nematophagous fungus *Verticillium chlamydosporium*, to control plant parasitic nematodes. Can J Bot 73:65–70

Kerry BR (2001) Exploitation of the nematophagous fungus *Verticillium chlamydosporium* Goddard for the biological control of root-knot nematodes (*Meloidogyne* spp.). In: Butt TM, Jackson C, Magan N (eds) Fungi as biocontrol agents. CAB International, Wallingford

Kerry BR, Crump DH (1998) The dynamics of the decline of the cereal cyst nematode, *Heterodera avenae*, in four soils under intensive cereal production. Fund Appl Nematol 21:617–625

Khan Z, Kim YH (2007) A review on the role of predatory soil nematodes in the biological control of plant parasitic nematodes. Appl Soil Ecol 35:370–379

Khan TA, Saxena SK (1997) Effect of root-dip treatment with culture filtrates of on root penetration, development and reproduction of *Meloidogyne javanica* on tomato. Int J Nematol 7:85–88

Khan A, Williams KL, Nevalainen HKM (2006) Control of plant-parasitic nematodes by *Paecilomyces lilacinus* and *Monacrosporium lysipagum* in pot trials. Biocontrol 51:643–658

Kiewnick S, Sikora RA (2006) Evaluation of *Paecilomyces lilacinus* strain 251 for the biological control of the northern root-knot nematode *Meloidogyne hapla* Chitwood. Nematology 8:69–78

Kladivko EJ (2001) Tillage systems and soil ecology. Soil Tillage Res 61:61–76

Kloepper JW, Rodriguez KR, McInroy JA et al (1992) Rhizosphere bacteria antagonistic to soybean cyst *Heterodera glycines* and root-knot *Meloidogyne incognita* nematodes identification by fatty acid analysis and frequency of biological control activity. Plant Soil 139:75–84

Kluepfel DA, McInnis TM, Zehr EI (1993) Involvement of root-colonizing bacteria in peach orchard soils suppressive of the nematode *Criconemella xenoplax*. Phytopathology 83:1240–1245

Kluepfel DA, Nyczepir AP, Lawrence JE et al (2002) Biological control of the phytoparasitic nematode *Mesocriconema xenoplax* on peach trees. J Nematol 34:120–123

Kok CJ, Papert A, Hok-A-Hin CH (2001) Microflora of *Meloidogyne* egg masses: species composition, population density and effect on the biocontrol agent *Verticillium chlamydosporium* (Goddard). Nematology 3:729–734

Kokalis-Burelle N, Dickson DW (2003) Effects of soil fumigants and BioYield™ on root-knot nematode incidence and yield of tomato. In: Proceedings of international research conference on methyl bromide: alternatives and emissions reductions, San Diego, pp 50.51–50.53

Kokalis-Burelle N, Vavrina CS, Rosskopf EN et al (2002) Field evaluation of plant growth-promoting Rhizobacteria amended transplant mixes and soil solarization for tomato and pepper production in Florida. Plant Soil 238:257–266

Kumar T, Wahla V, Pandey P et al (2009) Rhizosphere competent *Pseudomonas aeruginosa* in the management of *Heterodera cajani* on sesame. World J Microbiol Biotechnol 25:277–285

Lara Martez J, Acosta N, Betancourt C et al (1996) Biological control of *Meloidogyne incognita* in tomato in Puerto Rico. Nematropica 26:143–152

Li L, Mo MH, Qu Q et al (2007) Compounds inhibitory to nematophagous fungi produced by *Bacillus* sp strain H6 isolated from fungistatic soil. Eur J Plant Pathol 117:329–340

Madulu JD, Trudgill DL, Phillips MS (1994) Rotational management of *Meloidogyne javanica* and effects on *Pasteuria penetrans* and tomato and tobacco yields. Nematologica 40:438–455

Maehara N (2008) Reduction of *Bursaphelenchus xylophilus* (Nematoda: Parasitaphelenchidae) population by inoculating *Trichoderma* spp. into pine wilt-killed trees. Biol Control 44:61–66

Mankau R (1962) Soil fungistasis and nematophagous fungi. Phytopathology 52:611–615

Mankau R, Prasad N (1972) Possibilities and problems in the use of a sporozoan endoparasite for biological control of plant-parasitic nematodes. Nematropica 2:7–8

Manzanilla-López RH, Atkins SD, Clark IM et al (2009a) Measuring abundance, diversity and parasitic ability in two populations of the nematophagous fungus *Pochonia chlamydosporia* var. *chlamydosporia*. Biocontrol Sci Technol 19:391–406

Manzanilla-López RH, Clark IM, Atkins SD et al (2009b) Rapid and reliable DNA extraction and PCR fingerprinting methods to discriminate multiple biotypes of the nematophagous fungus *Pochonia chlamydosporia* isolated from plant rhizospheres. Lett Appl Microbiol 48:71–76

Mauchline TH, Kerry BR, Hirsch PR (2002) Quantification in soil and the rhizosphere of the nematophagous fungus *Verticillium chlamydospotium* by competitive PCR and comparison with selective plating. Appl Environ Microbiol 68:1846–1853

Mauchline TH, Kerry BR, Hirsch PR (2004) The biocontrol fungus *Pochonia chlamydosporia* shows nematode host preference at the infraspecific level. Mycol Res 108:161–169

Mazzola M (2007) Manipulation of rhizosphere bacterial communities to induce suppressive soils. J Nematol 39:213–220

McInnis TM, Jaffee BA (1989) An assay for *Hirsutella rhossiliensis* spores and the importance of phialides for nematode inoculation. J Nematol 21:229–234

McSorley R, Wang KH, Kokalis-Burelle N et al (2006) Effects of soil type and steam on nematode biological control potential of the rhizosphere community. Nematropica 36:197–214

Meyer SLF (2003) United States Department of Agriculture – Agricultural Research Service research programs on microbes for management of plant-parasitic nematodes. Pest Manag Sci 59:665–670

Meyer SLF, Roberts DP (2002) Combinations of biocontrol agents for management of plant-parasitic nematodes and soilborne plant-pathogenic fungi. J Nematol 34:1–8

Meyer SLF, Roberts DP, Chitwood DJ et al (2001) Application of *Burkholderia cepacia* and *Trichoderma virens*, alone and in combinations, against *Meloidogyne incognita* on bell pepper. Nematropica 31:75–86

Minton NA, Sayre RM (1989) Suppressive influence of *Pasteuria penetrans* in Georgia USA soils on reproduction of *Meloidogyne arenaria*. J Nematol 21:574–575

Monfort E, Lopez-Llorca LV, Jansson HB et al (2006) In vitro soil receptivity assays to egg-parasitic nematophagous fungi. Mycol Prog 5:18–23

Morton CO, Hirsch PR, Peberdy JP et al (2003) Cloning of and genetic variation in protease VCP1 from the nematophagous fungus *Pochonia chlamydosporia*. Mycol Res 107:38–46

Muller J (1985) The influence of two pesticides on fungal parasites of *Heterodera schachtii*. Colloques INRA 31:225–231

Noel GR (1990) Inability of a seed treatment with *Pseudomonas cepacia* to control *Heterodera glycines* on soybean. J Nematol 22:792–794

Notz R, Maurhofer M, Schnider-Keel U et al (2001) Biotic factors affecting expression of the 2, 4-diacetylphloroglucinol biosynthesis gene *phlA* in *Pseudomonas fluorescens* biocontrol strain CHA0 in the rhizosphere. Phytopathology 91:873–881

Notz R, Maurhofer M, Dubach H et al (2002) Fusaric acid-producing strains of *Fusarium oxysporum* alter 2, 4-diacetylphloroglucinol biosynthetic gene expression in *Pseudomonas fluorescens* CHA0 in vitro and in the rhizosphere of wheat. Appl Environ Microbiol 68:2229–2235

Nyczepir AP, Shapiro-Ilan DI, Lewis EE et al (2004) Effect of entomopathogenic nematodes on *Mesocriconema xenoplax* populations in peach and pecan. J Nematol 36:181–185

Okubara PA, Schroeder KL, Paulitz TC (2005) Real-time polymerase chain reaction: applications to studies on soilborne pathogens. Can J Plant Pathol 27:300–313

Olatinwo R, Becker JO, Borneman J (2006a) Suppression of *Heterodera schachtii* populations by *Dactylella oviparasitica* in four soils. J Nematol 38:345–348

Olatinwo R, Borneman J, Becker JO (2006b) Induction of beet-cyst nematode suppressiveness by the fungi *Dactylella oviparasitica* and *Fusarium oxysporum* in field microplots. Phytopathology 96:855–859

Olatinwo R, Yin B, Becker JO et al (2006c) Suppression of the plant-parasitic nematode *Heterodera schachtii* by the fungus *Dactylella oviparasitica*. Phytopathology 96:111–114

Oyekanmi EO, Coyne DL, Fagade OE et al (2007) Improving root-knot nematode management on two soybean genotypes through the application of *Bradyrhizobium japonicum*, *Trichoderma pseudokoningii* and *Glomus mosseae* in full factorial combinations. Crop Prot 26:1006–1012

Parvatha Reddy P, Rao MS, Nagesh M (1996) Management of the citrus nematode, *Tylenchulus semipenetrans*, by integration of *Trichoderma harzianum* with oil cakes. Nematol Medit 24:265–267

Perez EE, Lewis EE (2006) Use of entomopathogenic nematodes and thyme oil to suppress plant-parasitic nematodes on English boxwood. Plant Dis 90:471–475

Persmark L (1997) Ecology of nematophagous fungi in agricultural soils. Ph.D. thesis, Lund University, Lund

Pickett CH, Bugg RL (1998) Enhancing biological control. University of California Press, Berkeley

Pocasangre LE, zum Felde A, Canizares C et al (2007) Field evaluation of the antagonistic activity of endophytic fungi towards the burrowing nematode, *Rhadopholus similis*, in plantain in Latin America. In: ISHS/ProMusa symposium: Recent advances in banana crop protection for sustainable production and improved livelihoods, White River

Pyrowolakis A, Schuster RP, Sikora RA (1999) Effect of cropping pattern and green manure on the antagonistic potential and the diversity of egg pathogenic fungi in fields with *Heterodera schachtii* infection. Nematology 1:165–171

Pyrowolakis A, Westphal A, Sikora RA et al (2002) Identification of root-knot nematode suppressive soils. Appl Soil Ecol 19:51–56

Rao MS, Parvatha Reddy P, Nagesh M (1998) Evaluation of plant based formulations of *Trichoderma harzianum* for the management of *Meloidogyne incognita* on egg plant. Nematol Medit 26:59–62

Roberts PA (1993) The future of nematology: integration of new and improved management strategies. J Nematol 25:383–394

Robinson AF, Westphal A, Overstreet C et al (2008) Detection of suppressiveness against *Rotylenchulus reniformis* in soil from cotton (*Gossypium hirsutum*) fields in Texas and Louisiana. J Nematol 40:35–38

Ruberson JR, Nemoto H, Hirose Y (1998) Pesticides and conservation of natural enemies in pest management. In: Barbosa P (ed) Conservation biological control. Academic, San Diego

Rumbos CI, Kiewnick S (2006) Effect of plant species on persistence of *Paecilomyces lilacinus* strain 251 in soil and on root colonization by the fungus. Plant Soil 283:25–31

Samac DA, Kinkel LL (2001) Suppression of the root-lesion nematode (*Pratylenchus penetrans*) in alfalfa (*Medicago sativa*) by *Streptomyces* spp. Plant Soil 235:35–44

Santos MA, Ferraz S, Muchovej JJ (1992) Evaluation of 20 species of fungi from Brazil for biocontrol of *Meloidogyne incognita* race 3. Nematropica 22:183–192

Schmidt LM, Preston JF, Dickson DW et al (2003) Environmental quantification of *Pasteuria penetrans* endospores using in situ antigen extraction and immunodetection with a monoclonal antibody. FEMS Microbiol Ecol 44:17–26

Schmidt LM, Preston JF, Nong G et al (2004) Detection of *Pasteuria penetrans* infection in *Meloidogyne arenaria* race 1 in planta by polymerase chain reaction. FEMS Microbiol Ecol 48:457–464

Sharon E, Bar-Eyal M, Chet I et al (2001) Biological control of the root-knot nematode *Meloidogyne javanica* by *Trichoderma harzianum*. Phytopathology 91:687–693

Sharon E, Chet I, Viterbo A et al (2007) Parasitism of *Trichoderma* on *Meloidogyne javanica* and role of the gelatinous matrix. Eur J Plant Pathol 118:247–258

Siddiqui IA, Ehteshamul-Haque S (2001) Suppression of the root rot-root knot disease complex by *Pseudomonas aeruginosa* in tomato: the influence of inoculum density, nematode populations, moisture and other plant-associated bacteria. Plant Soil 237:81–89

Siddiqui ZA, Mahmood I (1996) Biological control of plant parasitic nematodes by fungi: a review. Bioresour Technol 58:229–239

Siddiqui MA, Shaukat SS (2002) Mixtures of plant disease suppressive bacteria enhance biological control of multiple tomato pathogens. Biol Fertil Soils 36:260–268

Siddiqui IA, Shaukat SS (2003a) Non-pathogenic *Fusarium solani* represses the biosynthesis of nematicidal compounds in vitro and reduces the biocontrol of *Meloidogyne javanica* by *Pseudomonas fluorescens* in tomato. Lett Appl Microbiol 37:109–114

Siddiqui IA, Shaukat SS (2003b) Plant species, host age and host genotype effects on *Meloidogyne incognita* biocontrol by *Pseudomonas fluorescens* strain CHA0 and its genetically-modified derivatives. J Phytopathol 151:231–238

Siddiqui IA, Shaukat SS (2003c) Suppression of root-knot disease by *Pseudomonas fluorescens* CHA0 in tomato: importance of bacterial secondary metabolite, 2, 4-diacetylpholoroglucinol. Soil Biol Biochem 35:1615–1623

Siddiqui IA, Shaukat SS (2004) *Trichoderma harzianum* enhances the production of nematicidal compounds in vitro and improves biocontrol of *Meloidogyne javanica* by *Pseudomonas fluorescens* in tomato. Lett Appl Microbiol 38:169–175

Siddiqui IA, Shaukat SS (2005) Phenylacetic acid-producing *Rhizoctonia solani* represses the biosynthesis of nematicidal compounds in vitro and influences biocontrol of *Meloidogyne incognita* in tomato by *Pseudomonas fluorescens* strain CHA0 and its GM derivatives. J Appl Microbiol 98:43–55

Siddiqui IA, Shaukat SS, Khan A (2004) Differential impact of some *Aspergillus species* on *Meloidogyne javanica* biocontrol by *Pseudomonas fluorescens* strain CHA0. Lett Appl Microbiol 39:74–83

Siddiqui IA, Atkins SD, Kerry BR (2009) Relationship between saprotrophic growth in soil of different biotypes of *Pochonia chlamydosporia* and the infection of nematode eggs. Ann Appl Biol 155:131–141

Smith ME, Jaffee BA (2009) PCR primers with enhanced specificity for nematode-trapping fungi (Orbiliales). Microb Ecol 58:117–128

Smith KS, Hewlett TE, Griswold S (2004) *Pasteuria* for nematode control: development of a commercial production process. In: Proceedings of the 2004 annual international research conference on methyl bromide alternatives and emissions reductions Orlando, #44. http://www.mbao.org/2004/Proceedings04/mbrpro04.html

Smitley DR, Warner FW, Bird GW (1992) Influence of irrigation and *Heterorhabditis bacteriophora* on plant-parasitic nematodes in turf. J Nematol 24:637–641

Sorribas FJ, Ornat C, Galeano M et al (2003) Evaluation of a native and introduced isolate of *Pochonia chlamydosporia* against *Meloidogyne javanica*. Biocontrol Sci Technol 13:707–714

Spaull VW (1984) Observations of *Bacillus penetrans* infecting *Meloidogyne* in sugarcane fields in South Africa. Rev Nematol 7:277–282

Starr JL, Ong KL, Huddleston M et al (2007) Control of *Meloidogyne marylandi* on bermudagrass. Nematropica 37:43–49

Stirling GR (1984) Biological control of *Meloidogyne javanica* with *Bacillus penetrans*. Phytopathology 74:55–60

Stirling GR (1991) Biological control of plant-parasitic nematodes: progress problems and prospects. C.A.B International, Wallingford

Stirling GR, Smith LJ (1998) Field tests of formulated products containing either *Verticillium chlamydosporium* or *Arthrobotrys dactyloides* for biological control of root-knot nematodes. Biol Control 11:231–239

Stirling GR, McKenry MV, Mankau R (1979) Biological control of root-knot nematodes (*Meloidogyne* spp.) on peach. Phytopathology 69:806–809

Stirling GR, Wilson EJ, Stirling AM et al (2005) Amendments of sugarcane trash induce suppressiveness to plant-parasitic nematodes in a sugarcane soil. Aust Plant Pathol 34:203–211

Talavera M, Mizukubo T, Ito K et al (2002) Effect of spore inoculum and agricultural practices on the vertical distribution of the biocontrol plant-growth-promoting bacterium *Pasteuria penetrans* and growth of *Meloidogyne incognita*-infected tomato. Biol Fertil Soils 35:435–440

Tedford EC, Jaffee BA, Muldoon AE et al (1993) Parasitism of *Heterodera schachtii* and *Meloidogyne javanica* by *Hirsutella rhossiliensis* in microplots over two growing seasons. J Nematol 25:427–433

Terefe M, Tefera T, Sakhuja PK (2009) Effect of a formulation of *Bacillus firmus* on root-knot nematode *Meloidogyne incognita* infestation and the growth of tomato plants in the greenhouse and nursery. J Invertebr Pathol 100:94–99

Timm L, Pearson D, Jaffee B (2001) Nematode-trapping fungi in conventionally and organically managed corn-tomato rotations. Mycologia 93:25–29

Timper P, Minton NA, Johnson AW et al (2001) Influence of cropping systems on stem rot (*Sclerotium rolfsii*), *Meloidogyne arenaria*, and the nematode antagonist *Pasteuria penetrans* in peanut. Plant Dis 85:767–772

Tobin JD, Haydock PPJ, Hare MC et al (2008a) The compatibility of the fungicide azoxystrobin with *Pochonia chlamydosporia*, a biological control agent for potato cyst nematodes (*Globodera* spp.). Ann Appl Biol 152:301–305

Tobin JD, Haydock PPJ, Hare MC et al (2008b) Effect of the fungus *Pochonia chlamydosporia* and fosthiazate on the multiplication rate of potato cyst nematodes (*Globodera pallida* and *G. rostochiensis*) in potato crops grown under UK field conditions. Biol Control 46:194–201

Trudgill DL, Bala G, Blok VC et al (2000) The importance of tropical root-knot nematodes (*Meloidogyne* spp.) and factors affecting the utility of *Pasteuria penetrans* as a biocontrol agent. Nematology 2:823–845

Tzortzakakis EA (2000) The effect of *Verticillium chlamydosporium* and oxamyl on the control of *Meloidogyne javanica* on tomatoes grown in a plastic house in Crete, Greece. Nematol Medit 28:249–254

Tzortzakakis EA, Gowen SR (1994) Evaluation of *Pasteuria penetrans* alone and in combination with oxamyl, plant resistance and solarization for control of *Meloidogyne* spp. on vegetables grown in greenhouses in Crete. Crop Prot 13:455–462

Tzortzakakis EA, Petsas SE (2003) Investigation of alternatives to methyl bromide for management of *Meloidogyne javanica* on greenhouse grown tomato. Pest Manag Sci 59:1311–1320

Tzortzakakis EA, Gowen SR, Goumas DE (1996) Decreased ability of *Pasteuria penetrans* spores to attach to successive generations of *Meloidogyne javanica*. Fund Appl Nematol 19:201–204

van den Boogert PHJF, Velvis H, Ettema CH et al (1994) The role of organic matter in the population dynamics of the endoparasitic nematophagous fungus *Drechmeria coniospora* in microcosms. Nematologica 40:249–257

Verdejo-Lucas S, Sorribas FJ, Ornat C et al (2003) Evaluating *Pochonia chlamydosporia* in a double-cropping system of lettuce and tomato in plastic houses infested with *Meloidogyne javanica*. Plant Pathol 52:521–528

Wang KH, Sipes BS, Schmitt DP (2002) Management of *Rotylenchulus reniformis* in pineapple, *Ananas comosus*, by intercycle cover crops. J Nematol 34:106–114

Wang KH, Sipes BS, Schmitt DP (2003) Enhancement of *Rotylenchulus reniformis* suppressiveness by *Crotalaria juncea* amendment in pineapple soils. Agric Ecosyst Environ 94:197–203

Wang KH, McSorley R, Gallaher RN (2004) Effect of *Crotalaria juncea* amendment on squash infected with *Meloidogyne incognita*. J Nematol 36:290–296

Wei BQ, Xue QY, Wei LH et al (2009) A novel screening strategy to identify biocontrol fungi using protease production or chitinase activity against *Meloidogyne* root-knot nematodes. Biocontrol Sci Technol 19:859–870

Weibelzahl-Fulton E, Dickson DW, Whitty EB (1996) Suppression of *Meloidogyne incognita* and *M. javanica* by *Pasteuria penetrans* in field soil. J Nematol 28:43–49

Weller DM, Landa BB, Mavrodi OV et al (2007) Role of 2, 4-diacetylphloroglucinol-producing fluorescent Pseudomonas spp. in the defense of plant roots. Plant Biol 9:4–20

Westphal A (2005) Detection and description of soils with specific nematode suppressiveness. J Nematol 37:121–130

Westphal A, Becker JO (1999) Biological suppression and natural population decline of *Heterodera schachtii* in a California field. Phytopathology 89:434–440

Westphal A, Becker JO (2000) Transfer of biological suppressiveness against *Heterodera schachtii*. Phytopathology 90:401–406

Westphal A, Becker JO (2001a) Components of soil suppressiveness against *Heterodera schachtii*. Soil Biol Biochem 33:9–16

Westphal A, Becker JO (2001b) Soil suppressiveness to *Heterodera schachtii* under different cropping sequences. Nematology 3:551–558

Yin B, Valinsky L, Gao XB et al (2003) Identification of fungal rDNA associated with soil suppressiveness against *Heterodera schachtii* using oligonucleotide fingerprinting. Phytopathology 93:1006–1013

Zhang L, Liu XZ, Zhu SF et al (2006) Detection of the nematophagous fungus *Hirsutella rhossiliensis* in soil by real-time PCR and parasitism bioassay. Biol Control 36:316–323

Zhu ML, Mo MH, Xia ZY et al (2006) Detection of two fungal biocontrol agents against root-knot nematodes by RAPD markers. Mycopathologia 161:307–316

Chapter 12
Root Patho-Systems Nematology and Biological Control

Keith G. Davies and Yitzhak Spiegel

12.1 Introduction

Since the publication of Stirling's book (1991), our knowledge has grown but the developments of robust strategies that exploit biological control have remained, in the main, anecdotal and illusive. A new era is now present in which the research and development of biological control is the result of several circumstances: (1) The inconsistent results, especially under field conditions, led to the necessity of an intensive understanding of the interaction mechanism(s) between the parasitic nematode and its natural enemy. Research here focused on a particular interaction between the nematode host and its parasite and was studied in detail. This approach was possibly naïve in thinking that this could be developed into a control strategy as it failed to acknowledge the more complicated multitrophic interactions between plant, parasite and the microorganism. Acknowledging these multitrophic aspects will result in an improvement of the efficacy of the biocontrol agent and perhaps will help with a better understanding of the interaction mechanism(s) between microorganisms, in general. (2) The maturation of molecular biology with all the new tools presently at its disposal, equips the researchers in the laboratory with approaches that hitherto have been impossible. (3) The acquaintance to environmental issues that brought research groups from other disciplines to deal with the topic within an 'ecological' perspective. (4) Constrains which were forced upon the industry following the restrictions in the registration process of biological control agents.

K.G. Davies (✉)
Department of Plant Pathology and Microbiology, Rothamsted Research,
Harpenden, Hertfordshire AL5 2JQ, UK
e-mail: keith.davies@rothamsted.ac.uk

Y. Spiegel
Division of Nematology, Institute of Plant Protection, ARO, The Volcani Center,
Bet Dagan 50250, Israel
e-mail: spiegely@volcani.agri.gov.il

K. Davies and Y. Spiegel (eds.), *Biological Control of Plant-Parasitic Nematodes: Building Coherence between Microbial Ecology and Molecular Mechanisms*,
Progress in Biological Control 11, DOI 10.1007/978-1-4020-9648-8_12,

12.2 Technological Developments

During the 1980s and the 90s most of the research groups had focused on the attempt to isolate soil microorganisms, mainly fungi, bacteria, mites, collembola and predatory nematodes (Chap. 1), without devoting any special concern to understanding the mode-of-action of the interaction between the microorganisms and the pest nematodes. Within the research groups that investigated more carefully the mechanistic interactions between the pest nematode and its natural enemy it is noteworthy to point out that research that initially started out as fundamentally morphological in nature and focused on the nematophagous fungi, performed by Nordbring-Hertz, H.–B. Jansson and their colleagues, was then 'developed' by this group to biochemical aspects of infection processes (Chap. 6) and ongoing today with emphasis on molecular and genomic aspects. A similar trend has been developed with the interaction between the fungus *Pochonia chlamydosporia* (previously defined as *Verticillium chlamydosporia*) and the sedentary cyst- and root-knot nematode species. These relations were also, in the beginning, investigated morphologically by using mainly light- and electron microscopy tools, followed by expansion to 'ecological' aspects based on observations made on microbiology of population measurements (Chap. 7), which then shifted to the usage of combination of several approaches (molecular biology, biochemistry, genetics/genomic and developmental biology) aimed to get a deeper understanding of such interactions (e.g. Chap. 9). The *Trichoderma* – phytonematode interactions (see Chap. 8) present an additional example of a thorough study where morphological, biochemistry, and molecular tools were assigned to understand the different nematicidal mechanisms in which the various *Trichoderma* isolates acted against plant-parasitic nematode species. Utilizing genomical and metabolomical means for the understanding of *Trichoderma*-plant diseases interactions (Shoresh and Harman 2008; Woo et al. 2009), can further enhance the efficacy of *Trichoderma* as a biocontrol agent against nematode pests. Modern day biocontrol scientists can employ the growing data bases using the readily available bioinformatic tools and use genomic approaches. Comparative genomics, therefore, can integrate the various scientific disciplines and build cohesion between disciplines which hitherto was impossible (Fig. 12.1).

For example, comparative genomics and maximum likelihood based methods showed that three different chitinase subgroups have expanded in copy number in *Trichoderma* species, proposing an important role of these chitinases during the mycoparasitism process (Ihrmark et al. 2010). Several regions and amino acid sequences have been identified in four chitinase genes from different *Trichoderma* species, that are likely to determine functional properties playing important roles in the interaction between plant/pathogens and mycoparasites, such as substrate-specificity, processivity or pH-optima (*ibid.*). These approaches can help lead to a better understanding of multitrophic nematode/plant/*Trichoderma* interactions and improvement of the biocontrol activity.

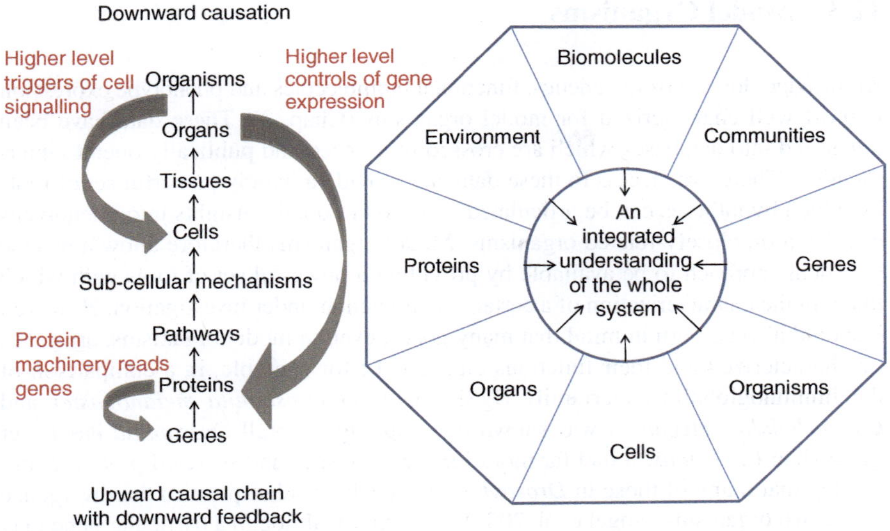

Fig. 12.1 The reductionist causal chain in which genes > proteins > sub-cellular mechanisms > cells > tissues > organs and organisms are under downward control through signaling pathways and other system level phenomena which ultimately also includes biotic and abiotic environmental factors (Adapted from Noble 2006)

Nematode/bacteria relationships are characterized by two eminent groups: (1) Bacteria that are directly related to phytonematode parasitism. (2) Bacteria that interact with the plant/nematode system, and thus, indirectly cause a nematicidal activity. The bacteria that directly bind to the nematode surface and then, at the end of the interaction process, cause the termination of the nematode life-cycle, are allocated within the first group. The most prominent bacteria related to this group are the *Pasteuria* isolates, which have been studied for many years by different groups in many aspects and shown to have associations with all the major groups of plant-parasitic nematodes (Chen et al. 2004). Again, the development in understanding the interactions' mechanisms resembles that of the nematode/fungi studies: at first morphological studies utilizing light, electron (transmission and scanning) initially by Mankau and Sayre and colleagues in the United States, and later, using confocal microscopy, accompanied by biochemical, molecular and more recently genomic tools (Chap. 4). Comparative genomics has and will in the future further illuminate the specific characteristics of these species (Chap. 4). The second group of bacteria include soil-borne bacteria that inhabitant root rhizosphere, naturally or deliberately, and do not necessarily exist in direct contact with the phytonematode, but can impose a biocontrol effect by interference with the host (plant) finding process, host recognition, or by metabolite secretion that paralyze or kill the nematode (Oka et al. 1993; Spiegel et al. 1991).

12.3 Model Organisms

Knowledge flowing from genetics, functional biomolecules and phenotype expression is most well characterized for model organisms (Chap. 3). These data have been integrated into databases which are crossed referenced and publically open to interrogation. Therefore, access to these data can provide a hugely powerful set of tools by which hypotheses can be articulated and used to obtain insights into phenotypes of related or, closely-related organisms. Model organisms therefore allow a reverse genetical approach to be available by providing a powerful set of tools with which to begin the characterization of a certain phenomenon under investigation. However, it should also be born in mind that many genes, even in model organisms, are yet to be characterized and their functions elucidated ; for example, in a comparison of the immunoglobulin superfamily (IgSF) between *Drosophila melanogaster* and *Caenorhabditis elegans* it was shown that, although overall *Drosophila* has fewer genes than *C. elegans*, it had far more IgSF cell-surface and secreted proteins, suggesting that many of those in *Drosophila* had evolved subsequent to the divergence of the two organisms (Vogel et al. 2003). In addition, although a particular gene may be nearly identical to one that has been characterized in a model organism, small differences in sequence can have large effects on biochemical function, especially when viewed from the perspective of biological control. An example of this was the characterization of a protease, VCP1, from *Pochonia chlamydosporia*, where a single change in an amino acid lead to enzymic specificity that enabled the proteolytic digestion of eggs of one group of nematodes but not another (Chap. 7).

Caenorhabditis elegans has over the last several years been developed as a model for investigating aspects of innate immunity. Much of the funding for this research has been driven by wishing to understand aspects of immunity that may be applicable to the medical or veterinarian scientist who desires to use this knowledge in order to protect a person or animal from a particular disease or pathogen. However, this generic science is also important for the biological control scientist, aiming to use a disease or pathogen to control a pest. *C. elegans* was originally isolated from decomposing plant debris, a habitat full of fungi and bacteria. Our knowledge about susceptibility and resistance mechanisms in this nematode is rudimentary, and our knowledge in plant-parasitic nematodes is even more elementary. However, the knowledge obtained from *C. elegans* as a model system can be a point of departure from which it will be possible to develop hypotheses about plant-parasitic nematodes and their hyperparasites. This knowledge will be of paramount importance if the application of biological control agents is ever to be successfully deployed in a robustly commercial way to control plant-parasitic nematodes.

12.4 Multitrophic Interactions

The growing interest in multitrophic interactions in soil has led to an increased amount of data provided by researchers from various disciplines. These interactions with the plant, or in-between microorganisms, which include pathogenic and non-pathogenic microorganisms, may enlighten interactions where nematodes are involved. Moreover, it has been shown by several groups that microbial mixtures can enhance the antagonistic effect(s) to the plant pathogen, e.g. by enhancing plant resistance to the pathogen (Woo et al. 2006), or by combining enzymatic activity of one organism with an antibiotic activity of another microorganism (Gazit-Fatal et al. 2009), or by combining two antibiotic activities of one organism against a plant pathogen (Moseri et al. 2009). Such synergistic activity should be carefully addressed, as sometimes an antagonistic activity may result by combining a microbial mixture in which one of them secrete a dominant enzyme and/or an antibiotic material which subjugates (or suppresses) the activity of the 'partner' organism.

The studies of tri-trophic interactions, between a plant, a parasitic nematode and a natural enemy, have tended to be studied in agricultural systems where any particular niche gets destroyed regularly by cultivation practices. The study of nematodes in natural systems has gained interest as nematodes are important in food webs that play an important part in nutrient cycling. A large proportion of carbon fixed by photosynthesis is exuded from roots, attracting a large rhizosphere microflora. Plant-parasitic nematodes infecting root-systems are likely to increase this loss of carbon into the rhizosphere, which in turn will further stimulate the rhizosphere microflora, and sequentially, free-living nematodes that feed on this microflora. Molecular based methods will certainly help identify key functional components of the rhizosphere communities, and the comparison of agricultural systems with natural systems will help to measure rhizosphere biodiversity (Chap. 2). Nematode suppressive soils rely on the microbial diversity to reduce the parasitic nematode populations (Chap. 10). Comparative studies between natural and agricultural soils, using molecular tools that can identify and characterize microbial population structures, will be important in assessing changes to microbial diversity due to agricultural practices.

Plant-associated microorganisms' secret proteins, and other small molecules, that have the ability to manipulate host-cell structure and function. These secreted molecules, which have been collectively defined as 'effectors', were attributed mainly to plant-pathogens (Hogenhout et al. 2009). Nematodes also have a large number of secreted proteins that may be acting as 'effectors' (Bird and Opperman 2009; Davis et al. 2008). The effectors can facilitate infection (triggering virulence or symbiotic factors, inhibitors, toxins) but may also activate defense responses such as elicitors or avirulence factors. Such dual, and contradictory, activity has been reported in various plant-microbe interactions. The effectors are delivered into host cells through a diversity of mechanisms, but can also act in the extracellular space at the plant-microbe interface, where they may interfere with apoplastic plant defenses (reviewed by Misas-Villamil and van der Hoorn 2008). Numerous effectors suppress plant immunity but other reveal other activities such as alter host plant behavior and morphology (Oldroyd and Downie 2008). Root-knot nematodes, for instance, are using their stylet to inject effector proteins inside their plant host, causing the giant cells formation (Davis et al. 2008). Microorganisms that interact 'intimately' with nematode-infected plants may introduce effectors such as proteins, peptides or secondary metabolites, which will suppress the plant host immunity (as in the case of the *Verticillium dahlia*/*Pratylenchus penetrans*/potato complex) or will enhance host protection to the nematode parasitism. In that last case (*Trichoderma*, endophytic bacteria? fungi? mycorrhiza?) the suppressive effect of microorganisms on phytonematodes can be due either via so-called 'local' interactions (direct parasitism or local induction of defense responses in the plant roots) or 'non-local' (systemic) effects, namely, by triggering an effect transferred through the plant to other distant organs.

Members of the *Bacillus* genus have been evaluated as candidates for bio control against plant diseases, including plant-parasitic nematodes (Niu et al. 2006; Oka et al. 1993; Sela et al. 1998). An immense range of biologically active molecules, capable of hampering pathogen growth, are constructed by this genus, and genetic markers associated with biological control activities of *B. subtilis* have been identified and characterized (Joshi and McSpadden Gardener 2006). *B. subtilis* produces a vast array of antibiotics, potentially more than 25 structurally diverse antimicrobial compounds (reviewed by Stein 2005). Cyclic lipopeptides (LPs) of the surfactin, iturin and fengycin families are the most well-known compounds (Ongena and Jacques 2008). These LPs' play a major role in the stimulation of host defense mechanisms in the plant either by performing a biofilm by the bacterium on the plant roots (Bais et al. 2004), secretion of inhibitory quantities of surfactin, iturins and fengycines that act as an anti-bacterial component against plant diseases (Houda et al. 2009), or as inducers of plant resistance (Ongena et al. 2007). So far no evidence for the involvement of lipopeptides in phytonematode/*Bacillus* spp./host-plant interaction has been published but such involvement cannot be ruled out.

12.5 Building Coherence Between Disciplines

The aim of biological control is to use knowledge about how soil organisms interact such that nematode pests can be managed to a soil density so that they do not have a detrimental effect on the yield of a particular crop. This involves an understanding of a complex web of multitrophic interactions. The last half of the twentieth century has seen the massive success for the reductionist approach applied to biological problems; structures of macromolecules have been unraveled and the chemical nature of DNA, the aperiodic crystal, determined and the elucidation of its role in heredity has been hugely instrumental. As discussed above, recent biological control has also, in the last decade or so, also taken advantage of this new knowledge and this will continue to happen. However, the success of this reductionist approach has also been applied to biological control, but perhaps not with the general success that was at first thought. The way in which biological control scientists have operated has in general tried to identify one causal agent, isolate it from the soil, grow it up in large quantities and then inundate the soil in the hope that it will control the pest nematode. Looking back over the last couple of decades this strategy has not produced the robust control that was originally hoped, and perhaps now seems a little naïve. However, successful though this reductionist approach has been in understanding cellular biology, it certainly has limitations and modern day biology has evolved a more subtle approach. The huge molecular databases and rise of computational biology has seen a development in which the integration of molecular information within the context of organizational hierarchies has seen the development of systems biology in which the upward causal chain - genes, proteins, pathways, cells, tissues, organs and organisms, - is as important as the downward causal chain (Noble 2006).

The soil ecologist has always been aware of the complexity of interacting organisms, and arguably, systems biology is the result of the molecular biologist having to take a leaf out of the ecologist's book and investigate the molecular ecology of the organism. While ecology has for several decades used models that are mathematically sophisticated, this approach, now being applied to physiology and biochemistry, is new. The power of any model is dependent in being able to identify the essential from the non-essential and thereby understand a biological phenomenon (Noble 2002). Similarly, the biological phenomenon of nematode suppressive soils might require the development of models that require the integration of two types of model: one that incorporates the various species of microbes in the soil and their special and temporal distribution, a population level model, and another, that includes biochemical information relating to proteins such as enzymes and adhesion factors that are important in the infection processes, a biochemical model. This biological systems approach would therefore link the reductionist approach, with its upward causal chain, with a downward causal chain that include the associated higher level controls of gene expression (Fig. 12.1) and therefore lead to an understanding of how a nematode suppressive soil is produced.

12.6 Commercialization

Plant-parasitic nematodes are ubiquitous in agricultural soils. In nematode suppressive soils, at the end of a growing season, only a very small fraction of the eggs that lie dormant, somewhat less than 10%, will fulfill their life-cycle and reproduce the next generation (Kerry and Crump 1977). This reduction in reproductive capacity of the plant-parasitic nematodes has been attributed to a whole diverse spectrum of microbial control agents such as *Dactylella oviparasitica, Hirsutella rhossiliensis, Fusarium* spp, *Pochonia chlamydosporia or Trichoderma*, and other groups of bacteria such as *Bacillus* spp., *Pasteuria penetrans* and *Pseudomonas aureofaciens.* Therefore, it is the phenomenon of nematode suppressive soils associated with many crops that has been the motivating force behind the research on biological control of nematodes (e.g. see Chap. 10). The goals of this research has been to understand the mechanisms of this phenomenon in order to develop environmentally benign strategies to manage these pests either through agronomic practices, or through the development of commercial products that can be applied. Biological control of plant-parasitic nematode pests would therefore seem, on the face of it, to be relatively straight forward and attainable. However, the reality has proven to be more difficult and if nematode suppression was easily understandable and applicable, biological control would already be a robust crop protection technology.

The wide range of different microbial entities associated with nematode suppression offer a spacious array of different approaches and control options. Each organism will have its advantages and disadvantages, for example candidates that can be cultured very easily in vitro have advantages over those that are obligate pathogens that cannot, and again when it comes to other aspects of development such as formulation, storage and application, each organism will have its advantages and disadvantages. To date, there are already a number of products on the market (Table 12.1) and even this small amount outnumbers the number of new commercial nematicides and there are other potential organisms that are being developed (Hallmann et al. 2009). Most biological control products for nematodes so far exist as a liquid or wettable powder formulations applied in furrow or through drip irrigation systems and one of the major drawbacks to these inundative practices is the volume of soil needed to be treated in order to protect root systems. It is important to protect plants from nematodes while they are establishing themselves in the soil and it has been estimated that around 2,500 t/ha of soil in the upper 25 cm usually needs to be treated to obtain effective control (Sikora et al. 2008). It is therefore perhaps not surprising that biological control of nematodes has concentrated on high value crops and niche markets, for example the development of Econem™, a formulation of *Pasteuria* usage by Pasteuria Biosciences LLC, to control sting nematode (*Belonolaimus longicaudatus*) which is a problem on the greens of golf courses.

12.7 Designer BCAs

The use of molecular tools by biological control scientists has grown since the publication of Stirling's book in 1991 when the impact of these techniques was in its infancy and is now routine. As our understanding grows, with respect to the key factors that are important in determining the mode of action of biological control agents and the biochemistry of each individual strains specificity, the application of such techniques to broaden the host range and aggressiveness of potential biological control agents becomes increasingly feasible and compelling. Transgenic approaches to improve biological control agents go back for over a decade and have been attempted for entomopathogenic nematodes (Hashmi et al. 1995) and fungi (Gressel 2001). The advent of synthetic biology and the possibility of developing a designer biological control agent for a particular nematode pest are technically now a possibility and perhaps where the future lies.

Table 12.1 Commercially available biological control products for control of plant-parasitic nematodes

Product name (Commercial trade-name)	Active antagonist	Formulation type	Treatment form	Treated-crop	Producer/country
Bioact WG	*Paecilomyces ilacinus*	Wettable powder		Vegetables,	Prophyta GmbH,
Melocon WG				Banana	Germany
Bio-Nemax		Liquid	Drench, spray	General	M.J. Exports, India
BioNem-WP	*Bacillus firmus*	Wettable powder	Drip irrigation	Vegetables	Bayer, Germany
BioSafe			Drench		AgroGreen, Israel
BioStart	*Bacillus chitinosporus*	Liquid	Irrigation	General use,	Microbial Solutions S Africa
RhizoBoost	*B. laterosporus* *B. licheniformis*		Drench	Vegetables, Fruits	Rincon Vilova; USA
Deny	*Burkholderia cepacia*				Rincon Vitova, USA
Blue Circle					Stine Microbial Products, USA
DiTera (natural product from a hyphomycete fungus)	*Myrothecium verrucaria*	Liquid		Vegetables	Valent, USA
Econem	*Pasteuria penetrans*	Liquid and powder	Drench	Vegetables	Pasteuria Bioscience, USA
			Drip irrigation	Turf, soybean	Nematech, Japan
KlamiC	*Pochonia chlamydosporia*	Granulate	Soil incorporation	Vegetables	Cuba
Nemix	*Bacillus* sp.	Powder	Drench/drip	Vegetables, fruit trees	AgriLife/Chr Hansen, Brazil
PlPlus	*Paecilomyces lilacinus*		Drench	Vegetables	BCP, South Africa
			Drip irrigation	Tobacco	
				Banana	
				Citrus	
Sudozome	*Pseudomonas fluorescence*	Liquid	Drench	General	Agriland Biotech Limited, India
Yorker	*Paecilomyces lilacinus*	Suspo-emulsion	Drench, drip irrigation	Vegetables, fruit trees	Agriland Biotech Limited, India

12.8 Society and Science

Science is not hermetically sealed from the rest of society and therefore takes place within a social milieu; biological control as a discipline cannot be removed from this general context. The release of biological control organisms, and especially 'designer' genetically engineered organisms, into the environment where they cannot be controlled, is controversial. The millennium development goals highlight the need to alleviate world hunger, while at the same time maintain global biodiversity. These two goals appear to be in conflict with one another and the control of plant-parasitic nematodes by the use of genetically modified organisms cannot escape from these generic issues around genetic modification. Although it would be inappropriate to review this growing literature here, it is necessary to discuss various aspects of the issues and how they impinge on biological control scientists in the context of the society in which they operate. These issues fall into three main areas and include (1) environmental safety, what effects any released organisms will have on the broader environment; (2) political concerns, such as *who owns the technology?* and *who benefits from the technology?* and (3) a so-called 'global-social view', namely, the appropriateness of altering the genetic constitution of an organism by human intervention.

Much effort has gone into assessing the possible environmental impacts on the release of genetically modified organisms into the environment (Dale et al. 2002; Hails and Morley 2005; Sanvido et al. 2007). Interestingly, farm scale field trials in which the effects of genetically modified herbicide tolerant crops on biodiversity was assessed, showed that the differences between the different individual non-transgenic crops had a larger effect on environmental biodiversity than did the fact of whether or not the crops were genetically modified (Firbank and Forcella 2000; Firbank 2003). However, in a recent review, concerns were articulated over the persistence and spread of feral herbicide resistant crops (Graef 2009) and similar concerns over the risks and potential problems associated with the ultimate destination of genes incorporated into potential microbiological control agents would need to be evaluated. Because of public concerns over genetically modified crops, regulatory systems have been developed which incorporate environmental risk as a function of hazard and exposure (Poppy and Wilkinson 2005; Pidgeon et al. 2007) and therefore such approaches could be evaluated for their appropriateness for dealing with genetically modified biological control organisms and it would not be required to initiate these assessment systems from the beginning.

The specter of genetic engineering, particularly as developed by private companies, who are primarily answerable to shareholders, has brought in its wake a public distrust of biotechnology (Jasanoff 2005). The reaction by publically funded research grant awarding bodies has been to push active engagement of the public into participation and into dialog with scientists, and ideally to engage in this dialog at an early stage in the scientific process (Wilsdon et al. 2005). However, these activities have not always proved fruitful, and the drive to see science in the form of a simple customer – contractor relationship has had the effect of politicizing science (Davies 2007). While it has been argued that the public needs to be involved in setting up the research priorities, scientists themselves have seen this as a loss of faith, but as is clear by the debates around food security and climate change, that there is clearly a conflict of interest. It is this conflict of interest that needs to be addressed and a new model developed. This new model needs to move away from the customer – contractor relationship and infuse science with a new set of social possibilities based on principles that are open and democratic such that the public can see who owns the technology and who benefits from the technology (Davies and Wolf-Phillips 2006; Wilsdon and Willis 2004; Wilsdon et al. 2005).

Acknowledging that the two issues described above are addressed successfully, when it comes to genetic engineering there is a third category that for some of the public will always deem genetically engineered biological control agents to be unacceptable. Those people holding this position have a *world view* in which they would wish to see the release of genetically engineered organisms prohibited, because they think their production as unnatural. In general, this group of people believes that moving genes around from one species to another is intuitively wrong, and therefore genetically engineered biological control agents would never be acceptable. It has been argued that such a world view comes from an essentialist position, dating back to Plato and Aristotle, who suggest that the everyday world of sense experience is not real but abstract, and that the real world consists of essences, such as cat and dog, which are immutable and good in their own right (Davies 2001; Ruse 2003). To this group of the public, moving genes around is unnatural because it violates these immutable essences. This view of the world was overthrown by Darwin who did not believe that the world of sense experience was abstract but very real, and therefore, also very mutable (Davies 2001; Ruse 2003). Although this view may be only held by a minority globally, it is a world view that dates back to the Greeks and upon which western civilization stands. Therefore, perhaps it is not surprising that the most ardent campaign against the development of genetically modified organisms is based in Europe with its very firm roots in early Greek philosophy.

12.9 Future Prospects

The aim of the book has been to integrate the current state of knowledge and build some bridges between ecological knowledge and molecular knowledge, and it is clear that substantial progress is currently being made. Understanding microbial diversity and the multitrophic interactions that are manifested in the rhizosphere will play an important role in managing plant-parasitic pest nematodes. Molecular biology and the tools this brings have had and will continue to have an important role. Perhaps its most important role will be in the development of tools in which it will be possible to reconcile agricultural food production with sustainable methods of crop protection while maintaining biodiversity.

References

Bais HP, Fall R, Vivancoet JM et al (2004) Biocontrol of *Bacillus subtilis* against infection of *Arabidopsis* roots by *Pseudomonas syringae* is facilitated by biofilm formation and surfactin production. Plant Physiol 134:307–319

Bird DMcK, Opperman CH (2009) The secret (ion) life of worms. Genome Biol 10:205. doi:doi:10.1186/gb-2009-10-1-205

Chen ZX, Chen SY, Dickson DW (eds) (2004) Nematology, advances and perspectives, vol 2: nematode management and utilization. CABI Publishing, Wallingford

Dale PJ, Clarke B, Fontes EMG (2002) Potential for the environmental impact of transgenic crops. Nat Biotechnol 20:567–575

Davies KG (2001) What makes genetically modified organisms so distasteful? Trends Biotechnol 19:424–427

Davies KG (2007) Bring back haldane. The british association, science and public affairs. p. 21, June 2007

Davies KG, Wolf-Phillips J (2006) Scientific citizenship and good governance: implications for biotechnology. Trends Biotechnol 24:57–61

Davis EL, Hussey RS, Mitchum MG et al (2008) Parasitism proteins in nematode-plant interactions. Curr Opin Plant Biol 11:360–366

Firbank LG (2003) The farm scale evaluations of spring sown genetically modified crops – introduction. Philos Trans R Soc Lond B Biol Sci 358:1777–1778

Firbank LG, Forcella F (2000) Agriculture – genetically modified crops and farmland biodiversity. Science 289:1481–1482

Gazit-Fatal I, Ovadis M, Gorovits R et al (2009) Introduction of the endochitinase ChiA gene from *Serratia plymuthica* enhances biocontrol activity of 2,4-DAPG-producing strains of *Pseudomonas fluorescens*. In: Elad Y, Maurhofer M, Keel C et al (eds) Biological control of fungal and bacterial plant pathogens. IOBC/WPRS Bulletin 43: 75–78

Graef F (2009) Agro-environmental effects due to altered cultivation practices with genetically modified herbicide-tolerant oilseed rape and implications for monitoring: a review. In: Lichtfouse E, Navarrete M, Debaeke P (eds) Sustainable agriculture. Springer, Netherlands, pp 229–242

Gressel J (2001) Potential failsafe mechanisms against the spread and introgression of transgenic hypervirulent biocontrol fungi. Trends Biotechnol 19:149–154

Hails RS, Morley K (2005) Genes invading new populations: risk assessment perspective. Trends Ecol Evol 20:245–252

Hallmann J, Davies KG, Sikora R (2009) Biological control using microbial pathogens, endophytes and antagonists. In: Perry RN, Moens M, Starr JL (eds) Root-knot nematodes. CABI, Wallingford, pp 380–411

Hashmi S, Hashmi G, Gaugler R (1995) Genetic transformation of an entomopathogenic nematode. J Invertebr Pathol 66:293–296

Hogenhout SA, Van der Hoorn RA, Terauchi R et al (2009) Emerging concepts in effector biology of plant-associated organisms. Mol Plant Microbe Interact 22:115–122

Houda Z, Romero D, García-Gutiérrez L et al (2009) Lipopeptides are key factors in the biocontrol activity of *Bacillus subtilis* towards powdery mildew and bacterial diseases of cucurbits. In: Elad Y, Maurhofer M, Keel C et al (eds) Biological control of fungal and bacterial plant pathogens. IOBC/WPRS Bulletin 43: 117–120

Ihrmark K, Asmail N, Ubhayasekera W et al (2010) Comparative molecular evolution of *Trichoderma* chitinases in response to mycoparasitic interaction. Evol Bioinform 6:1–26

Jasanoff S (2005) Designs on nature: science and democracy in Europe and the United States. Princeton University Press, Princeton, 344

Joshi R, McSpadden Gardener BB (2006) Identification and characterization of novel genetic markers associated with biological control activities in *Bacillus subtilis*. Phytopathology 96:145–154

Kerry BR, Crump DH (1977) Observations on fungal parasites of females and eggs of the cereal cyst-nematode, *Heterodera avenae*, and other cyst nematodes. Nematologica 23:193–201

Misas-Villamil JC, van der Hoorn RA (2008) Enzyme-inhibitor interactions at the plant-pathogen interface. Curr Opin Plant Biol 11:380–388

Moseri I, Bimerew M, Ovadis M et al (2009) Combining 2,-4-diacetylphloroglucinol and pyrrolnitrin activities in *Pseudomonas fluorescens* Q8r1-96 increases its ability to suppress *Rhizoctonia solani*. In: Elad Y, Maurhofer M, Keel C et al (eds) Biological control of fungal and bacterial plant pathogens. IOBC/WPRS Bulletin 43: 159–162

Niu Q, Huang X, Tian B et al (2006) *Bacillus* sp. B16 kills nematodes with a serine protease identified as a pathogenic factor. Appl Microbiol Biotechnol 69:722–730

Noble D (2002) The rise of computational biology. Nat Rev Mol Cell Biol 3:460–463

Noble D (2006) The music of life: biology beyond genes. Oxford University Press, New York, p 51

Oka Y, Chet I, Spiegel Y (1993) Control of the root-knot nematode *Meloidogyne javanica* by *Bacillus cereus*. Biocontrol Sci Technol 3:115–126

Oldroyd GE, Downie JA (2008) Coordinating nodule morphogenesis with rhizobial infection in legumes. Annu Rev Plant Boil 59:519–546

Ongena M, Jacques P (2008) *Bacillus* lipopeptides: versatile weapons for plant disease biocontrol. Trends Microbiol 16:115–125

Ongena M, Adam A, Jourdan E et al (2007) Surfactin and fengycin lipopeptides of *Bacillus subtilis* as elicitors of induced systemic resistance in plants. Environ Microbiol 9:1084–1090

Pidgeon JD, May MJ, Perry JN et al (2007) Mitigation of indirect environmental effects of GM crops. Proc R Soc B 274:1475–1479

Poppy GM, Wilkinson MJ (2005) Gene flow from GM plants – a manual for assessing, measuring and managing the risks. Blackwell Publishing, Oxford, 241pp

Ruse M (2003) Darwin and design: does evolution have a purpose? Harvard University Press, Cambridge/London, pp 17–19

Sanvido O, Romeis J, Bigler F (2007) Ecological impacts of genetically modified crops: ten years of field research and commercial cultivation. Adv Biochem Eng Biotechnol 107:235–278

Sela S, Schickler H, Chet I et al (1998) Purification and characterization of a *Bacillus cereus* collagenolytic/proteolytic enzyme and its effect on *Meloidogyne javanica* cuticular proteins. Eur J Plant Pathol 104:59–67

Shoresh M, Harman GE (2008) The molecular basis of shoot responses of maize seedlings to *Trichoderma harzianum* T22 inoculation of the root: a proteomic approach. Plant Physiol 147:2147–2163

Sikora RA, Pocasangre L, Felde A (2008) Mutualistic endophytic fungi and in-planta suppressiveness to plant parasitic nematodes. Biol Control 46:15–23

Spiegel Y, Cohn E, Galper S et al (1991) Evaluation of a newly isolated bacterium, *Pseudomonas chitinolytica* sp. nov. for controlling the root-knot nematode, *Meloidogyne javanica*. Biocontrol Sci Technol 1:115–125

Stein T (2005) *Bacillus subtilis* antibiotics: structures, syntheses, and specific functions. Mol Microbiol 56:845–857

Stirling GR (1991) Biological control of plant-parasitic nematodes: progress, problems and prospects. CAB International, Wallingford

Vogel C, Teichmann SA, Chothia C (2003) The immunoglobulin superfamily in *Drosophila melanogaster* and *Caenorhabditis elegans* and the evolution of complexity. Development 130:6317–6328, doi:10.1242/dev.00848

Wilsdon J, Willis R (2004) See-through science: why public engagement needs to move upstream. Demos, London, 69 pp

Wilsdon J, Wynne B, Stilgo J (2005) The public value of science: or how to ensure that science really matters. Demos, London

Woo SL, Scala F, Ruocco M (2006) The molecular biology of the interactions between *Trichoderma* spp., phytopathogenic fungi, and plants. Phytopathology 96:181–185

Woo SL, Ruocco M, Vinale F et al (2009) The *Trichoderma*-plant-pathogen interaction: understanding the mechanisms and improving biocontrol. In: Elad Y, Maurhofer M, Keel C et al (eds) Biological control of fungal and bacterial plant pathogens. IOBC/WPRS Bulletin 43:83–88

Index

A
Abiotic environment, 40, 52, 57, 146, 173, 236, 269–271, 293
Acanthocheilonema viteae, 128
Acremonium, 131, 215, 240
Acromonium strictum, 274
Actinobacteria, 4, 102, 103
Acyltransferase, 73
Adhesive, 106, 128–131, 133, 135, 146, 147, 152, 154, 156, 205, 206, 268, 280–281
Affinity chromatography, 150
Agaricus bisporus, 150
Agglutination, 132, 187, 190, 191
Agrobacterium
  A. tumefaciens, 71
Aldicarb, 272
Alkaline serine protease, 130, 134, 136
Alleles, 73, 75, 81, 93, 95
AMF. *See* Arbuscular-mycorrhizal fungi
Ammonia, 10, 11, 152, 217
Ammophila arenaria, 45–47, 236, 238
Amphid, 117, 121, 122, 124
Anguina, 135
Anguina funesta, 135
Animal manure, 9, 10, 14, 278
Antagonism, 5, 135, 184, 193, 210, 215, 216, 235, 241, 269, 273
Antagonist, 5, 8–10, 16–18, 50, 129, 135, 184, 187, 193, 198, 204, 216, 234, 236, 237, 243–247, 260–275, 277–281, 295, 299
Antibiosis, 4, 5, 184, 211, 214
Antibodies
  monoclonal (MAb) antibodies, 117, 118, 123, 125, 126, 134, 187–188, 190
  polyclonal (PAb) antibodies, 117, 187–188
Aphelenchoides fragariae, 266
Aphelenchoides sp., 269
Apoptotic corpse receptor CED–1, 82
Appressoria, 73, 131, 132, 156, 159, 162, 173, 179, 187, 205–208
Arbuscular-mycorrhizal fungi (AMF), 47–48, 198, 230–239
Arthrobotrys conoides, 148, 149
Arthrobotrys dactyloides, 148, 275–276, 278
Ascaris suum, 116, 117
Ascomycota, 147, 204–206
Aspergillus niger, 148, 152, 271
Attachment, 72, 98, 101, 106, 108, 109, 123, 129, 132–136, 147, 185–191, 197, 206, 265–266, 270, 271

B
Bacillus
  B. cereus, 101–103, 106, 107, 135, 266
  B. firmus, 245, 266, 273, 299
  B. halodurans, 101–103
  B. nematocida, 134
  B. subtilis, 101–103, 105, 107, 266, 296
  B. thuringiensis (Bt), 75, 101–103, 105–107, 266, 274
Bacteria, 2, 40, 67, 91, 117, 159, 184, 207–208, 233–234, 262, 292
Belonolaimus longicaudatus, 17, 98, 269, 298
Bioact®, 16, 299
Biocontrol agents, 16–19, 28–30, 49, 50, 91, 99, 108, 109, 133, 135, 145, 164, 171, 172, 177, 183–198, 204, 209, 214–218, 229, 245, 291, 292
Biofilm, 69–70, 72, 296
Biological control, 1, 49, 91, 133, 145, 176, 205, 229, 259, 291
Biotic environment, 2, 40, 50, 96, 293
Biotic interaction, 2, 4–6, 31
Blumeria graminis, 156, 240
Bottom-up controlled, 43

K. Davies and Y. Spiegel (eds.), *Biological Control of Plant-Parasitic Nematodes: Building Coherence between Microbial Ecology and Molecular Mechanisms*, Progress in Biological Control 11, DOI 10.1007/978-1-4020-9648-8,

Bradyrhizobium japonicum, 266, 273–274
Brevibacillus laterosporus, 134, 299
Brugia malayi, 56, 92, 94, 95, 116, 128
Burkholderia cepacia, 75, 266, 299
Bursaphelenchus xylophilus, 131, 134, 268

C

Caenacins (cnc), 83
Caenopores, 83
Caenorhabditis
  C. briggsae, 56, 94
  C. drosophilae, 73
  C. elegans, 56, 57, 65–84, 91–95, 108, 116–118, 120, 124, 129–131, 154, 155, 157–160, 294
  C. plicata, 73
Candida albicans, 71, 159
Captafol, 275
Carbofuran, 272
Carbohydrate-recognition domains (CRDs), 121, 190
Catalase, 77, 147, 248
Catenaria spp.
  C. anguillulae, 51–52, 132
cDNA libraries, 154
Cellulases, 4
Cell wall, 4, 57, 83–84, 133, 151, 162, 192, 193, 206, 217, 238–240, 248
CFs. *See* Culture filtrates
Chemosensation, 74, 124
Chemosensory, 76–77, 124, 131
Chitin, 11, 12, 116, 117, 185, 191–193, 206, 207, 209, 216–218, 238, 266, 274
Chitinase
  endochitinases, 184, 193–194, 208
  exochitinases *(Chitobiosidases)*, 193
Chitosan, 204, 216–218
Chloropicrin, 275
Citrus nematode *(Tylenchulus semipenetrans)*, 185, 274
Clonostachys rosea (syn. *Gliocladium roseum*), 131, 208
Coccidioides posadasii, 159
Coelomocytes, 76
Cofilin (actin-binding protein), 156
Collagen, 94, 101, 106–107, 118–120, 127, 134–136, 151, 179, 191–192, 206, 207
  genes, 119, 120
Collagenase, 107, 134–136, 208
Coloniser-persister strategy, 42
Competition, 5, 18, 19, 41, 42, 45–50, 52, 162, 177, 184, 197, 210, 218, 236, 237, 247, 265, 269–270, 273
Concanavalin A (Con A), 131
Conidia, 28, 73, 131, 132, 161–162, 177, 187–191, 194, 197, 205, 206, 213, 214, 217, 268, 270, 279–280
Conidial traps (CTs), 161, 162, 213
Cortical zone, 118–119
Cottonseed, 11
CRDs. *See* Carbohydrate-recognition domains
Criconemella, 236, 269
Crotolaria juncea, 276
Cryptococcus neoformans, 71, 74
CTs. *See* Conidial traps
Culture filtrates (CFs), 133, 135, 149, 151, 152, 185, 193, 195, 196, 246
Cunninghamella elegans, 136
Cuticle, 4, 57, 68, 70, 72–74, 76–77, 94, 98, 106–108, 115–116, 118–120, 122–125, 127–137, 147–148, 151, 152, 156–158, 192, 196, 205–209, 216, 218, 268, 271, 281
Cyclophilins, 156
Cysteine proteinases, 118, 128

D

Dactylella
  D. Oviparasitica, 262–264, 277, 279, 297
  D. shizishanna, 130, 208
Dactylellina candidum, 277, 280–281
DAPG. *See* 2,4-diacetylphloroglucinol
Defensin like ABF peptides (abf), 81, 83
Denaturing gradient gel electrophoresis (DGGE), 55, 56
Detoxification, 76, 77
DGGE. *See* Denaturing gradient gel electrophoresis
2,4-Diacetylphloroglucinol (DAPG), 4, 270, 271, 280, 281
1,3-Dichloropropene (1,3-D), 265, 272, 275, 276
Differential display reverse transcription (DDRT)-PCR, 241–242
Dihydroxy-phenylalanine (DOPA), 206–207
Dilophosphora alopecuri, 135
Diplogasterid nematodes, 269
Dirofilaria immitis, 116
Ditylenchus sp., 269
DMT. *See* Drug metabolite transporter
DOPA. *See* Dihydroxy-phenylalanine
Drechmeria coniospora, 72, 73, 81, 131, 160, 205
Drechslerella dactyloides, 205, 212
Drug metabolite transporter (DMT), 103, 104
Duddingtonia flagrans, 72–73, 146, 148

E

Ecosystem, 2, 4, 6, 14, 31, 39–57, 184, 197, 210, 216
Effector, 66–67, 77, 78, 81, 83, 84, 137, 159–160, 228, 295
Encapsulation, 76
Endoglycosidase, 134
Endogone gigantea, 230
Endophyte, 6, 18, 19, 30, 209–216, 228, 229, 240–249
Endosymbiont, 69, 70, 93
Energy channels, 23, 43–45, 51, 52
Enterobacter cloacae, 270, 271
Enterobacteriacae, 69
Enterococcus faecalis, 71
Entomopathogenic nematodes, 53, 269, 298
Environment, 2, 4–6, 8, 10–11, 13, 16–19, 23–30, 40–42, 50, 52, 54–55, 57, 66, 68, 76, 77, 83, 96, 101, 104, 105, 115–116, 122–124, 127, 129, 136, 137, 145–147, 161, 163–164, 184–185, 191, 194, 206, 215, 229, 236, 241, 260, 262, 264, 266, 269, 278, 279, 291, 293, 297, 300
Eosinophils, 128
Erwinia carotovora, 71
Erwinia chrysanthemi, 71
Escherichia coli, 70, 71, 75, 134
Esterases, 174
ESTs. *See* Expressed sequence tags
Ethylene, 195–196, 228
Excretory cell, 122
Exosporium, 101, 106, 107, 134, 271
Expressed sequence tags (ESTs), 92, 154–155, 157, 163
Extracellular enzymes, 4, 174, 194, 207

F

Fatty acid binding protein (GpFAR1), 126
Fenamiphos, 12
Fibronectin, 107, 134
FLN. *See* Free-living nematode
Fluorescent pseudomonads, 4, 18
Food chains, 3, 5, 25, 42–45
Food web, 2–8, 14, 20, 22, 23, 25, 27, 28, 42–46, 49, 52, 55, 295
Fosthiazate, 272
Free-living nematode (FLN), 3, 13, 24, 56, 66, 68, 91–92, 94, 123, 124, 129–130, 146, 152, 179, 196, 209, 246, 295
Fucose, 75, 123, 190–191
Fucosyltransferases, 57
Functional genomics, 56–57, 146, 154, 160
Fungal channel, 23, 43–44
Fungitoxic effect, 217
Fusarium spp.
  F. oxysporum, 15, 19, 213, 216, 229, 243–249, 263, 280

G

Gaeumannomyces graminis var. *tritic*, 4, 216
Galactosyltransferases, 73–74
Galβ1–3GalNAcα, 150
Gall, 11, 12, 16, 18, 41–42, 48, 96, 131, 175–176, 178, 185, 186, 231, 238, 244, 247, 265–268, 270–275, 281
Gelatin, 134
Gelatinous matrix (gm), 116–117, 123, 132, 135, 178–179, 185–192, 194
Genomics, 56–57, 91–109, 145–165, 292, 293
Giant cells, 125, 185, 295
Gigaspora margarita, 232–235, 238
Gliocladium, 16, 131
Globodera pallida, 108, 122, 126, 178, 209, 234–235, 245, 267, 280
Glomus etunicatum
  G. fasciculatum, 232, 235–237
  G. intraradices, 232–236, 238
  G. macrocarpum, 232, 234
  G. mossae, 230, 274
  G. tenue, 232
Glucanase, 4, 184, 196, 248
Glucose, 131, 156, 190, 192–194
Glutathione peroxidase, 126, 128
Glycocalyx, 71
Glycolipid, 75, 120, 196
Glyconjugates, 72
Glycosylation, 73, 74
Glycosyltransferase, 73–75, 107
gm. *See* Gelatinous matrix
GpFAR1. *See* Fatty acid binding protein
G-protein, 94, 187

H

Haemonchus contortus, 95, 118, 126, 127
Helicotylenchus sp.
  H. dihystera, 236, 241, 267
Heterodera spp.
  H. arenaria, 46, 47, 49, 50
  H. avenae, 14–15, 98, 171–172, 262, 274
  H. glycines, 11, 15, 98, 108, 117, 134, 172, 234, 261, 266, 277
  H. schachtii, 15, 131, 136, 171–172, 176–177, 240, 245, 263, 264, 268, 276, 277
  H. trifolii, 52

Heterorhabditis
H. bacteriophora, 69, 269
Hirsutella rhossiliensis, 28, 176, 177, 205, 206, 212–213, 268, 275, 276, 279, 280, 297
Hoplolaimus sp., 269
Horizontal gene transfer, 41–42, 48, 57, 162
Host-recognition, 57, 93, 217, 248
Hydrogen peroxide, 75, 126
Hypersensitive, 67, 82, 97
Hypodermis, 115, 119, 120, 122

I
IAA. *See* Indole-acetic acid
Immunodetection, 51
Indole-acetic acid (IAA), 124
Induced-resistance, 48, 184, 195–196, 238, 248, 249
Innate immune, 66, 67, 76–84, 128

J
Jasmonic acid, 126, 195, 196, 228

K
Kinetin, 124
Knock-out mutant, 161

L
Lecanicillium lecanii, 205, 209, 214
Lectins
ABL, 150, 151, 153
AOL, 150, 151, 164
C-type, 67, 83, 84, 128, 130, 160
C-type lectin domain containing proteins (CTLD), 48
legume-rhizobia symbiotic interaction, 48
limulin, 131
UEA-I, 190
Leucine-rich repeat (LRR), 81
Leucobacter chromiireducens subsp. *solipictus*, 72–73
Linoleic acid, 126, 152
Lipase, 67, 76, 83, 117, 136, 174
Lipophilicity, 123, 124, 127
Lipoproteins, 120, 191, 207
Lolium perenne, 52, 240
Longidorus, 70
LRR. *See* Leucine-rich repeat
Lysis, 4, 184
Lysozymes, 67, 76, 77, 83, 84

M
Macrophage, 76
Magnaporthe grisea, 156, 159
Mannose, 75, 131, 190
Marram grass, 45–49, 51
Medicago truncatula, 48, 233
Melancon, 16
Meloidogyne spp.
M. artiellia, 116, 120
M. graminis, 241
M. hapla, 56, 84, 92–97, 126, 231, 232, 237, 241, 266, 271, 274
M. incognita, 16, 19, 56, 57, 84, 92–97, 117, 118, 120, 122–127, 173, 178, 185, 186, 190, 191, 193, 209, 212, 232–234, 237–241, 244–248, 261, 264, 266–268, 270–271, 273, 274, 279
M. javanica, 11, 12, 14, 48, 97, 116, 117, 119, 121–123, 132, 135, 185–188, 190, 192, 195, 196, 232, 233, 245–246, 262, 266–268, 271–274, 278, 281
M. maritima, 47, 49, 50
M. marylandi, 241–243, 267
M. naasi, 241
Meristacrum sp., 278
Mesocriconema xenoplax, 262, 263, 269, 273, 276
Metalloprotease, 151
Metallothionein, 77, 156
Metridin-like ShK toxin domain, 83
Microarray technology, 81, 83, 130, 154, 157, 160–162, 164, 165, 241–242, 280, 281
Microarthropod, 4–5, 20, 50, 270
Microbabacteriaceae, 73
Microbacterium nematophilum, 72–74, 76–77, 82
Monacrosporium ellipsosporum, 148, 268
Monacrosporium gephyropagum, 148, 268
Monacrosporium haptotylum (syn. *Dactylaria candida*), 72–73, 130, 148, 153–161, 163, 280–281
Monacrosporium lysipagum, 161, 274
Monochamus, 268
Mononchoides gaugleri, 19, 269
Mutagenesis, 67, 72
Mycelium, 16, 147, 149–151, 154–156, 161, 173, 233

N
N-acetylglucosamine (GlucNAc), 238
N-acetylglucosaminidases, 193
Natural enemies, 23, 43, 45, 48, 51–53, 57, 97, 174, 176, 260, 292, 295

Neem cake, 185, 274
Nematicide, 7, 12, 21, 30, 95, 117, 216, 259, 260, 272, 275, 276, 298
Nematoctonus leiosporus
N. pachysporus, 205, 212–213
N. robustus, 205, 212
Nematode-trapping fungi, 18, 24, 28–30, 52, 129, 145–165, 205, 209, 212, 213, 276, 278, 279
Nematophagous fungi, 23, 25, 28–29, 51, 136, 145, 146, 152, 160, 171, 176, 184, 203–210, 212–214, 217, 218, 249, 270, 272, 275–278, 280, 292
Nematophthora
N. gynophila, 14–15, 27, 262
Nemin, 149
Neotyphodium, 241–243
Neotyphodium coenophialum, 241, 242
Neotyphodium lolii, 241, 242
Neuropeptide-like proteins (nlp), 83, 160
Nuclear receptors (NR), 94, 108

O
Oesophageal glands, 57, 117
Oesophagus, 70, 128
Oil-cake, 10
Oligonucleotide fingerprinting, 15, 263, 264
Onchocerca volvulus, 118
Operons, 72, 92–94, 107
Orbiliales, 28–29, 147, 162–164, 279
Organic amendments, 7, 9–13, 20–22, 24, 177, 216–217, 274–278
Organic matter, 2, 3, 5, 7, 9–10, 12–14, 20, 22–25, 27, 28, 41, 43, 44, 52, 68, 228, 247, 264, 266, 269, 275–278
Orthologue, 79–81, 94, 104–105, 108
Over-expressing mutant, 161
Oxamyl, 267, 272

P
PAE. *See* Pectin acetylesterase
Paecilomyces
P. chlamydosporia, 206, 208, 209, 212, 213, 216–218
P. lilacinus, 15, 16, 29, 30, 146, 206, 208, 245, 260, 263, 267, 268, 271, 273, 274, 277, 279, 280, 299
P. marquandii, 274
PAL. *See* Phenylalanine ammonia lyase
Panagrellus redivivus, 129–131, 134, 149, 152, 245–246
Parasitism, 5, 17, 23, 26–28, 43, 50, 52, 56, 57, 68, 71–72, 92, 95, 123, 129, 132, 135, 137, 162, 163, 173, 176, 177, 179, 184–198, 210–212, 214, 247, 265–266, 268, 270–273, 275–277, 280, 293, 295, 296
Paratrychodorus, 70
Pasteuria penetrans, 14–15, 17, 27, 30, 49–50, 91–109, 133, 134, 136, 235–236, 238, 262, 265, 266, 270–273, 275, 277–278, 295, 297, 299
Pathogenesis-related (PR) proteins, 57, 83, 195, 196
PC. *See* Phosphorylcholine
Pectin acetylesterase (PAE), 248, 249
Penetration tube, 130, 131, 147, 148, 268
Penicillium, 131, 133
Peptaibol, 194–195
Peptidyl-prolyl cis-trans isomerases, 156
Peroxiredoxin (thioredoxin peroxidise), 126
Phagocytosis, 76
Pharyngeal glands, 117, 122
Phenoloxidase, 76
Phenylalanine ammonia lyase (PAL), 248, 249
Phoma, 131
Phospholipase C (PLC), 74
Phospholipid fatty acid analyses (PLFA), 52, 54, 55
Phosphorylcholine (PC), 128
Photorhabdus
P. luminescens, 69
P. rubescens, 131, 179, 205, 208
Phylloplane microbials, 204
Phytohormones, 124, 229
Piriformospora indica, 212, 239–240
Plant growth promoting rhizobacteria, 6, 19, 30, 239–240
PLC. *See* Phospholipase C
Plectosphaerella cucumerin, 279
PLFA. *See* Phospholipid fatty acid analyses
Pochonia (Verticillium)
P. chlamydosporia, 14–16, 18, 27, 30, 51, 131, 171–179, 205, 206, 208, 209, 212, 213, 216–218, 262, 264, 267, 268, 270–273, 275–277, 279, 292, 294, 297, 299
Polysaccharide, 72
Polythienyl, 48
Pratylenchus
P. coffeae, 235–236, 238
P. jordanensis, 12
P. penetrans, 49–50, 237, 238, 266, 272, 295
P. zeae, 12, 14, 21, 242, 245–246, 277

Predation, 4–5, 12–13, 17, 23, 26, 42, 43, 46, 265
Profilin (actin-binding protein), 156
Protease, 76, 127, 128, 130, 131, 134–136, 149, 151–154, 157, 174, 179, 184, 192–195, 207–209, 267–268, 280, 281, 294
Protozoans, 4–5, 23, 50
Pseudomonas
  P. aeruginosa, 71, 74, 75, 78, 81–82, 266, 273–274
  P. fluorescens, 266, 270, 271, 273–274, 280, 281
  P. synxantha, 273
PTS. *See* Sugar-specific phosphotransferase family
Pyocyanin-phenazine, 75
454 Pyrosequencing, 163
Pythium ultimum, 271, 280

Q
Quantitative PCR (QPCR), 15, 164, 175–176, 216, 279–280

R
Radopholus
  R. citrophilus, 235
  R. similis, 18–19, 235–236, 238, 244–249, 268
Rathaybacter (formerly referred as *Clavibacter* and *Corynebacterium*)
  R. toxicus, 135
RBC. *See* Red blood cells
Receptor, 76–83, 94, 108, 128, 130, 134, 151
Recognition, 57, 76, 78–79, 81, 83, 108, 121, 124, 129–131, 134, 149–151, 160, 187, 190, 207, 217, 248, 293
Red blood cells (RBC), 122, 150
Repetitive/transposable elements (TEs), 93
Resistance, 17–19, 46, 52, 67, 74, 75, 77, 78, 81, 84, 96, 97, 106, 179, 184–186, 211, 239–244, 246, 249, 259, 260, 272, 294, 295
Retinol, 126
Rhizoplane, 6, 212
Rhizosphere, 5, 6, 17–18, 41, 43, 46–49, 51–53, 55, 102, 136, 161, 164, 171–179, 184–185, 194, 197, 204, 212, 213, 216, 229, 234, 238, 247–248, 262, 263, 266, 267, 271, 272, 275, 277, 278, 280, 293, 295, 301
Ribosomal RNA (rRNA), 15, 16, 55–56, 94, 263, 264
Rickettsia-like alpha-proteobacteria, 69, 103
RNAi. *See* RNA interference
RNA interference (RNAi), 67, 70, 81, 95, 108, 116–117
Root channel, 43–44, 51
Root exudates, 3, 5, 123–124, 184–185, 213, 234, 238, 242, 246, 247
Rotation crop, 13, 14, 20–23, 177, 259, 260, 272, 277, 278
Rotylenchulus reniformis, 16, 25, 261, 267, 274, 276
rRNA. *See* Ribosomal RNA

S
Saccharomyces cerevisiae, 148, 159
Salicylic acid, 195, 228
Salmonella typhimurium, 71
Saposin, 76, 77, 83–84
Saposin-like proteins (ssp), 83
SC. *See* Surface coat
Schizosaccharomyces pombe, 159
Sequence-selective PCR primers, 263, 264
Serine proteinase, 173, 179, 192
Serratia marcescens, 71, 74, 76
Serrawettin, 76
Shedding, 70, 95–96, 108, 120–125, 127, 128, 191
Signal transduction, 66, 78–79, 84, 155, 160–161, 187, 196, 240, 248
Soil ecology, 2–8, 40, 57
Soil environment, 2, 19, 29, 30, 40–42, 104, 146
Soil solarization, 273
Solanum tuberosum, 48–49
Solexa, 163
Split-root, 48, 185–186, 196, 238, 239, 247, 248
Sporogenesis, 99
ssp. *See* Saposin-like proteins
Staphylococcus aureus
  S. epidermis, 71
Steinernema
  S. carpocapsae, 242, 243, 269
  S. riobrave, 269
Streptomyces costaricanus, 266, 274
Streptoverticillium albireticuli, 72–73
Stylet, 3, 19, 68, 69, 295
Subtilisin, 130, 152–154, 157, 207, 208
Sugar-specific phosphotransferase family (PTS), 103, 104
Superoxidase dismutase, 77

Suppressive soil, 8–16, 25–26, 28, 172–173, 177, 209, 218, 245, 261–264, 267, 277, 295, 297
Surface coat (SC), 108, 115, 118–128, 133–136, 187, 190–191, 207
Symbionts, 41–42, 69, 184, 197, 233, 242
Systemic resistance, 19, 184–186, 195–197, 211, 240, 247, 269

T

Tagetes erecta, 48–49
Tagetes patula, 48–49
Telluria chitinolytica (Pseudomonas chitinolytica), 135
Terminal restriction fragment length polymorphism (T-RFLP), 56
TEs. *See* Repetitive/transposable elements
Thaumatins (thm), 77, 83
Thermosensation, 74
Thioredoxins, 126, 156
thm. *See* Thaumatins
Thomsen-Friedenreich/T antigen, 150
Tillage, 13, 14, 20–23, 25, 30, 277–278
Top-down controlled, 43, 47, 50–53, 57
Toxin, 4, 71, 74–75, 81–84, 101–102, 105, 133, 135, 197, 205, 206, 211, 212, 246, 249, 260, 280, 295
Toxocara canis, 128
Transcription factors, 77, 79–83, 105, 160
T-RFLP. *See* Terminal restriction fragment length polymorphism
Trichinella spiralis, 126–127
Trichoderma
  T. asperellum, 132, 186–189, 191–196
  T. atroviride, 132, 133, 136, 186–189, 192–196, 268, 281
  T. hamatum 382, 186, 196
  T. harzianum (T–12), 185
  T. koningii (T–8), 185
  T. pseudokoningii, 268, 274
  T. virens, 185, 195
Trichodorus, 70
Trichostrongylus colubriformis, 134
Trifolium repens, 52
Tritrophic interactions, 46, 53, 136, 173, 295
Trypsin, 134, 193, 208
Tylenchorhynchus
  T. annulatus, 277
  T. vulgaris, 236
Tylenchus sp., 269
Tyrosine kinase, 77

U

Ubiquitin, 156
Unfolded protein response (UPR), 79, 82

V

Vesicular-arbuscular mycorrhizal fungi (VAM), 198, 230
Virus, 5, 70, 184, 195
Vitelline, 116, 117, 173, 179, 191

X

Xenorhabdus
  X. nematophila, 69–70, 243
Xiphinema, 70

Y

Yersinia pestis, 71, 72

9781402096471